Readings in Environmental Psychology

Series Editor

David Canter

The Child's Environment

Edited by

Christopher Spencer
Department of Psychology, University of Sheffield, Sheffield, U.K.

ACADEMIC PRESS
Harcourt Brace & Company, Publishers
London   San Diego   New York   Boston   Sydney   Tokyo   Toronto

ACADEMIC PRESS LIMITED
24–28 Oval Road
London NW1 7DX

United States Edition published by
ACADEMIC PRESS INC.
San Diego, CA 92101

A catalogue record for this book is available from the British Library

ISBN 0-12-656640-2 ✓

Typeset by Galliard (Printers) Ltd, Great Yarmouth, Norfolk
Printed and bound in Great Britain by Hartnolls Ltd, Bodmin, Cornwall

CONTENTS

PREFACE

In an age of summarizing textbooks and popular distillations of scientific research it is easy to forget that new concepts and discoveries emerge out of the painstaking work of individuals. Usually this work is presented directly, in detail, in the pages of learned journals. Since its inception more than 12 years ago the *Journal of Environmental Psychology* has demonstrated its pre-eminence through publishing original, innovative papers. By bringing them together in one volume, ready access has been provided to the first-hand accounts of a range of explorations that are central to the growth and development of environmental psychology itself.

My day-to-day involvement in editing JEP had somewhat inured me to the range and quality of material with which I had to deal, so that when I set about planning the volumes to be drawn from the existing JEP papers I was pleasantly surprised to be reminded of how many interesting and important papers we had published. Indeed, it was clear that the variety of papers we had published was so great that it was no longer possible to keep track of all the strands and developments within environmental psychology. By identifying areas of study that had generated a critical mass of papers it was possible to provide some guide through the maze of environmental psychology research.

A further great value of these volumes is that it helps new generations not to forget the discoveries of earlier generations. Environmental psychology is not a field in which to-day's discoveries cause yesterday's research to become irrelevant and obsolete. Quite the reverse is true. As our understanding of the human experience of the environment deepens it becomes more important to have a full awareness of the earlier research on which current understanding is based. That research is best examined in the form of the original journal publications.

This first volume of papers drawn from the *Journal of Environmental Psychology* brings together a number of studies that explore children's knowledge and experience of the environment. By bringing together these papers with an introduction from their editor, Christopher Spencer, the reader is provided with a valuable overview of some of the most significant studies in this important area of environmental psychology.

The study of children's development has always been central to psychology. Through the examination of the changes throughout childhood it is possible to understand more fully the essence of many psychological processes. It is therefore no surprise that environmental psychology has always had a strong stream of explorations of children's understanding and experience of their surroundings. In many senses these are the most fundamental studies in environmental psychology, both because they reveal the base lines of children's environmental capabilities against which studies of adults can be compared and because they inform us of the experiences of people who are environmentally vulnerable.

However, I believe that this volume of papers does more than facilitate the examination and comparison of earlier studies of children and their environment. It provides detailed examples of how the community of scholars are engaged in a constant debate with each other through the pages of learned journals. It also shows how the accumulation of individual papers does lead to an increase in our knowledge that is greater than the sum of those papers. The environmental psychology of children has moved from the importing phase, considering the applicability of Piagetian and other theories to the environmental arena. It is now in an exporting phase of developing theories and methods that have illuminated issues beyond those that are strictly environmental.

Of particular importance has been the consideration of the environments that are especially, sometimes uniquely, the domain of children. Before environmental psychologists, responding to questions from architects, started to study schools and playgrounds, children's cognitions and experiences were examined in developmental psychology as if they had no actual setting at all. They floated freely in an interpersonal matrix. Now not only is there more awareness of the importance of looking at the physical context of children's behaviour but children's understanding of that context has been seen as an important route into their understanding of many aspects of the world.

The present volume will therefore be of value to everyone who wishes to enhance their knowledge of children's experiences of the places in which they live. People who have not looked at environmental psychological studies before will find it a useful introduction to that growing field. For those who know environmental psychology and JEP it will provide a valuable compendium of key papers that have laid the basis for future research into children and their environment.

David Canter, Managing Editor
University of Liverpool

INTRODUCTION: THE CHILD'S ENVIRONMENT — A CHALLENGE FOR PSYCHOLOGISTS AND PLANNERS ALIKE

CHRISTOPHER SPENCER

Department of Psychology, University of Sheffield, U.K.

The two main strands of environmental psychology research have been the individual's understanding of the environment, and the shaping of individual behaviour by the physico-social environment: both of these strands are represented in this volume on the child's environment.

Environmental psychology has joined forces with geography and with developmental and cognitive psychologies in studying the development of the child's environmental skills and cognitions. The past 15 years have seen a radical reassessment of the young child's competence in understanding spatial and environmental information: the first reading in the volume, by Christopher Spencer and Zhra Darvizeh, should be read as an early call to recognize this competence; and summarizes a range of empirical evidence from the 1970s in support. Much of the rest of the readings represent more recent, systematic studies of a whole range of childhood (and adolescent) competencies: for example, Alison Conning and Richard Byrne's study of three- and four-year-olds' ability to point to out-of-sight targets, as a demonstration that they already have a vector map of space. Reg Golledge and his colleagues, in their 1985 paper, offer another good example: this multidisciplinary team of authors offer both a conceptual model of the child's knowledge structures used in learning a novel environment; and, in support of the model, an in-depth case study of one child learning a particular novel route. From modelling exercises such as this, it is now possible to present formal computational process models, as the paper by Nathan Gale, Reg Golledge and their colleagues indicates. Interaction with the environment and understanding of that environment must be seen as mutually supportive—knowledge arises from and then guides action.

The development of route knowledge has proved a particularly good way of investigating the child's emerging environmental competencies, as the next reading, by Glen Waller, also illustrates. As he argues, in the real world, children manifest clear wayfinding skills; yet early developmental psychology research had tended to stress the child's *deficiencies* with relation to adult performance. A sensitive approach to children's actual performance indicates a range of strategies employed by children at different ages, and for different wayfinding tasks. Hugh Matthews' paper continues this theme, in offering a systematic comparison of techniques for eliciting young children's representations of the environment: three widely used techniques are free-recall mapping, verbal descriptions and the interpretation of large scale plans and aerial photographs. Clearly, the method of stimulus presentation is a strong influence upon representation and recall of the child's environment. Further techniques are illustrated in the remaining papers in this section of the book. Mark Blades and Louise Medlicott have compared the ability of children from four age groups to give route descriptions from a map; and describe an unfolding developmental sequence, with, by 12 years, of age an adult style of narrative emerging.

By adolescence, the individual may well have developed their capacity to parse the grammar of environments such that newly encountered, complex areas are quickly and efficiently learned. The case study by Wood and Beck gives us just such an example: an American teenager on holiday in London for a week produces a series of sketch maps which document her rapid extending cognitive map of the city. But, as the paper forcibly reminds us, what we learn and store are not just facts but also feelings about places; and Wood's mapping language, Environmental A, offers a unique and accessible way for people to show us their *individual* 'affective maps'.

Knowledge and feelings about place both derive from and then also influence patterns of behaviour in the environment. This is clearly seen in Paul Webley's study, which demonstrates that, at eight years of age, boys have a more extensive cognitive map than do girls of the same age; and that this is attributable to their more extensive home range. When the children were given the same controlled experience of a novel area, then the sex difference in recall disappeared.

Interaction with the environment enables the individual to acquire knowledge of spatial locations. What is particularly striking about Anders Biel's study is that, by the age of six, children have developed the ability to combine perspectives of spatial layouts in their own neighbourhoods—something which conventional Piagetian developmental psychology would not predict.

The next two papers shift the focus to adolescence; but clearly there are strong continuities with some of the childhood patterns we have already seen. Thus, for example, in Fred van Staden's study of 11- to 13-year-olds in New York City, we again find males as being more free to explore a wider neighbourhood than females. van Staden's main focus is on the experience of and the coping with high density living by this age group. Using a whole range of measures of everyday adolescent activities, he attempts to assess the impact of crowding on cognitive functioning and style.

How would adolescents choose to use and identify with a less crowded environment? Cotterell chose to study the emergence of what he describes as 'adolescent territories' within the grounds of a major world fair in Australia; and analyses the patterns of usage of the area in terms of the degree of fit between adolescents' leisure and experience needs, and the facilities afforded by the different pavilions of the fair.

So far, in environmental psychology, there has not been sufficient research on the particular needs of children with relation to the world around them. An early indicator of the kind of work needed is offered by Fernando Bernáldez and his colleagues: they have used a photo-preference test to examine 11- and 16-year olds' aesthetic responses to landscape. Three dimensions of preference response were found; and younger and older children differ subtly on these dimensions. All the children may agree in their rating of a particular landscape on a blandness to harshness dimension, but only the older children will then show a more positive preference for the harsher region's challenge.

After the home, the single most used behaviour setting for children is the school. We include in the concluding section of this volume five papers concerned with the school as meeting various of the child's needs. Carole Weinstein devised an experimental method of assessing the need for privacy during the elementary school day. What individual differences might be found? Would these relate to the opportunities for privacy back at the child's home? And would children who expressed a desire for a private place consistently use it once it was offered?

Carole Weinstein, with Leanne Rivlin, reminds us in the next paper that we can construe schools not only as places for learning, but also as places for socialization and for psychological development. Privacy, discussed in the earlier paper, is only one aspect of the school setting reviewed by Rivlin and Weinstein as factors which can contribute to the improvement of education and well-being in schools: topics included range from sex role stereotyping to the school as a haven.

Going *away* to school can, however, be anything but a haven, as Shirley Fisher's series of studies of homesickness in boarding schools reminds us: this is apparently a high incidence phenomenon among children sent away for the first time. Perhaps not surprisingly, problems at school exacerbate feelings of loss of valued features of the home.

The final paper in this volume brings us full circle, to link the theme of schools with that of the child's capacity to represent and understand familiar environments. Giovanna Axia and her colleagues offer a study of eight-year-olds recalling their school, which study not only acts as a useful comparison of investigation techniques, but also serves to show us that, in effect, pupils and teachers see and remember rather different aspects of the familiar school environment.

These 18 papers can only give a taste of the extensive empirical work that is now available linking children and the environment; and the remainder of this introductory chapter will attempt to indicate why work with children has played such a significant part in environmental psychology.

Several themes will be developed:

—Much of this work has been, whether intentionally or otherwise, a challenge to the orthodoxy of the developmental psychology tradition following Piaget.

—Environmental psychology has often already been at work in the child's everyday environment, while developmental psychology has stayed in the laboratory, worrying about how to make its research more 'ecologically valid'.

—Children have a discernibly different stake and interest in the environment from that of adolescents, who in their turn differ from adults in their transactions with the environment. Within this observation, one could well go further in subdividing childhood into different periods of interests and needs.

—This in turn has implications for design—at individual room level through buildings to the scale of neighbourhoods and whole settlements—and the natural areas in between. As will be argued below, the planning agenda is usually so adult-centred that it remains in ignorance of the research on children's environmental attachments and valued places (on which there is a growing literature, which is aggregating to something of a consensus).

—Children's different needs in and use of the environment perforce leads to a conceptualization of the environment which is distinctive: areas which have high affordances for exploration, play and sociability, areas of exclusion, because the adult world denies children access, areas lacking meaning for the child.

—Leading on from this point is the concentration that much research on children's environments has had on the most identifiable and (for the experiments) most accessible elements of the child's world: the school and home.

—Future research must thus address the more hidden worlds of childhood.

—In doing so, the current wide range of research techniques with children will need further extension, and children will have to join the ranks of research collaborators/informants, rather than being seen as the 'subjects' of research.

Environmental Psychology and the Piagetian Legacy within Developmental Psychology

Within many of the papers in this volume, there is an acknowledgement of the practical and conceptual debt that anyone conducting empirical work on children's epistemology must pay to the Swiss psychologists, Jean Piaget (1896—1980). Piaget showed how it was possible to frame enquiries into the child's developing conceptual world: by a mixture of practical testing, observation and questioning of the child, he showed an orderliness in development which he described in terms of stages.

Much of environmental psychological research with children has (whether intentionally or not) served to challenge the orthodoxy of the developmental psychology tradition which has followed on from Piaget; and it has tried to contribute to the general critique of his methods and conceptual approach.

At one level, environmental psychologists could simply be seen as challenging Piaget's findings about the child's capabilities at particular ages and stages—stressing the early competencies rather than incompetencies. Thus, in the first paper in these readings, empirical evidence is offered to challenge Piaget's account of the egocentricity of the young child in perspective taking. Young children's competence in distance estimation, in route following, and in integrating this information into a useful spatial frame of reference, would all seem to be much greater than the Piagetian position would indicate likely (see Conning and Byrne's paper). Understanding the conventions of a map, or interpreting an aerial photograph, are tasks which, again, seem to be within the competence of children at ages earlier than Piaget would have predicted (see the paper by Blades and Medlicott). However, the challenge to Piagetian orthodoxy goes beyond the proving of earlier and earlier competencies; the debate is also about the very processes by which the child comes to develop these competencies (Blades and Spencer, 1994).

Take the debate back to the earliest weeks of life: how does the world appear to the neonate? Is there a coherence for the young, or does the individual have to develop cognitive structures to discover such coherence by interactions with the world? Gibson (1979) takes the former position, arguing that the perceptual system is tuned from birth to relational information about objects in space; in contrast to Piaget (e.g. Piaget and Inhelder, 1956) who takes the constructionist position: young infants perceive unconnected images well into their second year, when abstract schemes for perceptual construction are forming. At first, objects have no permanence to the infant, whose universe consists of moving tableaux which appear and disappear with no fixed location. When the infant begins to coordinate actions, then a 'schema of the permanent object' develops, and the infant now searches for an object after it disappears. During this sensorimotor period, Piaget shows how the infant is able to coordinate positions and changes of position, which allows him or her the possibility of a return to their point of departure, either by directly retracing or via an alternative route.

The Piagetian concept which has perhaps caused most controversy is that of *egocentricity*. Piaget argues that in the period of preoperational thinking (roughly ages two to seven years) the necessary condition for operational development is a decentring for the child's own actions and own point of view. By egocentricity, Piaget meant nothing more than the absence of cognitive decentring; but this has been taken to indicate the spatial and conceptual limitedness of the young, child, citing Piaget's now-classic three mountains experiment. In this, the young child is faced with a table top model and asked to predict what a toy figure could see from different locations within the model.

If you conduct the experiment with the instructions and kind of materials Piaget originally used, the young child indeed does appear unable to take any other perspective than their own; but changing the mode of presentation can considerably reduce the chances of an egocentric response (Liben, 1982). Children may actually have very good representations of their spatial world, but still remain unable to cope with the extraneous demands of the tasks used by Piaget. They may be puzzled by the instructions in the experiment, and try and make 'human sense' of the testing situation; or they may simply lack the expressive (rather than conceptual) skills to respond to the experimenter. Papers in the first section of the book are acutely aware of the need to make sense to the child, to engage him or her in the activity under study, and to devise appropriate ways in which the child can fully demonstrate his or her skills.

One respect in which Piaget was a pioneer who was insufficiently followed within developmental psychology, and where environmental psychology is beginning to fill the gap, is in considering the child's understanding of and appreciation of the natural world. This topic was central to some of his earliest books. In *The Child's Conception of the World* (1929), Piaget reported his verbal investigations with children on how the child views the environment and what he or she believes to be the origins of the sun and moon, of trees and of dreams. This line of enquiry is extended in *The Child's Conception of Physical Causality* (1930) to the child's ideas about, for example, the causes of everyday phenomena: the movement of rivers, the passage of clouds, what shadows are, how water is displaced by an object being immersed. Only towards adolescence do the conceptualizations resemble those held by most adults: prior to this are a series of qualitatively different stages of explanations through which the child will travel. Thus, to take as an example the child's changing conceptualization of the sun and moon:

Stage One is characterized by *animism* (the sun and moon are alive in the same sense as people are alive); by *artificialism* (some agency, rather than material processes, formed them) and by the idea of *participation* (a continuing connexion between human activities and those of things).

Stage Two is transitional: artificialism and animism continue, but are mixed with some natural explanations for the origins of things.

Stage Three: attainment is marked by the dominance of natural processes as the origins of physical objects, without appealing to agencies (human, divine or mysterious). Whereas the actual explanations that the child gives at this stage may be crude and erroneous, they are qualitatively similar to the astronomical, geomorphological or biological accounts which they will develop into. Piaget's enterprise is principally to investigate cognitive development; yet in many of the examples of verbal protocols from children talking about the world comes a strength of affective feeling for the natural world which has been little commented upon by developmental psychologists, but which has become a focus for environmental psychology (e.g. Hart, 1979; Moore, 1986; and papers in this volume by Biel, Bernáldez and by Axia).

Ecological Validity of Research with Children

It would be a caricature of developmental psychology, either in Piaget's time or subsequently, to characterize it as wholly laboratory bound; yet there has been such a considerable reliance on tightly-controlled and often novel situations such as the laboratory, that major thinkers, including Bronfenbrenner (1979) and Garbarino (1985) have called for an ecologically valid developmental psychology. Environmental psychology has, from its much more recent inception in the 1970s, been acutely aware of this same need, and many of the studies cited in this volume are 'real-world' based.

Laboratory-based studies can, of course, be valuable. We can, for example, learn much of value about the development of cognitive processes involved in spatial understanding from the sandbox modelling of the study by Piaget et al. (1960). From this experiment, we can see how important to four- to seven-year-olds is the serial order of landmark information: children between these stages, Piaget et al. argue, tend to construct each portion of a route serially, with each subsection only loosely related. Yet the caution always remains: would we find the same performance at the same ages in an environment rich in cues and associations? Studies subsequent to Piaget's, using real world travel, have cast doubt. Over-reliance on laboratory studies of the development of cognitive mapping skills may indeed divert our thinking away from the question: what functions does the cognitive map perform for the child? And are there different functions at different ages?

> Looking at children in the rooms of an experimental laboratory, and in the room of a school, researchers have most frequently interpreted age changes in learning, memory, cognitive mapping and the like as reflecting mechanisms of cognitive growth and quasi-embryological structural development. But our everyday knowledge...is that children travel from place to place. (Siegal, 1982, p. 84).

If, following Siegal, we focus on the functional needs of the child in travelling and exploiting the world, then we have need of good, analytical natural histories of childhood. The early pioneers of the environmental psychology tradition, travelling under the banner of 'ecological psychology' (e.g. Barker, 1978) offer us some excellent examples of the everyday challenges to the child's cognitive skills. What are the behaviour settings within the home, the neighbourhood, the school, the shops, and the community as a whole that the child must learn? The great challenge offered by Barker's concept of the behaviour setting is to describe how individuals deal with both the spatial–geographical knowledge *and* the social expectancies and opportunities of places.

Other excellent, detailed studies of the child's everyday world include those by Hart (1979), Moore (1986), Torrell (1990) and Matthews (1992). Drawing on European and North American settings, rural suburban and big city, these researchers have shown how knowledge acquisition and its structuring are based on the 'functional significance of places and structures in the environment, and their potential and actual usefulness' (Torrell, 1990). These studies have provided strong evidence of the critical role played by the child's self-directed activities in the establishment of cognitive representatives of the environment.

Because children's everyday behaviour has context and purpose, their environment is rich with values and feelings. Heft and Wohlwill (1987), in their major review of the literature, have pleaded for researchers to incorporate the role of affective processes into our models for the analysis of environmental knowledge (p. 198); and Matthews (1992) has called for further work on the child's effective responses to place, especially given children's highly personalized views about place and space (p. 237).

The Changing Activity Patterns of the Growing Child: Cognitive Consequences

It does not take the findings of a psychologist to bring home the ever widening sphere of experience that generally comes with age—from the restricted range of the non-mobile neonate, through exploration of the home as a toddler, to the neighbourhood range and wider of the remainder of childhood and adolescence. What the psychologist can do is to indicate the cognitive and affective implications of this increase in environmental experience; and to note that the increase is neither uniform nor universal: culture, gender, social position, rural vs. urban cultural settings have all been indicated as important differentiators. Let us first consider the effects of gender.

For many years, gender differences in spatial skills have been noted—often using micro rather than macro spatial tests as their criterion; and the explanations have ranged from the purely neurological to the largely experiential. What is the evidence at the macro-geographical level? At the level of simple spatial information, Hart (1979), Torrell (1990) and other similar studies have shown that, not surprisingly, the child's extent of knowledge of the environment maps closely onto the activity patterns of the child, and that this in turn is closely related to age and to gender. Younger children's home range *and* their cognitive maps are more restricted than those of older children; with boys at any particular age having wider experience than girls of that same age. As Hart indicates, boys' wider range only partly reflects their parents' willingness to allow them greater freedom to travel in the neighbourhood: boys also admit to being more likely to transgress these bounds and explore still further!

Field evidence such as this needs to be available for the debate on the origins of gender differences in spatial abilities. The review by Self *et al.* (1992) of this debate distinguishes between the proponents of a deficiency theory (primarily biologically based); a difference theory (based on socio-cultural expectancies and early experience) and an inefficiency theory (spatial abilities are actually approximately the same, but that performance is culturally encouraged or limited). If children who have wider ranges of environmental experiences develop richer skills, then presumably the environmental psychology evidence cited above argues for a difference theory, a conclusion which would be in line with much cross-cultural evidence (e.g. Berry, 1971; Williams & Best, 1982).

Ranging behaviour differences between boys and girls is a noticeable, outward difference, whose relationship to environmental cognition was intuitively plausible well before we had supportive evidence. Can we extend the logic, and examine how differences in activity pattern across the life span might relate to other, perhaps less obvious, cognitive outcomes?

Adopting a life span approach enables us to contextualize the child's cognitive development: at each stage, we should ask what earlier stages have so far equipped the individual to do and to understand, what the demands of the current situation are upon the individual, and how experiences at the present stage can affect later development (see the review in Spencer, 1992). The approach allows one to emphasize processes of change, the adaptive nature of the cognitive schema for the particular life stage reached by the individual, while at the same time emphasizing the individual continuities throughout life.

Much mainstream 'developmental' research lacks this sense of continuity; often what we get is a series of snapshots of the typical child at different ages or stages. In contrast, if we follow an individual through their early life span, we are more aware of personal (and indeed cultural) life events occurring during the period, and the individual's personal travel through the roles and settings of childhood.

The Changing Activity Patterns of the Growing Child: Implications for Planning

It is the subtle, social, contextualized development of the individual through childhood and adolescence into adulthood which demands a more subtle review of the corresponding changes in environmental needs than we normally get. Both planners and developmental psychologists would broadly acknowledge that children, adolescents and adults have different stakes in the home, and in the environment beyond (even if this acknowledgement does not always then lead to appropriate provisions). We should be able to go beyond this broad acknowledgement and offer a usefully detailed account of the individual's developing needs.

From birth to about 18 months of age (all ages given here are offered as guidelines only) the *tasks* the child faces include: development of body awareness; location of objects in space; relation of objects to self, both whilst stationary and in motion; and an early understanding that there are social rules governing the use of space. *Environmental needs* during this period include, in addition to basic safety, shelter and sustenance, an environment which is stimulating, offers a range of visual, textual and relational experiences, and a variety of behaviour settings.

From approximately 18 months to six years, the child progresses through a set of cognitive achievements from a seemingly egocentric perspective to a relatively sophisticated articulate spatial reference system; from early observations on object and place properties to a more complex world view (albeit one which is limited by the range of direct and media experience). The socio-physical environment *should* expand to match—and facilitate—these cognitive expansions. The child also needs to develop not only a range of social competencies but also general rules for place-understanding and adaptability for novel settings. The range from environmental deprivation to enrichment has been much documented at the extremes (from institutionalized children's sensory deprivation to programmes of educational enrichment for both gifted and disadvantaged children): we need to have an account of the majority who will be mid-range. Between these ages, the child should have a massively extended sensory input from an ever-extending exploration and use of space. Accompanied and independent travel play their part in this explosion of experiences, as do television, books and, latterly, the school.

The early school years' tasks include the integration of spatial information into broad representation of areas (the child will become increasingly accurate in direction and distance estimates); as well as the development of a knowledge about areas not directly experienced. Independent travel, at first on prescribed routes, and later with more freedom, characterizes this period; and the child comes to have a more idiosyncratic and personal structuring of places, in ways which link to the emerging self. (For an excellent developmental account of place attachments among children see the volume edited by Altman and Low, 1992 and, especially the chapters therein by Chawla and by Cooper Marcus.)

The years leading to adolescence see a continuation of these trends—a greater detail and precision in cognitive maps, with a huge individual variation, reflecting ranging and interest differences. By this age, the 'support services' of the neighbourhood become almost as important as those of the home in facilitating the older child's social and personal needs. Robin Moore's studies (e.g. 1986, 1989) of the valued places of childhood are particularly instructive here; he stresses the 'unofficial' uses of adult-dominated areas, and children's adoption of the places left over after planning as 'private places'. Nostalgia for the rural or for past landscapes of childhood often overcome writers' rationality at this point: as a corrective, I would recommend Colin Ward's essays comparing the potential richness of cityscapes with the ambivalent affordances of the country, in his *The Child in the City* (1977) and *The Child in the Country* (1988).

The environment has, so far, been discussed as if wholly positive: but the dangers to children from traffic (e.g. Sandels, 1975), density (Wohlwill & van Vliet, 1985) and social isolation (Weinstein & David, 1987; Fisher *et al.*, this volume) are now receiving much research attention, as also are parents' views on the risks to their children (Gärling & Gärling, 1990).

The evidence is now accumulating on children's environmental needs and a consensus would seem to be emerging, on the importance of certain desiderata for full and healthy development of the child and adolescent: yet we have to ask whether these stakeholders are having their interests considered alongside those of adults by planners, politicians and 'space managers'. Those who manage the child's space range from parents and siblings through school teachers, playground adults and shopkeepers, to the police and other guardians of 'the commons'. One must acknowledge that space planning often represents balancing potentially conflicting interests: for example, privacy and sociability, and of places for the old and play areas for the young (see, e.g. Coulson, 1980).

How can their interests be promulgated? Can the young be adequately consulted, or included in the planning process itself? Or should we, rather, be talking about advocacy for the young?

Children, Adolescents and Involvement in Planning

User participation in planning was one of the 'causes' of the 1970s (e.g. Appleyard, 1970; Kaplan, 1978); which, during the 1980s, was extended to include the rights of children to be involved (e.g. Moore & Hart, 1980, 1981, 1982; Stea, 1985; Sutton, 1985).

Two points arise: however articulate the users are, their participation in the design process is no guarantee that their full range of needs will be expressed, or can be incorporated in the eventual design. (Architect–user communications are a field of study to themselves: see, for example, McKechnie, 1977.) And children need further facilitation even than do adults in the process of participation—extra game-like activities to involve them and help their design visualization.

Secondly, in many of the trial projects with children reported above, their participation has been limited to the design of settings specifically labelled as for children—typically the school yard, and the public playground—and not in the wider, shared public settings (e.g. housing estates, shopping malls, town centres) in which children tend to be the excluded stakeholders. Children are not short of ideas about what they would wish to see in such public spaces, as I can attest from a marvellous (unpublished) exercise carried out by my home city's planning department in conjunction with a score of local primary schools. Each school was visited by the planners, and after introductory meetings, it was allocated a sector of the city centre to replan as an exercise. Site visits followed, with modelling and map-drawing back in the schools, before everybody came together for a plenary session in which the sectors were brought together as a complete model of the city centre, renewed according to the children's suggestions. Few of these ideas for replanning were exclusively child-centred: their idealized city centre was a more humane area for all age groups—more greenery, more pedestrianization (including play areas), more visual appeal. One school group was particularly concerned to build old people's flats close to the centre 'so they would be close to the shops'. Part of the 'contract' with the planners was that, by the end of the plenary day, the city architects and planners would discuss with the children which ideas they, the planners, could bring into the immediate budget, which might be realized within two years, and which they felt would be impractical. A year on, I can report that some of the ideas *have* been implemented, to the benefit of all; and that others are on the future agenda.

Sheat & Beer (1989) have reviewed the philosophies underlying the various participation exercises which have been published. These include the general principle of giving children a sense of responsibility in society (as exemplified in my Sheffield example); and more specific aims, often using the children's expertise in planning their own particular settings. Eriksen (1984), for example, said that children could and should be involved in designing their own places, e.g. school playgrounds, because 'they have an intuitive understanding of the nature of play', in a way that the adult observer could not have.

Does it work? Advocates of user participation often argue from the moral and ideological position, without offering empirical evidence on the various outcomes. Sheat & Beer (1989) set up an empirical study with secondary schools rather than the more usual primary school children. Their aims included asking:

Which stages of design are most suited to user participation?
Which of the many techniques used works best?
How far does the process also result in general learning about the environment?

The study compared the success of a range of approaches, and found, amongst other things, that the participatory techniques which facilitated most thinking and involvement tended to occur when the design professionals had minimal influence on the pupils' output; and that working 'on site' rather than away from it produced the best results. Many detailed lessons can be learned from this and from other well written up case studies (see the above references). Children's participation in planning also opens up the whole related area of environmental awareness and education (Adams & Ward, 1982; Spencer & Bishop, 1989), itself an area which is characterized by pro-children political/advocacy values.

Largely Missing from the Research Agenda

Supposing one were to conduct a research study on the behaviour settings of the developmental researcher; one would, as already noted, discover that, for the majority of developmental psychologists, the laboratory was the preferred habitat (despite their protestations about ecological validity); whereas many environmental psychologists and planners concerned with children's needs are likely to be conducting field research. But even here—and this is the main point of this section—these researchers still tend to play it easy: the most accessible sites are, not surprisingly, the most studied. Schools and playgroups are places where—permission granted—the researcher is assured of finding his 'subjects'. (Changing away from regarding people as 'subjects' to research collaborators and experts is another theme to be developed here.) Access to the home, often initially via recruitment through school or clinic, is several stages more difficult; working with children beyond school and home, in their everyday free play and travel; something that only a few persistent and ingenious researchers have attempted. The work of Hart (1979), Moore (1989) and Torrell (1990) has already been mentioned as participatory observation of children's play, leisure activities, informal clubs, etc. The reader can imagine the lengths the researcher must go to in order to establish their credentials for harmless eccentricity, not only to the children but also to their protective community.

One of the pioneers of, and still strongest advocates for, this kind of approach within the social sciences, is Whyte (e.g. 1991), perhaps familiar to readers for his early studies of 'street corner society'. Whyte now talks of 'participatory action research' (PAR) as his aim—effecting change through a joint investigation by researcher and those whose activities are under study. Typically, in his studies, the researcher comes to the behaviour setting, discusses and observes, writes up a preliminary set of observations, circulates these as the basis for the next round of discussions. Whyte argues that one can achieve a rigour with PAR in a way never considered by conventional experimental or survey research: in these usual studies, the subjects have no opportunities to check the 'facts' as the researcher sees them. In contrast, the PA researcher checks the observations with those who have first-hand experience, before any reports are written; and the eventual results come from a sharing of findings and interpretations; and from a mutual development of underlying theory.

Should we be doing this with children, as the experts on environmental use? Could we? Robin Moore's studies (e.g. 1986, 1989), in essence, have adopted part of the approach. Activity observation and participation is backed up with discussions about the observations. Take as a sample topic the role of adults as 'gate-keepers' to places and resources:

> Most field trips gave the impression that children were treated with passive toleration by the large majority of unrelated adults they encountered. There were isolated examples of where adults and children had developed a strong rapport ... Sometimes, adult objections to children's play seemed justifiable, especially with relation to private property. But his was not always the case. (1986, p. 199).

Moore's discussion here of neighbours, nice and nasty, is full of supportive examples, some of which he directly observed. some of which were reported by 'his' children; but all of which bear the mark of discussion before inclusion in the final published record.

There are indeed some aspects of children's lives which are, by definition, so private and personal that the writer could only know of them through such a process of sharing. (Poignantly, Moore records a succession of children vouch-safing to him the whereabouts of their own favourite, private, even secret places: the *same* bush, bank or hideaway that other children in his study also regarded as their own secret place.)

These are the places which, for younger children, are usually missing from the re-search agenda. The daily routines and identifications of *adolescents* are, if anything, even less well documented. Natural landscapes are still used and valued, often as places to 'be alone', 'be oneself' (Owens, 1988). These, and more formal leisure settings (e.g. Noack & Silbereisen, 1988), are the frequently used settings for sociable and, later, dating activities. As Noack and Silbereisen's longitudinal studies show, reported place preferences during adolescence are clearly related to what they describe as 'progress in partnership development'. Participant observers definitely not welcome!

Comparative studies across a range of cities in different continents have been carried out by Lynch (1977) and his associates, in their 'growing up in cities' project. Riddick (1988) has studied the activity patterns of homeless 'street kids' in Hollywood (of whom she estimates there are some several thousand in any one year); and Anthony (1985) has described the important role shopping malls can play in the social and personal lives of city teenagers. These studies range, methodologically, from standard surveys to, in the case of the last, an ethnographical approach involving observation, detailed interviews, participation and subsequent interpretation. Anthony lays stress on the alternative, unofficial functions of the shopping mall as the main attraction to teenagers. Adults primarily visit the malls to make purchases; teenagers in her study regularly spend up to five hours at a time 'people watching', cruising around with friends, only occasionally using commercial facilities such as cafes and video game arcades. She suggested that

> One of the major attractions...is the lack of organised activities, structures and schedules. For many teenagers, the shopping mall may well serve as an antidote to the regimentation of school and home life. (p. 312).

The ethnographic approach used here surely offers us much more profound in-sights into adolescent life in general, and their use of this type of behaviour setting in particular than does, for example, the conventional attitude surveys of geograph-ical and market researchers (e.g. Smith *et al.*, 1979) whose discussion of use of shopping malls omits all consideration of the unofficial adolescent scripts, concen-trating only on 'official' economic behaviour. An ethnographic analysis of teenagers' places may well indicate that the familiar adolescent lament 'There's nothing to do' is in reality the complaint that in many neighbourhoods 'There's nowhere to do it'. It takes a stretch of the imagination to realize that 'hanging around' and 'doing nothing' need environmental support just as much as specific leisure activities such as swimming and bowling. Wood (1984) has even suggested the slogan 'kids hanging around doing nothing *make* a neighbourhood'.

Conclusion

Research on children and adolescents in the environment has been, as the following papers indicate, a challenge to environmental psychology. A challenge to its conceptualization of needs, and how these may be discerned and described in terms usable by the planner. A challenge also to its research methodology, in developing tools and approaches appropriate to the expressive capacities of the particular age group under study.

And the findings of environmental psychology in addition one hopes to being *useful* to planner and child alike, may also be seen as a challenge to some orthodox thinking within education and within developmental psychology about the competencies of the child, who manifestly is a more skilled user of the environment than is often acknowledged.

References

Adams, G. & Ward, C. (1982). *Art and the Built Environment.* Burnt Mill, Essex: Longman for the Schools Council.

Altman, I. & Low, S. M. (Ed.). (1992). *Place Attachment.* New York: Plenum.

Anthony, K. H. (1985). The shopping mall: a teenage hangout. *Adolescence, 20,* 307–312.

Appleyard, D. (1970). Elitists versus the public's cry for help. *Landscape Architect, 61* (24–25).

Barker, R. G. (1978). *Habitats, Environments and Human Behaviour.* San Francisco: Jossey-Bass.

Berry, J. W. (1971). Ecological and cultural factors in spatial perceptual development. *Canadian Journal of Behavioural Science, 3,* 324–336.

Blades, M. & Spencer C. P. (1994). The development of children's ability to use spatial representations. In H. M. Reese, Ed., *Advances in Child Development and Behaviour.* New York: Academic Press.

Bronfenbrenner, U. (1979). *The Ecology of Human Development.* Cambridge: Harvard.

Chawla, L. (1992). Childhood place attachments. In I. Altman & S. M. Low, Eds., *Place Attachments.* New York: Plenum.

Cooper Marcus, C. (1992). Environmental memories. In I. Altman & S. M. Low, Eds., *Place Attachments.* New York: Plenum.

Cornell, E. H. & Hay, D. H. (1984). Children's acquisition of a route via different media. *Environment and Behaviour, 16,* 627–641.

Coulson, T. (1980). Space around the home. *Architect's Journal,* (24 December), 1245–1260.

Eriksen, A. (1984). The play's the thing. *Landscape Architect, 74,* 72–77.

Garbarino, J. (1985). Habitats for children: an ecological perspective. In J. F. Wohlwill & W. v. Vliet, Eds., *Habitats for Children.* Hillsdale, NJ: Erlbaum.

Gärling, A. & Gärling, T. (1990). Parents' residential satisfaction and perception of children's accident risk. *Journal of Environmental Psychology, 10,* 27–36.

Gibson, J. J. (1979). *The Ecological Approach to Visual Perception.* Boston: Houghton Mifflin.

Hart, R. A. (1979). *Children's Experience of Place: a developmental study.* New York: Irvington Press.

Heft, H. & Wohlwill, J. F. (1987). Environmental cognition in children. In D. Stokols & I. Altman, Eds., *Handbook of Environmental Psychology.* New York: Wiley.

Kaplan, R. (1978). Participation in environmental design. In S. Kaplan & R. Kaplan, Eds., *Humanscape.* Belmont: Wadsworth.

Liben, L. S. (1982). Children's large-scale spatial cognition. In R. Cohen, Eds., *New Directions for Child Development: Children's Concept of Spatial Relationships.* New York: Academic Press.

Matthews, M. H. (1992). *Making Sense of Place: children's understanding of large-scale environments.* Hemel Hempstead: Harvester.

McKechnie, G. E. (1977). Simulation techniques in environmental psychology. In D. Stokols, Eds., *Perspectives on Environment and Behaviour.* New York: Plenum.

Moore, R. (1986). *Childhood's Domain*. London: Croom Helm.

Moore, R. (1989). Plants as play props. *Children's Environments Quarterly*, **6**, 3–6.

Moore, R. & Hart, R. (Issue editors) (1980). Participation. *Childhood City Newsletter*, (22), entire issue.

Moore, R. & Hart, N. (Issue editors) (1981). Participation 2: a survey of projects, programs and organisations. *Childhood City Newsletter*, (23), entire issue.

Moore, R. & Hart, R. (Issue editors) (1982). Participation 3: Techniques. *Childhood City Newsletter*, (Double Issue 9(4) & 10(1)).

Noack, P. & Silbereisen, R. K. (1988). Adolescent development and the choice of leisure settings. *Children's Environments Quarterly*, **5**, 25–33.

Owens, P. E. (1988). Natural landscapes, gathering places and prospect refuges: characteristics of outdoor places valued by teens. *Children's Environments Quarterly*, **5**, 17–24.

Piaget, J. (1929). *The Child's Conception of the World*. London: Kegan Paul.

Piaget, J. (1930). *The Child's Conception of Physical Causality*. London: Kegan Paul.

Piaget, J. & Inhelder, B. (1956). *The Child's Conception of Space*. London: Routledge & Kegan Paul.

Piaget, J., Inhelder, B. & Szeminska, A. (1960). *The Child's Conception of Geometry*. London: Routledge & Kegan Paul.

Riddick, S. (1988). Debunking the dream machine: the case of street kids in Hollywood, California. *Children's Environments Quarterly*, **5**, 8–16.

Sandels, S. (1975). *Children in Traffic*, London: Elek.

Self, C. M., Gopal, S., Golledge, R. G. & Fenstermaker, S. (1992). Gender-related differences in spatial abilities. *Progress in Human Geography*, **16**, 315–342.

Sheat, L. G. & Beer, A. (1989). User participation—a design methodology for school grounds design and environmental learning. *Children's Environments Quarterly*, **6**, 15–30.

Siegal, A. W. (1982). Towards a social ecology of cognitive mapping. In R. Cohen, Ed., *New Directions in Child Development: Children's Conceptions of Spatial Relationships*. San Francisco: Jossey Bass.

Smith, G. C., Shaw, D. J. B. & Huckle, P. R. (1979). Children's perceptions of a downtown shopping centre. *Professional Geographer*, **31**, 157–164.

Spencer, C. P. (1992). Life span changes in activities, and consequent changes in the cognition and assessment of the environment. In T. Gärling & G. W. Evans, Eds., *Environment Cognition and Action: an integrated approach*. New York: Oxford University Press.

Spencer, C. P. & Bishop, J. (Eds.) (1989). Environmental Education and Children's Environmental Learning. *Children's Environments Quarterly*, **6** (Special Double Issue 2/3).

Spencer, C. P., Blades, M. & Morsley, K. (1989). *Children in the Physical Environment*. Chichester: Wiley.

Stea, D. (1985). From environmental cognition to environmental design. *Children's Environments Quarterly*, **2**, 22–26.

Sutton, S. E. (1985). *Learning through the built environment*, New York: Irvington.

Torrell, G. (1990). *Children's Conception of Large Scale Environments*. Göteberg: Göteberg University Press.

Ward, C. (1977). *The Child in the City*, London: Architectural Press.

Ward, C. (1988). *The Child in the Country*, London: Robert Hale.

Weinstein, C. S. & David, T. G. (Ed.). (1987). *Space for Children: the built environment and child development*. New York: Plenum Press.

Whyte, W. F. (1991). *Participatory Action Research*. Newbury Park: Sage.

Williams, J. & Best, D. (1982). *Measuring Sex Stereotypes: a thirty nation study*, Beverly Hills, CA: Sage.

Wohlwill, J. F. & van Vliet, W. (Eds). (1985). *Habitats for Children: the impacts of density*. Hillsdale, NJ: Erlbaum.

Wood, D. (1984). A neighbourhood is to hang around. *Children's Environments Quarterly*, **1**, 29–35.

THE CASE FOR DEVELOPING A COGNITIVE ENVIRONMENTAL PSYCHOLOGY THAT DOES NOT UNDERESTIMATE THE ABILITIES OF YOUNG CHILDREN

CHRISTOPHER SPENCER AND ZHRA DARVIZEH

Department of Psychology, University of Sheffield, Sheffield S10 2TN, U.K.

Abstract

Both developmental and environmental psychology, where influenced by the work of Piaget, tend to have underestimated the environmental skills and potential of young children. This paper argues that what is needed is a cognitive environmental psychology which derives from observations of young children's real-life behaviour in their everyday environment, as well as from the laboratory-based studies which predominate in the field. Much of the present evidence on young children's capabilities in cognitive mapping and other locational skills comes from studies by geographers and educationalists; whereas psychologists have concentrated in their small-scale laboratory studies upon the processes whereby the child learns and uses information about the physical environment. Combining findings from both laboratory and large-scale settings, it is now possible to describe a developmental sequence; although many more naturalistic as well as experimental studies are needed.

Although empirical studies of environmental cognition must now be numbered in hundreds (Moore, 1979) and some proportion of these have been devoted to onto-genetic developmental changes in environmental cognition, it is the argument of the present paper that, despite the existence of these studies, there is a considerable under-estimation of the environmental skills and potential of young children and expecially those below five years of age. What is needed is a cognitive environmental psychology geared to such an age group and its real-life behaviour in the everyday environment; and not just a predominantly laboratory-based one which treats young children as imperfect apprentice adults.

Theoretical expectations of what young children are capable of have been influenced by Piagetian developmental psychology, and, in particular, by Piaget and Inhelder's (1967) use of the 'three mountains experiment' to demonstrate the young child's ego-centricity in perspective taking. Piaget and Inhelder further suggest that the reproduction of a model landscape layout is dependent upon the co-ordination of perspectives which they find is usually achieved at about nine to ten years old (their sub-stage IIIb); and that formal mapping operations are not fully attainable until about 11–12 years (their Formal Operations stage). However, as Hart (1979) and others have argued, much of the younger child's apparent difficulty is occasioned by the mode of presentation of the test material; and a number of developmental psychologists broadly within the Piagetian tradition have, by using altered methodologies, shown that young children do not necessarily view the world as egocentrically as Piaget and Inhelder have stated (e.g. Huttenlocher and Presson, 1973; Liben, 1978; Kurdek, 1978).

There has been rather less tendency to underestimate the young child's capacity to structure environmental information on the part of some psychologists from other traditions, and by some behavioural geographers. The cognitive psychologist, Neisser (1976) for example, argues that very young children and even babies show by their actions in everyday environments that they possess what he calls 'orienting schema': cognitive maps of those portions of the local environment through which they travel, and containing the permanent and recent locations of objects important to them. Neisser is clear about the methodological problems of working with young children:

> 'Cognitive maps are defined by information pick-up and action, not by verbal description. Travel is one thing, travelogue another. A child can find his way long before he can give an adequate account of where he has been or how he got there.'

Thus, although the cognitive maps of adults have been investigated by getting subjects to externalize them by drawing sketch maps ever since Lynch's (1960) pioneering research, such exercises are much less appropriate for children who lack the drawing skills to produce a map of sufficient clarity and accuracy to represent their environmental knowledge. [The extent to which such 'graphicacy' is an intervening factor in the mapping exercises of *adults* has often been conveniently ignored by investigators, but can quite easily be demonstrated (see Murray and Spencer, 1979).] Considerable ingenuity and patience are needed in developing and working with alternative means of investigating the environmental cognitions of young children; and the remainder of the paper will discuss studies which have evolved a great variety of such techniques.

Have We Underestimated Young Children's Environmental Cognition?

Some of the strongest claims for young children's possessing relatively complex environmental cognitions have been made by geographers (e.g. Blaut *et al.*, 1970; Piché, 1977; Catling, 1979).

Blaut and Stea (1971) showed children of five years and older aerial photographs, and, finding them able to interpret several geographical features in each, have argued that children are able to engage in the fundamental processes needed for map making and map reading (namely, rotation from horizontal to vertical views, reduction in scale, and abstraction to semi-iconic signs) long before they are exposed to traditional maps. Children who have never interacted with large spaces can nevertheless interpret the aerial properties of such spaces, and solve mapping problems concerning them. Blaut and Stea suggest that 'we can explain these abilities only if we assume that a very highly evolved cognitive map has already been formed in many children by the age of five'.

Experiences contributing to such a map are likely to include not only the child's active movement through space (the mobile observer in a stationary environment) but also the stationary spectator in the 'mobile environment' afforded by film and television. (Note, however, that Blaut and Stea were able to show a level of interpretation of aerial photographs by Puerto Rican children which was comparable to that of their mainland U.S.A. children, although the exposure to the visual media of the former group was reported to be limited.) The child's toy play is also claimed to be an important experience, enabling as it does active control over a miniature environment. The child chooses the positions of toy versions of objects relative to each other, and himself moves through and above this world, continually changing perspectives. Blaut and Stea stress the role of toy play: it gives the child experimental control over what would otherwise be, to him, impossible environments and view-points. Clearly, such play is characteristic of children much younger than five years of age; in most cultures, adults provide some kind of toy models of everyday objects for children, or children improvise them for themselves; and children from an early age can be observed delighting in organizing and moving toys, and in repeatedly changing perspectives. One pattern we have frequently observed is the child, having played with, for example, a model car by kneeling over it to push it, then lies down flat to observe the car moving past at eye level. An accumulation of many such switches from plan to elevation, and from distant to close-up viewpoints must surely develop interpretative skills which could be translated to the full-scale environment, and representations of it. We need further empirical work to test whether such experience manipulating self and *objects* to get different perspectives does actually transfer to interpreting relations about largely non-manipulable spatial layouts.

It is indeed possible to demonstrate the ability to interpret aerial photographs in children as young as three years of age (Spencer *et al.*, 1980): children who in Piagetian terms are clearly at a pre-operational stage, yet who found few problems in inter-preting photographs comparable with those used by Blaut and Stea. In this study, all 30 children in the sample, without hesitation or need for prompting interpreted the photographs as being in some way geographical ('There are roads and houses'; 'It's a town', etc.) rather than as, for example, a pattern of lines or shapes. The children used the context then to interpret further: e.g. 'If these are roads, then are these cars?' None of the children claimed having seen such photographs before, or could explain how they could have been made (in contrast with primary school children, many of whom would imply an aerial perspective by mentioning an aeroplane). Revealingly, many of the three-year-old children in the Spencer *et al.* study referred to railways in one of the photographs by the proprietory name of the toy train tracks that all regularly played with in the nursery school they all attended: further indication that Blaut and Stea may well be right to stress the importance of toy play in the construction of geographical images.

How significant are such findings? Piché (1977) has argued that such decoding and recognition is a relatively easy task, based upon the principles of object perception, compared with the kind of encoding and reconstruction involved in the mapping exercises, which are dependent upon higher logical operations. She is more cautious than Blaut and Stea when describing the child's abilities *vis-à-vis* the large scale environment:

> That children can find their way in the neighbourhood does not mean that they have completed their representation of it . . . it is possible that they only imagine the roads very concretely, like corridors, and that their displacements from them are guided by perceptual cues.

Piché's own experimental work with primary school children would seem to bear this out: she describes a series of cognitive stages through which the child passes before he develops a continuous, spatial image of the neighbourhood. These stages, starting with particular routes, then linking routes, and eventually building a neighbourhood map upon this network of routes, is very reminiscent of several accounts which have been given of the stages an adult may go through when rapidly developing his cognitive map of a novel area (e.g. Appleyard, 1970; Gittins, 1969; Spencer and Weetman, in press): a case of ontogenesis prefiguring microgenesis?

Even adopting the cautious theoretical position advanced by Piché, the performance of her youngest (i.e. five-year old) subjects demonstrates a degree of geographical competence which would not be predicted from an orthodox Piagetian view.

A further strong piece of evidence that young children's ability has been under-estimated is given by Bluestein and Acredolo (1977). Three- to five-year old children were asked to identify and find items in a room using a simple map under a variety of conditions. Most three-year olds were able to perform these rudimentary map reading exercises when the map was correctly aligned inside the room; by four-years old the child could typically retain and use the map image when he had been shown it, correctly aligned, before entering the room; and that five-year olds were not even dependent upon a correct alignment of the map. Thus, it seems that young children are able to comprehend simple drawn maps, and to use an image of them to problem solve in at least the immediate situation.

Small-scale, Analytical Studies of Young Children's Abilities

Many of the studies which have investigated age-changes in children's locational and geographical competence have deliberately placed their subjects in limited environ-ments—a room, a corridor—and studied one selected aspect of the child's capabilities. Acredolo et al. (1975), for example, had children retrace their route along a corridor, and asked them to locate the point at which the experimenter had dropped her key ring on their previous walk together. They demonstrated that the presence of dis-tinguishable landmarks in the corridor considerably aided recognition; and conclude that 'four-year old children are sensitive to the topological relationships of a neighbourhood, to the point that, given landmarks, their performance is as good as that of eight-year olds'.

In another series of such studies, Kosslyn et al. (1974) showed that four- and five-year old children were able to make systematic distance judgements from every location to every other location within an experimental room, although they had previously had direct experience with only a few of the possible relations. These authors argue that their data provided further support for the existence of usable cognitive maps in young children. Hardwick et al. (1976) set their subjects a triangulation task within a room familiar to the children. They showed that their youngest subjects, six-year olds, possessed very accurate, coherent cognitive maps of this familiar, limited space as conventionally experienced; but found that the six-year olds performed less well than 11-year olds when the experimenter asked them to imagine either the room rotating (mental rotation) or themselves moving within the room (perspective taking). Their conclusion was that, at least on such tasks, there was a considerable increase with age in the accuracy and completeness of the mental manipulation. The six-year olds, unlike the 11-year olds, found especial difficulty with the transformation of ordinal spatial relationships.

Hardwick et al., having presented their experimental findings, then speculate about the implications for the child's use of the large scale environment. They think that, in the broader environment, young children may be unable to abstract what Hardwick et al. describe as 'general level representations'—summaries of the spatial information contained in all the separate specific perspectives with which the individual is familiar.

> If this is the case, then young children would be very limited in what they could do with their cognitive maps ... and (this) would severely limit their ability to plan spatial routes, to communicate directions to others, or to take an imagined perspective within the environment.

Further evidence to suggest that the ability to shift between spatial frames of reference develops considerably between 6 and 11 years of age comes from Cohen *et al.* (1979). In yet another limited-environment and novel-task experiment, children had to estimate the distances between, in the novel condition, six pieces of furniture in an unfamiliar room, or, in the familiar condition, six points in their school library. The younger children could as easily reconstruct the setting—whether familiar or novel—as could the older; but they were significantly less accurate when the tasks involved the rescaling or re-orientating of their representation of the room they had just seen.

It should be noted, however, that such tasks involve more than just the spatial abilities that the authors focus on. Herman (1980) has suggested that the speed of acquisition and the storage of spatial information—both of which can be demonstrated to be age-related—might explain some of the improvement in performance in such supposedly 'cognitive mapping tasks'. In his experiment, children of five- and six-years, and of eight- and nine-years, saw a very simple, eight item, model town; and were then asked to reconstruct it from memory. The older children were more accurate in these reconstructions, regardless of whether the child had viewed the model by walking round its perimeter, by walking along its road layout, or by wandering backwards and forward at will. (It was, incidentally, found that all children developed more accurate maps when their movement and attention were directed by their experimenter than when they wandered at will.)

Herman deduces that the superiority in task performance of the older children must be a function of the speed of acquisition and storage of spatial information, because in other respects older and younger children behaved similarly during the learning phase of the experiment: 'When left to explore the town freely, both groups of children took about the same amount of time and entered about the same number of quadrants.' He admits, however, that the children may have differed in terms of their awareness of the task demands; and that the older children might have developed better strategies for externalizing their mental maps for the experimenter. [That older and younger children use rather different strategies for *searching* their environment has been shown by Wellman *et al.* (1979).]

Laboratory Tests may Underestimate Performance by Young Children in the Familiar, Large-scale Environment

We suggest that laboratory-based and other small-scale tests, such as those reported above, although invaluable in investigating aspects of the development of spatial cognition, tend to underestimate quite how efficient are young children as users of the 'real-world', familiar environment. It should be noted, however, that laboratory-based investigations tend to have as their aim the analysis of the *process* by which some spatial behaviour is carried out, rather than the investigation of the *absolute competencies* of children. Characteristic of such studies are tasks dependent upon relatively complex verbal instructions, or some relatively novel or unusual behaviour, such as learning to align a sighting tube (as in Hardwick *et al.*, 1976); and younger children's performance on the tasks, although showing some degree of spatial or locational competence, may well be depressed relative to that of older children because of intervening factors such as comprehension of the task, attention span, etc.

Furthermore, most of these studies fail to set the child's performance in the test situation against the context of the everyday environmental competence typically exhibited by children of the same age. More than anecdotal evidence suggests that, in some cases at least, such everyday competence is impressive; and that we are in danger of underestimating it:

> Most mornings, thousands of 6-year-old children will find their way to school. There will be protections along the way, but no-one is particularly worried that the children will get lost. They will find their way to school (often taking varying routes), within the school to their classrooms, after school, perhaps to a friend's house, and then, home. However, most 3-year olds do not typically, or are not allowed to, engage in the same process. Is it possible that such very young children can recognize landmarks in a large-scale terrain, but have limited knowledge of routes, and that adults are aware of this? (Siegal, 1977)

Siegal, in answer to his own question, finds that the available data is sparse; but suggests that the evidence indicates the following sequence:

(1) Initially, the young child notes and remembers *landmarks*: the core of spatial representation is the visual recognition of memory landmarks; and from this core, memory-in-context is derived (e.g. Smothergill, 1973).

(2) Once landmarks are established, the child's acts are registered and assessed with relation to them (e.g. Zaphorzets, 1965): they provide a basis for a *temporal sequence of routes and landmarks* (e.g. Piaget and Inhelder, 1967).

(3) Next, the child forms clusters of landmarks, 'minimaps': internally coherent clusters of elements, which may well lack any correct relationship with other clusters (e.g. Siegal and Schadler, 1977).

(4) The formation of some kind of *objective frame of reference* organizing the separate perspectives into a system in space. [Siegal here cites literature indicating this may have happened by 8 years of age: e.g. Coie *et al.* (1973).]

(5) Full *survey representations* appear only after the establishment both of routes and an objective frame of reference (e.g. Shemyakin, 1972).

It is significant that, although Siegal speculates on the abilities of three-year olds, the literature which he reviews, and bases the above sequence upon, makes the presumption that efficient spatial cognition only develops after about six years of age; and has therefore not bothered to investigate how younger children cope with their world.

Indeed, many of the researchers take what might be called an adulto-morphic view of spatial cognition: how far does the child approximate to an adult's knowledge about structuring of and strategies for using the environment? But, as Rapoport (1976) has argued with relation to cross-cultural differences, there are more ways than one of giving meaning to the world: thus, for example:

> It may well be that the cognitive categories and orientation system used by Australian Aborigines are as valid and *potentially* useful as ours, and their way of defining place

through purely symbolic and conceptual means may be even more useful: at the very least, it shows the extreme range of place definition of which people are capable.

For one discussion of these aboriginal systems, and the high accuracy of orientation they facilitate [see Lewis (1976)]. The investigator working with very young children needs to be as sensitive as the anthropologist to the possibility that other 'cultures' have their own systems for cognizing the environment. Moore (1979), in a major review of the environmental cognition literature, found the role of cultural variables to be 'grossly understudied'; and we would wish to extend his observation to cover the ontogenesis of spatial cognition.

What Studies are Needed?

What are needed are techniques of investigation which are appropriate to the interests and real-life activities of young children. Furthermore, to look at the *potential* of younger children, as opposed to their average achievements, we suggest that some attention be given to those groups of young children whose local spatial exploration has been extensive, and not hedged in by adult constraints. (The writers have in mind, for example, children living in village communities, or the populations of young children who live in and move freely around certain pedestrianized housing areas in cities.) Another way of investigating potential is to explore the local environment in as far as it is known to the child and systematically extend it by taking him on trips to novel places or using unfamiliar routes, in order to study the child's expanding environmental knowledge. (Our own current investigations with three- to five-year olds, to be reported in future papers, indicate the fruitfulness of such techniques, and support our conviction that average children of this age can quickly learn and describe novel routes, can construct maps of familiar and newly learned areas, and are eccentrically observant of the city, whether directly experienced or presented as a filmed walk through the city.)

It is the argument of the present paper that young children have been underestimated in their cognitive abilities—or potentialities—with relation to the environment both by developmental psychologists and by educationalists (e.g. Satterly, 1964; D.E.S., 1978; but also see Catling, 1979). Such underestimation, we have argued, stems partly from the theoretical perspectives some investigators have brought to their experiments; and partly from the modes of investigation used. Although laboratory-based experiments may be attractive because they are apparently more rigorous, we feel that the sheer convenience of the laboratory has also been a factor in its prevalance in the literature. Compare the time investment required to put a score of children through an experimental task with that which would be needed to record and analyse the same group of children's analogous natural exploration and use of their local environment. One of the classic books of the ecological psychology tradition (Barker and Wright, 1966), comprises the meticulously detailed and interpreted account of the activities of one eight-year old boy in one day of his life: and took several years to analyse and write-up. Whilst we are not arguing that the ecological psychologists' *methodology* would be appropriate for the kind of *cognitive* environmental psychology of children advocated, we are suggesting that their degree of concern and awareness of the child's naturally occurring behaviour should be its starting point. Such observations, akin to the species descriptions of early ethology,

would then generate hypotheses about process which could be tested in the everyday
environment by carefully contrived experiments, akin to the modifications Tinbergen
(1951) wrought in the natural habitat of his studied species.

Which groups of children would provide the best subjects for such a programme
of research? As already indicated, within any particular age group, one should expect
a wide variation to exist in the amount and nature of environmental experience gained:
indeed, such variety might well form the basis of a series of experimental comparisons.
Not only do young children vary considerably in the amount of freedom to explore
accorded them by parents; but also one must consider the variety of 'habitats' that they
could potentially experience. Some, like Goodman (1961), feel that the modern city
child effectively inhabits an impoverished environment:

> The city, under inevitable modern conditions, can no longer be dealt with practically
> by children, because concealed technology, family mobility, loss of the country, loss
> of neighbourhood tradition, and eating up of the play space have taken away the
> real environment.

Against this point of view, one can set the powerful essay by Ward (1977) which argues
that the city as experienced is not such an inhuman or impoverished place for
children. Nonetheless, Ward does admit that most city children have considerably less
freedom—and thus potential for environmental learning—than have the children from
smaller communities. The recent cross-national studies by Lynch et al. (1977) on
adolescents growing up in cities also point to the same conclusion.

Thus, we have argued that to study the *potential* of young children to learn and
structure their environment, and to navigate successfully through it, studies should
include populations of children who have considerable freedom and range extensively
around their locality.

In this respect, Hart (1979) chose well when he selected the younger children of a
small New England community, 'Innavale', to study children's natural experience of
place. In contrast to the laboratory studies cited earlier, in which each child typically
spends a matter of minutes with a strange experimenter performing an unusual task,
Hart spent two years in day to day contact with his subjects, in child-guided field trips
around the children's usual range: a kind of tracking, in which the investigator
enters into close collusion with the child as he goes about his daily play, errands and
other journeys throughout the town and surrounding country. A study such as this
indicates the great extent to which children, following their own concerns, become
environmental experts, developing very effective mental maps of their area, and an
accompanying sub-cultural (and often personal) account and evaluation of the various
'places' scattered throughout the area. His natural history of the children's spatial
behaviour indicates that young children are highly selective users of neighbourhood
space, some areas being practically unvisited by any children, others being their own
private places, and the focus of much activity.

Hart argues that when children develop their ranges, what in essence they do is to
find or make a series of places, and connect them with a network of pathways, some
of almost a ritual nature in their importance to children. (Even quite a small modification can serve to create a 'place' for children, although Hart notes that the younger
children tended to use found rather than created places.)

Even in such a community, the investigator quickly notes that the child's home
range is one delimited by parental controls (although Hart also notes the regular trans-

gressions of these limits made, especially by the boys in his sample, which are tacitly acknowledged by children and parents alike). Nonetheless, we argue that in such a setting, the cognitive environmental psychologist is much more likely to appreciate the potential of the young child in comprehending and using his environment than he is in his laboratory.

What is of importance to the child may well, as Hart's study shows, include aspects which are simply not reproducible in the laboratory: for example, the importance of natural objects to the child. One other study which has illustrated this point is by Moore (1973), who investigated the elements of place significant to children; asked his subjects to map or draw 'all your favourite places'; and categorized their responses in terms of frequency of mention. Homesite was, not unexpectedly, the most frequently mentioned, followed by references to people and to vegetation, with particular pathways receiving many mentions. Altogether, natural features accounted for just over a quarter of the aggregate rate of mentions.

Young children, because they lack the verbal and graphical skills that are relied upon in many cognitive environmental psychology investigations, tend to be an understudied and, worse, an underestimated population. Yet, as the present paper has argued, competence in handling and using environmental information does not suddenly emerge in those years of middle childhood which many investigators have found easiest to concentrate upon: the young child, either from surrogates such as toy play or from direct experience with the environment, is developing a useful but often very personal and idiosyncratic set of schema for handling further information about the environment. As Neisser observed (1976) the young child has developed much environmental competence long before he can give an account of where he has been or how he gets there; and it is up to us to discover ways in which he can manifest his skills and the processes underlying them.

References

Acredolo, L. P., Pick, H. L. and Olsen, M. G. (1975). Environmental differentiation and familiarity as determinants of children's memory for spatial location. *Developmental Psychology*, **11**, 495–501.

Appleyard, D. (1970). Styles and methods of structuring a city. *Environment and Behavior*, **2**, 100–118.

Barker, R. G. and Wright, H. F. (1966). *One Boy's Day*. New York: Harper and Row.

Blaut, J. M., McCleary, G. S. and Blaut, A. S. (1970). Environmental mapping in young children. *Environment and Behavior*, **2**, 335–349.

Blaut, J. M. and Stea, D. (1971). Studies in geographical learning. *Annals of the Association of American Geographers* **61**, 387–393.

Bluestein, N. and Acredolo, L. P. (1977). Developmental changes in map reading skills. E.R.I.C. Document No. ED 154902.

Catling, S. (1979). Maps and cognitive maps: the young child's perception. *Geography*, **64**, 288–296.

Cohen, R., Weatherford, D. L., Lomenick, T. and Koeller, K. (1979). Development of spatial representations: role of task demands and familiarity with the environment. *Child Development*, **50**, 1257–1260.

Coie, J. D., Costanzo, P. R. and Farnhill, D. (1973). Specific transitions in the development of spatial perspective-taking ability. *Developmental Psychology*, **9**, 167–177.

Department of Education and Science (1978). *Primary Education in England*. London: H.M.S.O., paragraphs 5.136 and 5.14.

Gittins, J. S. (1969). Forming impressions of an unfamiliar city: a comparative study of aesthetic and scientific knowing. Unpublished M.A. Thesis, Clark University, Worcester, Mass.

Goodman, P. (1960). *Growing Up Absurd*. New York: Random House.

Hardwick, D. A., McIntyre, C. W. and Pick, H. L. (1976). The content and manipulation of cognitive maps in children and adults. *Monographs of the Society for Research in Child Development*, **41(3)**, Serial No. 166.

Hart, R. (1979). *Children's Experience of Place*. New York: Irvington.

Herman, J. F. (1980). Children's cognitive maps of large-scale spaces: effects of exploration, direction and repeated experience. *Journal of Experimental Child Psychology*, **29**, 126–143.

Huttenlocher, J. and Presson, C. C. (1973). Mental rotation and the perspective problem. *Cognitive Psychology*, **4**, 277–299.

Kosslyn, S. M., Pick, H. L. and Fariello, G. R. (1974). Cognitive maps in children and men. *Child Development*, **45**, 707–716.

Kurdek, L. A. (1978). Perspective-taking as the cognitive basis of children's moral development: a review of the literature. *Merrill-Palmer Quarterly*, **24**, 3–28.

Lewis, D. H. (1976). Observations on route finding and spatial orientation among the aboriginal peoples of the Western Desert Region of Central Australia. *Oceania*, **56**, 249–282.

Liben, L. S. (1978). Perspective-taking skills in young children: seeing the world through rose-coloured glasses. *Developmental Psychology*, **14**, 87–92.

Lynch, K. (1960). *The Image of the City*. Cambridge, Mass.: M.I.T. Press.

Lynch, K. (ed.) (1977). *Growing Up in Cities: Studies of the spatial environment of adolescence in Cracow, Melbourne, Mexico City, Salta, Toluca and Ivarszawa*. Cambridge, Mass.: M.I.T. Press, with UNESCO.

Moore, G. T. (1973). Developmental differences in environmental cognition. In W. F. E. Prieser (ed.) *Environmental Design and Research*, Volume 2. Stroudsburg, Pa: Dowden, Hutchinson and Ross.

Moore, G. T. (1979). Knowing about environmental knowing: the current state of theory and research on environmental cognition. *Environment and Behavior* **11**, 33–70.

Murray, D. and Spencer, C. P. (1979). Individual differences in the drawing of cognitive maps: the effects of geographical mobility, strength of mental imagery and basic graphic ability. *Transactions of the Institute of British Geographers*, New Series, **4**, 385–391.

Neisser, U. (1976). *Cognition and Reality: Principles and Implications of Cognitive Psychology*. San Francisco: Freeman.

Piaget, J. and Inhelder, B. (1967). *The Child's Conception of Space*. London: Routledge.

Piché, D. (1977). *The Geographical Understanding of Children Aged 5 to 8 years*. London: London School of Economics, Unpublished Ph.D. thesis.

Rapoport, A. (1976). Environmental cognition in cross-cultural perspective. In G. T. Moore and R. G. Golledge (eds.), *Environmental Knowing: Theories, Research and Methods*. Stroudsburg, Pa.: Dowden, Hutchinson and Ross.

Satterly, D. J. (1964). Skills and concepts involved in map drawing and map interpretation. *New Era*, **45**, 263.

Shemyakin, F. N. (1962). Orientation in space. In B. G. Ananyev (ed.) *Pychological Science in the USSR*. Volume 1, Part 1, U.S. Office of Technical Reports, No. 11466, 186–255.

Siegal, A. W. (1977). Finding one's way around the large-scale environment: the development of spatial representations. In H. McGurk (ed.), *Ecological Factors in Human Development*. Amsterdam: North Holland.

Siegal, A. W., Kirasic, K. C. and Kail, R. V. (1978). Stalking the elusive cognitive map. In I. Altman and J. F. Wohlwill (eds.) *Children and the Environment*. New York: Plenum Press.

Siegal, A. W. and Schadler, M. (1977). Young children's cognitive maps of their classroom. *Child Development*, **48**, 388–394.

Smothergill, D. W. (1973). Accuracy and variability in the localization of spatial targets at three age levels. *Developmental Psychology*, **8**, 62–6.

Spencer, C. P., Harrison, N. and Darvizeh, Z. (1980). The development of iconic mapping ability in young children. *International Journal of Early Childhood*, **12(2)**, 57–64.

Spencer, C. P. and Weetman, M. (In Press). The microgenesis of cognitive maps: a longitudinal study of new residents of an urban area. *Transactions of the Institute of British Geographers*.

Tinbergen, N. (1951). *The Study of Instinct*. Oxford: Oxford University Press.

Ward, C. (1977). *The Child in the City*. London: Architectural Press.

Wellman, H. M., Somerville, S. C. and Haake, R. J. (1979). Development of search procedures in real-life spatial environments. *Developmental Psychology*, **15**, 530–542.

Zaphrozets, A. V. (1965). The development of perception in the pre-school child. *Monographs of the Society for Research in Child Development*, **30**, Serial No. 100, 82–101.

POINTING TO PRESCHOOL CHILDREN'S SPATIAL COMPETENCE: A STUDY IN NATURAL SETTINGS

ALISON M. CONNING and RICHARD W. BYRNE

Psychological Laboratory, University of St Andrew's, St Andrew's, Fife, KY16 9JU, Scotland

Abstract

This paper describes a series of experiments which explore the extent to which children aged three years five months to four years seven months can build a Euclidean mental representation of the environment. It was found that the children showed a higher level of spatial competence than previous research had indicated, and that their spatial ability was dependent upon the nature of the environment. Euclidean knowledge was found at all ages in a familiar environment, but especially when familiarity was brought about through self-exploration. Euclidean knowledge was least likely in a novel environment. Previous experience with the test in the familiar environments did not lead to an increase in Euclidean knowledge in the novel environment. The results are interpreted in terms of Byrne's 'network-map'/'vector-map' theory of spatial knowledge.

Introduction

Until recently, work on preschool children's spatial knowledge has largely been based upon the elaboration of Piaget's developmental sequence (Piaget *et al.*, 1960; Piaget and Inhelder, 1967) by Siegel and White (1975). According to this, the child first acquired knowledge of unbranched strings of locations which are encountered along a route ('route knowledge'). Later, clusters of objects whose locations are known relative to each other are built up ('minimaps'); relative positions of minimaps are not known. Finally, a co-ordinated frame or 'survey map' is acquired, linking many minimaps. This theory is vague in various places. Does route knowledge include vector information or is it only topologically equivalent to actual routes? Are minimaps any different from small-scale survey maps? Indeed, are survey maps to be taken as more than branched and connected route knowledge or are they intended to be Euclidean in nature? Siegel and White (1975) use the example of a subway map, which is largely topological (but has some Euclidean properties), so this is unclear. Perhaps the difficulty is that their theory is largely a redescription of the categories of maps which children *draw* at different ages, and as such says more about their drawing abilities or lack of them than their mental representation of large-scale space.

The present authors prefer to use a different categorization, which concentrates on possible mental represenations and their intrinsic properties. This has two main divisions.

(a) 'Network-maps' (Byrne, 1979, 1982): these are branching networks, each string of which is a linear sequence of locations. Each branch-point corresponds to a junction, and other locations are coded along the strings. These maps do not encode knowledge about the distance between locations nor the precise angle at which routes join, only the order of location and branches, and are thus entirely topological. Each string of a network-map is conceptualized as a program for action, whose execution would enable travel along the particular route: each step is in the form 'continue until location X, when do Y'. As such, steps would not be expected to be reversible, so a route *back* would entail another string to be added to the network. A network-map encoding information about only *one* unbranched route would partly resemble the route knowledge of Siegel and White. However, no vector information could be included, and it may therefore be safer to use a different term: 'string-map'. One question which follows is whether two or more intersecting string-maps may remain independently stored in memory, or whether, as in Byrne's theory, they are automatically integrated into one network-map.

(b) 'Vector-maps' (Byrne, 1979, 1982): these are representations which include vector information, that is, knowledge about the distance between locations, and their relative bearings. They therefore show isomorphism to the layout of the real world, although may of course be distorted or inaccurate.

The present experiment examines preschool children's spatial knowledge in the light of Byrne's network-map/vector-map theory of representation. Since the presence or absence of vector information is crucial to this theory, the study concentrates on children's ability to point to out of sight locations. If the subjects respond by pointing along the path they had just walked down, this is indicative of a string-map/network-map form of encoding. In contrast, pointing the crow-flight direction to an object implies vector-map knowledge, the accuracy of which can then be ascertained.

For other reasons, direction estimates have recently been used as a tool for investigating spatial knowledge (for example, Hardwick *et al.*, 1976; Piché, 1977; Anooshian and Young, 1981; Lockman and Pick (cited in Pick and Rieser, 1982)). Direction estimates have the advantage of overcoming the execution problems of the verbal, drawing and modelling tasks which have previously been used to assess spatial knowledge (for criticisms of such methods see, for example, Herman and Siegel, 1978; Byrne, 1979; Evans, 1980; Downs and Siegel, 1981; Spencer and Darvizeh, 1981*a*, *b*), and are appropriate for testing young children in the real world.

Direction estimates, of course, only provide one measure of spatial knowledge. Unfortunately, requiring distance estimates, as well as direction estimates, in a single experiment places a large memory demand on preschool children which may detract from their performance, so this experiment asks three- and four-year-old children to make direction estimates only. Every effort was made to ensure that the task was comprehensible to the preschool subjects, and that the equipment used was easily transportable, so the mounted 'sighting tubes' and 'telescopes' of previous researchers (for example, Hardwick *et al.*, 1976; Anooshian and Young, 1981) were abandoned for a wooden arrow and compass.

This experiment aims to answer three main questions. First, to what extent do preschool children show evidence of vector-map representation? The work of several authors (for example, Biel and Torell, 1979; Biel, 1979; Anooshian and Young, 1981; Hart, 1981; Biel, 1982*a*) has hinted that young children may be able to show quite advanced spatial knowledge in their home area, whilst other experiments with older children (for example, Schouela *et al.*, 1980; Cousins *et al.*, 1983) have shown that spatial knowledge is affected by familiarity with the environment. The following experiment uses the children's homes and places round about the home: if young children can show vector knowledge, it would be expected here. Second, does the type and quantity of knowledge shown differ between a small self-explored environment and a larger familiar environment which is necessarily passively explored? Self-exploration has enhanced young children's chances of success in other spatial tasks (Feldman and Acredolo, 1979; Hazen, 1982; Biel, 1982b). Third, does experience with the procedure of testing itself affect the nature of the children's responses? Experience with the same task in a familiar environment, where the likelihood of vector responses is maximized, may enable the children to then make vector responses in a novel environment or experience with the task *per se* in a novel environment may affect the children's responses on future tests.

Method

Subjects
Twenty-four children (12 boys and 12 girls) who attended playgroups in St Andrews, Fife, were used. They were divided into four groups, matched on sex, intellectual ability and, as far as possible, age. The children in each group ranged from three years five months old to four years seven months old, with a mean age of three years and eight months.

Materials
(1) Large wooden arrow, (2) Silva compass, (3) Scale drawings of the ground floor of each subject's home, (4) Ordnance Survey maps (scale 1:2500) of each child's home area and of the novel environment and (5) English Picture Vocabulary Test (Brimer and Dunn, 1973).

General procedure
Before testing began the experimenter played with each child individually for several hours in order to overcome the child's shyness and establish a relationship. Each child was tested on the English Picture Vocabulary test (EPV test), which is a measure of comprehension of spoken words. This provided a means of matching groups on intellectual capacity. The subjects' scores ranged from 88 to 133, the means for each group being 110·7, 111·3, 109·5 and 113·2 respectively.

In all experimental environments, the children were taken on a walk with the experimenter, and were asked to point to out-of-sight targets using the wooden arrow, which they had previously learned to hold very still and level, whilst the experimenter took the bearing of their response by placing the compass on top of the arrow. This bearing was then corrected for magnetic variation and compared with the true bearing as obtained from a scale drawing (in the case of the children's houses) or the Ordnance Survey map. The children were encouraged to point the crow-flight direction of the target by being asked to imagine that 'everything between them and the target had fallen down so that they could see it and point right to it', or that 'they were superman/wonderwoman/a ghost who could walk through walls', and to point the way they would walk to the target. The more anxious children took a very popular glove puppet with them on the walks. After each test, regardless of accuracy of response, the subject was rewarded with two small sweets or a balloon.

Environments and test locations

The children were tested in two familiar environments: the ground floor of their own house and the area round their home in which they frequently went on walks with their parents. In each of the familiar environments, the child helped the experimenter to choose four targets; but to avoid ambiguity, the experimenter ensured that the locations were such that the angle between the walked-direction and crow-flight direction to the target were as large as possible.

In the children's homes, the targets were in different rooms and were such things as a lamp, or a chair. The subject and the experimenter visited each target to make sure that the child knew its location and was in agreement as to which was the chosen target. The location of each target was also used as a test location for the bearings to the other targets. In the area around the home the targets chosen were out of sight of each other and they also acted as test locations for all other locations. One of the targets had to be the home and the others were such things as a swing in the park or the front door of a friend's house. The child was asked to lead the experimenter to each of the targets, to test his or her knowledge of how to get there and to check that the child and experimenter were in agreement over what constituted each target. The children were tested on one novel environment. This was a simple outdoor route with two approximately right-angled corners. The target was a memorable archway at the start of the route, and the test locations were specified by the experimenter along the route.

Design

All the children were tested in the novel and the two familiar environments; half the subjects received the Home test first and half the Around Home test first, to avoid order effects. The Novel test was repeated three times to test for practice effects. In order to detect any changes in responding caused by experience of the task in a familiar environment, half the subjects were tested in the novel environment *before* either familiar environment, then between the two, and finally after both. The other half received the Novel test only *after* one of the two familiar environments (and thus were tested on it twice at the end of the series of tests) in order to (a) describe preschoolers' naive ability in each environment and (b) detect changes in responding in the familiar environment caused by experience of the same task in the novel environment. Thus, there were four experimental groups as follows.

Group 1: Novel test, Home test, Novel test, Around Home test, Novel test.
Group 2: Home test, Novel test, Around Home test, Novel test, Novel test.
Group 3: Novel test, Around Home test, Novel test, Home test, Novel test.
Group 4: Around Home test, Novel test, Home test, Novel test, Novel test.

Results

Scoring

The bearings obtained from the children's pointing responses were converted to errors from the correct bearings; whether the direction of pointing was clockwise or anticlockwise of the correct bearing (that is, the sign) was not taken into account, only the difference between the two. Inspection of the results indicated a bimodal pattern of responding, so analysis of variance (ANOVA) could not be used simply on errors, and a more qualitative analysis was preferred. The children's responses were categorized into 'path responses' when the children pointed in the walked direction to the target, and 'crow-flight responses' when the children pointed directly towards the target. A pilot study indicated that a suitable definition for 'path responses' is those errors which fall within a range of $\pm$ 22·5 degrees from the true path bearing, unless the angle between the path and correct bearing is less than 45°, when the response is categorized according to which of the path or correct bearing is closest; and a suitable definition of crow-flight responses was $\pm$ 30° from the true bearing to the target.

Qualitative analysis
To test whether the children's tendency to make path responses was affected by the nature of the environment the proportion of path responses given in each test was analysed using ANOVA (four groups × sex × subjects × five tests). There were no significant effects of group or sex, but a highly significant effect of test (F = 18·87, 64 df, $P < 0.0001$). Tukey's HSD tests gave the significant differences shown in Table 1. There were significantly less path responses in the Home test than in any of the other tests; significantly less path responses in the Around Home test than in any of the novel environment tests; and significantly less path responses in Novel test 3 than in Novel test 1. These results are consistent with a hypothesis that increased familiarity with an environment leads to decreased path responding.

To provide further verification for the effect of the environment upon the children's tendency to make path responses all the children's responses in the Home, Around Home, and the Novel tests were analysed into patterns of responses according to whether they made path responses or not. This, of course, meant pooling data over different orders of testing, but the lack of significant differences between groups, and interactions between groups and tests made this possible. Six patterns emerged, as shown in Table 2, and form a perfect 'cumulative scale' often called a Guttman Scale (Guttman, 1941, 1944). The probability of all 24 subjects giving responses which fall only into these six patterns is 6/32 to the twenty-fourth: a chance which is not significant. Path responding is lost in the home first, then in the area around home, then in the Novel tests in decreasing order of familiarity. However, when the data were analysed into patterns of response according to whether the children made crow-flight responses or other responses, 11 patterns emerged as shown in Table 3. Unlike, path responses, there is no neat change in response accord-

TABLE 1
Differences between the proportion of path responses given in each environment

More path responses							
Home	Around home	Novel 3	Novel 2	Novel 1			
—	18·4[a]	31·2[b]	39·6[b]	50·0[b]	Home		
		12·8[b]	21·2[a]	31·6[b]	Around Home	Less path responses	
		—	8·4	18·8[a]	Novel 3		
			—	10·4	Novel 2		
				—	Novel 1		

[a]HSD = 17·8, 64 df, $P < 0.05$.
[b]HSD = 21·6, 64 df, $P < 0.01$.

TABLE 2

Patterns of responses defined by the most frequent response type

Pattern (No. of subjects)	Test				
	Home	Around Home	Novel 3	Novel 3	Novel 1
1 (4)	×^a	×	×	×	×
2 (6)	√^b	×	×	×	×
3 (4)	√	√	×	×	×
4 (2)	√	√	√	×	×
5 (3)	√	√	√	√	×
6 (5)	√	√	√	√	√

[a] × Indicates crow-flight and wild responses.
[b] √ Indicates path responses.

TABLE 3

Pattern of responses defined by the most frequent response type

Pattern (No of subjects)	Test				
	Home	Around Home	Novel 3	Novel 2	Novel 1
1 (5)	×^a	×	×	×	×
2 (4)	√^b	×	×	×	×
3 (2)	√	√	×	×	×
4 (1)	√	×	√	×	×
5 (1)	√	×	×	√	√
6 (1)	√	×	√	×	√
7 (1)	√	×	√	√	×
8 (1)	√	×	√	√	√
9 (2)	√	√	×	√	×
10 (4)	√	√	√	√	×
11 (2)	√	√	√	√	√

[a] × Indicates crow-flight responses.
[b] √ Indicates wild and path responses.

ing to the environment. The only consistent finding is that crow-flight responses develop first in the home. The two results taken together show that although change from path responding to making other responses happens in a predictable pattern across the environments, whether these non-path responses will be accurate enough to be called crow-flight responses is not predictable, except for the home environment. This suggests that children abandon path responding before they are able to make crow-flight responses.

In order to ascertain whether the children's tendency to make crow-flight responses was affected by the nature of the environment, an ANOVA (four groups × sex × five environments) was carried out on the proportion of crow-flight responses made in each environment. There was a significant effect of environment only ($F = 19.84$, 64 df, $P < 0.001$), and there were no significant interactions. Tukey's HSD tests showed that significantly more crow-flight responses were made in the home than in any other environment, and significantly more around home than in the first Novel test (all HSD = 19.5, $P < 0.01$).

Quantitative analysis: effect of previous testing
An ANOVA (four groups × subjects × three Novel tests) was carried out on the error scores produced in each Novel test to detect changes in responding in the Novel environment due to previous experience in a familiar environment. However, contrary to expectation, no significant interaction was found between group and Novel test. In view of the danger of artifactual results from using ANOVA on bimodal data, further tests were performed. Preselected comparisons were made on the error scores produced in different Novel tests, using t-tests, to ascertain (1) whether experience in a familiar environment transfers to a novel one (Groups 2 and 4, Novel test 1 versus Groups 1 and 3, Novel test 1), (2) whether the type of familiar environment matters or not (Group 2 Novel test 1 plus Group 1 Novel test 2 versus Group 4 Novel test 1 plus Group 3 Novel test 2). Neither of the t-tests was significant: (1) $t = -1.24$, 22 df; (2) $t = -0.28$, 22 df), suggesting that response accuracy is determined by qualities of a particular environment, and does not thereafter transfer to environments with different qualities. There was a nonsignificant tendency for accuracy to increase over the three trials.

Quantitave analysis: comparison of Home and Around Home tests
To compare accuracy of response in the two environments, the children's unsigned errors in pointing *to* each target, and *from* each target were analysed using ANOVAs (four groups × sex × two environments × four targets). Both errors 'from' targets and 'to' targets gave significant effects of group and environment (group: $F = 3.81$, 16 df, $P < 0.05$; $F = 4.17$, 16 df, $P < 0.05$, respectively. Environment: $F = 36.67$, 16 df, $P < 0.001$; $F = 34.53$, 16 df, $P < 0.0001$, respectively). Both 'to' and 'from' targets, errors were significantly larger in the around home test than in the home ($P < 0.001$). Tukey's HSD tests show that 'to' and 'from' targets, errors were significantly greater in Group 1 than in Group 2 (both $P < 0.05$), and that errors 'to' targets were just significantly greater in Group 1 than Group 3 ($P < 0.05$). Errors 'to' targets gave a just significant effect of sex, with boys producing smaller errors than girls ($F = 4.52$, 16 df, $P < 0.05$).

If the children's spatial knowledge is Euclidean, regardless of how distorted, one would expect their responses to be commutative; that is, the bearing pointed from target A to target B should be the 180° reversal of the bearing pointed from target B to target A. An ANOVA on each child's average error from 180° (four groups × sex × subjects × two environments) gave a significant effect of environment only. Children's responses were less commutative around home than in the home ($F = 24.55$, 16 df, $P < 0.001$). In both the Home and Around Home tests, the more path responses a child made, the less commutative were his or her responses, but this correlation between average divergence from commutativity and number of path responses given by each child only reached significance in the Around Home test (Home; $r = 0.401$, 22 df: Around Home; $r = 0.815$, 22 df, $P < 0.01$. Moreover, the mean divergence from commutativity is even larger than the mean errors: home, error 23°, commutativity 48°; around home, error 45°, commutativity 57°.

Quantitative analysis: Home test
As each child's home was different, it is possible that their responses were affected by the size and shape of their house. Therefore, the data were analysed to see whether each child's error scores were correlated with the crow-flight distance to

the target, the number of walls through which the crow-flight direction passed, and the number of corners on the route walked between the two targets. One child (out of 24) produced errors which correlate significantly ($r = 0.717$, 10 df, $P < 0.01$) with crow-flight direction; there were no correlations between average errors and number of walls, and three children out of 24 produced significant correlations between average errors and number of corners ($r = 0.6$, 10 df, $P < 0.05$; $r = 0.71$, 10 df, $P < 0.01$; $r = 0.58$, 10 df, $P < 0.05$). No firm conclusions can be drawn from this, but some children, who are not necessarily the youngest, may be working out the way to point by thinking of the twists and turns on the route they would walk between targets (compare Biel, 1979).

Average errors within the home ranged from $8.0°$ to $45.5°$, with those who made the smallest average errors making the most commutative responses ($r = 0.67$, 22 df, $P < 0.01$). Sixteen children made few or no path responses, and made small average errors (less than $30°$), that is, most of their responses were vector-map responses. The mean age of these 16 children was 4.0 with a standard deviation of 0.45. They possessed knowledge of the layout of their home which contained fairly accurate Euclidean information, that is, knowledge of the direction between locations.

There were no significant correlations between either age of the child, or scores on the EPV test, and either average error scores, or average number of path responses per child.

Quantitative analysis: Around Home test

The children's unsigned average errors ranged from $12.6°$ to $79.4°$; those who made the smallest average errors made the least path responses ($r = 0.82$, 22 df, $P < 0.01$), and the most commutative responses ($r = 0.86$, 22 df, $P < 0.01$). Seven children made few or no path responses, and made small average errors, that is, the majority of their responses were vector-map responses. These children possessed knowledge of the environment around their home which was neither route-like and poorly integrated (Siegel and White, 1975), not topological, static, and egocentric (Piaget *et al.*, 1960; Piaget and Inhelder, 1967) but was apparently Euclidean and integrated with accurate knowledge of the position of all four targets relative to each other. These seven children had a mean age of 4.12 with a standard deviation of 0.31.

There was a significant negative correlation between the average number of path responses per child and age ($r = -0.479$, 22 df, $P < 0.05$). There were no significant correlations between average number of path responses and scores on the EPV test or between average error scores and either age or scores on the EPV test.

Discussion

The major findings of the experiment are as follows. First, some children aged between three years five months and four years seven months can show consistent vector-map/Euclidean knowledge. Second, network-map knowledge is lost in the home first (an environment which is familiar and self explored), then in the area around the home (familiar, but passively explored) and lastly in a novel environment (passively explored). Vector-map knowledge is most likely in the home, then in the area around the home and lastly in the novel environment. Sixteen out of

24 children had accurate bearing knowledge of target locations within their own homes, and seven of those had integrated and accurate knowledge of the position of locations within their home neighbourhood. Of the others, some seemed to understand the notion of crow-flight direction since they avoided path responses. In the home, such children often showed clear vector-map knowledge; however, in other environments they showed vector-map knowledge only erratically and with no pattern of predictable increase across environments. Third, previous testing in either familiar environment had no effect on the accuracy of responses in the novel environment. There is a slight indication that previous experience with the experimental task *per se* in a novel environment may improve later responding but this finding was not consistent. Fourth, within the age range and educational status tested, the ability to make directional estimates and accuracy of responses, is dependent upon the qualities of the test environment, rather than intellectual capacity or age.

According to Piaget (Piaget *et al.*, 1960; Piaget and Inhelder, 1967) preschool children should have no knowledge of projective and Euclidean relationships, yet many of the children here can show accurate knowledge of direction in some or all of the test situations. Vector-map/Euclidean knowledge appears to be expressed first in the most familiar environment, and last in the least familiar environment. This finding, that the type of knowledge expressed is dependent upon the nature of the environment, questions the generality of the concept of a stage. It is consistent with a view of development which suggests that development does not take place in stages across *all* domains of knowledge, but that a child's ability is unevenly distributed across tasks (Fodor, 1972; Feldman, 1980; Fischer, 1980) and that the response shown in dependent both on the experience of the child and on the nature of the test situation (in this case the size and familiarity of the environment). This is not to say that no cognitive development is taking place within the child, however what develops is the ability to build up vector-map/Euclidean knowledge in more and more situations. The finding, that path responses derived from network-maps are abandoned before any accurate vector-map knowledge is available, resembles the 'U-shaped' patterns of development found in other domains of knowledge (e.g. language, Ervin, 1964).

The few significant differences found between groups of subjects show that the familiarity and size of the test environment had more effect on the children's responses than the order of presentation. The one sex difference found—that girls make larger errors than boys in pointing to locations around the home—is consistent with the general trend of the literature (for example, Keogh, 1971; Harris, 1981; Spencer and Weetman, 1981). However, this was the only significant difference between the sexes possibly because the majority of sex differences have been found on tasks which required mental manipulation of the spatial representation (for example, Anooshian and Young, 1981). It could also be due to the small number of subjects per group. The lack of significant findings due to differences in EPV scores again suggests that, within the range of intellectual capacity tested, qualities of the environment had a greater effect upon the children's responses.

Those children who were unable to express vector-map/Euclidean knowledge in some situations nevertheless had network-map/topological knowledge, as they were able to lead the experimenter from one target to another. Some children did not rely on network-map knowledge, but their direction estimates lacked accuracy, suggesting that they had not yet built up accurate vector-map knowledge. The fact that errors of commutativity are even higher than errors of bearings argues against the idea that inaccurate pointing is sometimes due to distorted but Euclidean mental representations; rather it is due to use of network-map knowledge or none at all.

The results found in this paper were interpreted in terms of Byrne's (1979, 1982) network-map/vector-map theory of spatial knowledge because, unlike Piaget's theory, it carries no implications about development in stages, and it defines more explicitly the responses expected by each type of knowledge. However, Byrne's categorization is otherwise close to Piaget's distinction between topological and Euclidean knowledge, and we merely use it because it does not open itself to misinterpretation by those who cannot separate Piaget's theory from its interpretation.

Acknowledgement

This research was carried out whilst the first author was in receipt of an award from the Economic and Social Research Council.

References

Anooshian, L. J. and Young, D. (1981). Developmental changes in cognitive maps of a familiar neighbourhood. *Child Development*, **52**, 341–8.

Biel, A. (1979). Accuracy and stability in children's representation of their home environment. *Göteborg Psychological Reports*, **9**, No. 2.

Biel, A. (1982a). Children's spatial representation of their neighbourhood: A step towards a general spatial competence *Göteborg Psychological Reports*, **12**, No. 3.

Biel, A. (1982b). Children's spatial knowledge of their home environment. *Göteborg Psychological Reports*, **12**, No. 10.

Biel, A. and Torell, G. (1977). The mapped environment: Cognitive aspects of children's drawings. *Göteborg Psychological Reports*, **7**, No. 16.

Brimer, M. A. and Dunn, L. M. (1973). *English Picture Vocabulary Test Full Range Edition*. Bristol: Educational Evaluation Enterprises.

Byrne, R. W. (1979). Memory for urban geography. *Quarterly Journal of Experimental Psychology*, **31**, 1–8.

Byrne, R. W. (1982). Geographical knowledge and orientation. In A. W. Ellis (ed.), *Normality and Pathology in Cognitive Functions*. London: Academic Press, chapter 8.

Cousins, J. H., Siegel, A. W., and Maxwell, S. E. (1983). Way finding and cognitive mapping in large-scale environments: A test of a developmental model. *Journal of Experimental Child Psychology*, **35**, 1–20.

Downs, R. M. and Siegel, A. W. (1981). On mapping researchers mapping children mapping space. In L. S. Liben, A. H. Patterson and N. Newcombe (eds.), *Spatial Representation and Behaviour Across the Life Span*. New York: Academic Press, chapter 9.

Ervin, S. (1964). Imitation and structural change in children's language. In E. H. Lenneberg (ed.), *New Directions in the Study of Language*. Cambridge, Massachusetts: M.I.T. Press.

Evans, G. W. (1980). Environmental congnition. *Pyschological Bulletin*, **88**, 259–87.

Feldman, D. H. (1980). *Beyond Universals in Cognitive Development*. Norwood, New Jersey: Ablex.

Feldman, A. and Acredolo, L. (1979). The effect of active versus passive exploration on memory for spatial location in children. *Child Development*, **50**, 698–704.

Fischer, K. W. (1980). A theory of cognitive development: the control and construction of hierarchies of skills. *Psychological Review*, **87**, 477–531.

Fodor, J. (1972). Some reflections on L. S. Vygotsky's *Thought and Language, Cognition*, **1**, 83–95.

Guttman, L. (1941). The quantification of a class of attributes: A theory and method of scale construction. In P. Horst, with the collaboration of P. Wallin and L. Guttman, assisted by F. Wallin, J. Clausen, R. Reed and E. Rosenthal (eds.), *The Prediction of Personal Adjustment*, Bulletin No. 48. New York: Social Science Research Council.

Guttman, L. (1944). A basis for scaling qualitative data. *American Sociological Review*, **9**, 139–50.

Hardwick, D. A., McIntyre, C. W., and Pick, H. L. (1976). The content and manipulation of cognitive maps in children and adults. *Monographs of the Society for Research in Child Development*, Serial No. 166, Volume **41**, No. 3.

Harris, L. J. (1981). Sex related variations in spatial skills. In L. S. Liben, A. H. Patterson, and N. Newcombe (eds.), *Spatial Representation and Behaviour Across the Life Span*. New York: Academic Press, pp. 83–125.

Hart, R. A. (1981). Children's spatial representation of the landscape: lessons and questions from a field study. In L. S. Liben, A. H. Patterson, and N. Newcombe (eds.), *Spatial Representation and Behaviour Across the Life Span*. New York: Academic Press, pp. 195–233.

Hazen, N. (1982). Spatial exploration and spatial knowledge: individual and developmental differences in very young children. *Child Development*, **53**, 826–33.

Herman, J. and Siegel, A. (1978). The development of cognitive mapping of large-scale environments. *Journal of Experimental Child Psychology*, **26**, 389–406.

Keogh, B. K. (1971). Pattern copying under three conditions of an expanded visual field. *Developmental Psychology*, **4**, 25–31.

Piaget, J. and Inhelder, B. (1967). *The Child's Conception of Space*. New York: Norton.

Piaget, J., Inhelder, B. and Szeminska, A. (1960). *The Child's Conception of Geometry*. New York: Basic Books.

Piché, D. (1977). *The Geographical Understanding of Children Aged Five to Eight Years*. Unpublished Ph.D. Thesis, London School of Economics.

Pick, H. L. and Rieser, J. J. (1982). Children's cognitive mapping. In M. Potegal (ed.), *Spatial Abilities: Development and Physiological Foundations*. New York: Academic Press.

Schouela, D. A., Steinberg, L. M., Leveton, L. B., and Wapner, S. (1980). Development of the cognitive organization of an environment. *Canadian Journal of Behavioural Science*, **12**, 1–18.

Siegel, A. W. and White, S. H. (1975). The development of spatial representations of large-scale environments. In H. W. Reese (ed.), *Advances in Child Development and Behaviour*, **10**. New York: Academic Press.

Spencer, C. and Darvizeh, Z. (1981a). Young children's descriptions of their local environment: A comparison of information elicited by recall, recognition and performance techniques of investigation. *Environmental Education*, **1**, 275–84.

Spencer, C. and Darvizeh, Z. (1981b). The case for developing a cognitive environmental psychology that does not underestimate the abilities of young children. *Journal of Environmental Psychology*, **1**, 21–31.

Spencer, C. and Weetman, M. (1981). The microgenesis of cognitive maps: a longitudinal study of new residents of an urban area. *Transactions of the Institute of British Geographers*, New Series, **6**, 375–84.

A CONCEPTUAL MODEL AND EMPIRICAL ANALYSIS OF CHILDREN'S ACQUISITION OF SPATIAL KNOWLEDGE

REGINALD G. GOLLEDGE*, TERENCE R. SMITH*,
JAMES W. PELLEGRINO†, SALLY DOHERTY†
and SANDRA P. MARSHALL‡

*Department of Geography, †Graduate School of Education and ‡Department of Psychology, University of California, Santa Barbara, U.S.A.

Abstract

How adults and children come to understand, represent and behave within their spatial environment are topics of great interest to geographers, psychologists, environmental planners and laypeople. Considerable research and theory has been published on these and related topics. In this paper, we will review some of what is known and theorized about spatial cognition and then consider elements of our research program on the acquisition of spatial knowledge. We focus on two intimately related topics. The first is the development of a conceptual model of the knowledge structures and processes associated with acquiring, representing and accessing knowledge of a given environment. The conceptual model forms the basis for a formal computational process model intended as a simulation of actual knowledge and performance in way finding tasks. The second emphasis is an in-depth case study of the acquisition of spatial knowledge. The case study focuses on a single child acquiring knowledge of a lengthy route through an unfamiliar suburban neighborhood. It is presented as an empirical test of certain assumptions embodied within the conceptual model.

Before introducing the conceptual model and the case study, we first review the state of current theory and data on spatial cognition and identify four central issues confronting researchers in this field. This review provides a necessary context for describing and evaluating our program of research. The second section of this paper discusses elements of the conceptual model and its relationship to other formal computational models. The third section considers specific hypotheses about the acquisition and representation of spatial knowledge and tests of these hypotheses from the single in-depth case study. The final discussion section of this paper is a reconsideration of the four issues raised in the first section and necessary and proposed extensions of the current research.

Overview of Theory and Research on Spatial Cognition

Modeling spatial knowledge

Research and theory in cognitive science have increasingly focused on how to represent the knowledge structures and processes that underlie human behavior in a wide variety of complex domains. The general assumption is that an individual's permanent knowledge structures provide the basis for interpreting objects, actions and events in the external environment. They guide the decisions and actions of the individual in response to perceptions and interpretations of self and environment. Critical issues within this approach include the types of knowledge that exist, how

Funds for this research were made available through NSF Grant No. SES-82-09621.

such knowledge is represented and organized, the mechanisms by which it is activated
and the elementary and higher level cognitive processes that operate upon the
knowledge base to produce new knowledge, inferences, evaluations and external
behaviors.

Most of the knowledge domains that have been explored using computational
process models involve well-defined and abstract systems having minimal corre-
spondences to an external physical reality. Thus, they tend to include non-spatial
forms of cognition. Spatial cognition, however, has become an increasingly important
area of study since it represents a major type of human knowledge with considerable
practical significance. It has long been proposed that such knowledge is represented
in memory as a 'cognitive map' (Tolman, 1948; Trowbridge, 1913), that is, as an
organized body of information obtained through perceptual experience and, to some
extent, through inference. Despite considerable attention focused on cognitive maps
(e.g. Downs and Stea, 1973; Moore and Golledge, 1976), little is known about their
structural organization, their origin, their relationship to cognitive mapping as a
process, and their similarity to the structures of nonspatial knowledge representations.

Even without extensive proof it is commonly assumed that people's decisions and
behaviors within the spatial environment are some function of the cognitive representation of that environment. Both everyday spatial behavior (e.g. journey to work)
and episodic behaviors with less frequency (e.g. recreational behavior), are assumed
to be integrally related to knowledge about both general and specific features of
particular spatial environments. In general, geographers distinguish between spatial
behavior (i.e. spatially overt activity or spatial manifestations of the decision process)
and the *a priori* decision-making process that culminate in spatial activity. In
cognitive science a related distinction exists. The initial procedure consists of conceptualizing a problem, i.e. constructing an internal representation of it. Understanding is taken to be the ability of a problem solver to construct an adequate
representation of the problem and depends on the nature and organization of
knowledge in memory. The solution process then operates on the problem-represen-
tation to produce the overt behavior.

Cognitive mapping
'Cognitive representation' is a hypothetical construct referring to a person's knowledge or thought concerning a segment of the external world. The process of extracting
information from an external environment and storing it in the mind is generally
called cognitive mapping. As defined by Downs and Stea (1973, p. 9) this process
is 'composed of a series of psychological transformations by which an individual
acquires, codes, stores, recalls and decodes information about the relative locations
and attributes of phenomena in the everyday spatial environment.' For the most
part the term 'map' as used in cognitive mapping is more a metaphor than a strict
analogy (Downs, 1981). Throughout the development of research in the cognitive
mapping area, there has existed controversy over the way information is stored in
memory. In a recent review of interdisciplinary work on images, Lloyd (1982) classified
various approaches to the internal representation of spatial phenomena as radical
image theory (e.g. Kosslyn *et al.*, 1974); conceptual–propositional approaches
(Anderson and Bower, 1973); or the dual coding approach (Paivio, 1969). Basic
differences in theory relate to how information is coded and stored. Obviously,

one only accepts the existence of these alternate approaches if one denies the possibility that the coding of information may span several layers of cognitive processes each of which absorbs and codes information in its own unique way (see Kosslyn, 1981; Anderson, 1983).

While one segment of the literature in this area continues to argue about the way that information is stored, represented and accessed, another segment assumes that spatial information is stored 'in some way' and concentrates on externally representing it. In other words, the main problem is to obtain directly or indirectly information about spatial knowledge from individuals and represent that information in an appropriate form. To date different representational forms have included sketch maps, verbal description, pseudo-cartographic representations, configurations obtained from multi-dimensional scaling procedures, and interactive computer mapping procedures (Lynch, 1960; Appleyard, 1970; Golledge *et al.*, 1982; MacKay *et al.*, 1975; Golledge and Spector, 1978). In each of these cases a critical feature is the attempt to assess the accuracy of the representation by comparing it with an objective reality. The objective reality usually chosen is a two-dimensional representation of the objective phenomena (typically a map).

The question of errors in representation is still very much open and tied for the most part to specific representational formats. There appears to be substantial evidence that error varies differentially across a cognized area and that one may expect holes, folds, cracks, tears and other disruptions to occur in the knowledge fabric of a place (Golledge and Rayner, 1976; Golledge *et al.*, 1979). Such interruptions appear to be almost a logical outcome from things such as asymmetric distance judgments (Sadalla and Staplin, 1980a), spatial applications of the power function for distance distortion (Briggs, 1972) and translation invariance and triangle equality violations (Cadwallader, 1979; Burroughs and Sadalla, 1979). In addition there appears to be some conflict over the preservation of fully metric properties of cognitive information (Kosslyn and Pomerantz, 1977; Golledge and Spector, 1978; Richardson, 1981) with varying evidence supporting hypotheses of Euclidean properties, other Minkowskian properties (e.g. city block metric), hyperbolic spaces and various non-metric spaces (Tobler, 1976; Golledge and Hubert, 1982).

Components of spatial knowledge
A variety of spatial knowledge is acquired by individuals as a function of their experience within a given environment. At the simplest level, an individual must have knowledge of important objects and/or places and this is generally referred to as landmark knowledge. Such knowledge includes the ability to state with certainty that an object or place exists and to be able to recognize it when it is within the perceptual field. Spatial knowledge also includes information about the relationships among objects. In its simplest form, the individual possesses information about topological relationships such as proximity. A more detailed knowledge of relationships among objects or places involves Euclidean or metric properties such as distance and direction within a coordinate space. Knowledge of an environment also includes how to move from a given location to another location and the routes that are available to do so.

Empirical studies of spatial knowledge have focused on several issues related to landmark, route and configurational knowledge. One issue concerns the features of objects and locations that cause them to be selected as landmarks. Appleyard

(1970) provided evidence that the most easily recalled and recognized buildings are those with high use and important symbolic functions. Physical features such as sharp contours, bright surfaces and noticeable size contrasts relative to the immediate surroundings also influence memory. These attributes of buildings, as well as attributes such as location near an intersection, have also been shown to affect the ability to accurately locate buildings in a small-scale model reconstruction task (Herman, 1980; Herman and Siegel, 1978; Pezdek and Evans, 1979).

A deficiency of research on landmark identification and placement is that the importance of landmarks as components of routes is ignored. More recent studies of route knowledge have tended to focus on issues associated with distance knowledge within routes and the importance of landmarks in making such judgments. The general finding is that a power function captures the relationship between actual and judged distance (e.g. Allen *et al.*, 1978; Briggs, 1972; Erickson, 1975; Thorndyke, 1981). There also appear to be asymmetries in distance judgments depending upon choice of the reference point (e.g. Siegel *et al.*, 1979). Golledge *et al.* (1969), Briggs (1972) and Ericksen (1975) found that distances toward a central business district were overestimated while distances away from such an area were foreshortened.

Explanations of distance judgments have been linked to assumptions about the nature of route knowledge. At a minimum, route knowledge is a series of procedural descriptions involving a sequential record of the starting point or anchor point (e.g. one's home), subsequent landmarks, and destination of each path. The procedural knowledge must contain the decision-point landmarks where a change in direction takes place. More detailed route knowledge includes information about secondary and tertiary landmarks along a route and distances between landmarks. Routes of equivalent physical distance may differ substantially in the amount of stored information necessary to complete the route, i.e. the number of landmarks, intersections and locations where changes in direction are required. Estimates of distance between two points along a route may be a function of the amount of information stored in memory that must be processed to 'mentally' traverse the route. According to this hypothesis, an individual's estimate of distance between two points is based upon information pertaining to the effort involved in traversing the route. Effort can refer to directional changes and travel time and the two are correlated. Empirical evidence has been obtained to support such a conjecture (Sadalla and Staplin, 1980*a,b*; MacEachren, 1980).

Recent studies by Allen and his colleagues have examined several components of route knowledge including landmark selection, distance judgments and subdivisional processing by different age groups. The general paradigm involved a simulated walk along a complex route through the use of a series of successive photographic slides. Allen *et al.* (1978) showed that distance judgments for high landmark potential scenes were more accurate than corresponding judgments for low landmark potential scenes and this was the case after one or two presentations of the original route (see also Allen *et al.*, 1979). Allen (1981) has also explored the importance of landmarks with respect to their role in subdividing routes into separate segments or chunks of information. The basis for his study was the assumption that route information may be organized as a series of subdivisions bounded by distinct environmental features rather than being organized strictly as a sequence of landmarks connected by pathways. Thus, route knowledge may involve organization in terms of higher level units such as subdivisions and discrete loci within units. The data from Allen's (1981) forced-choice distance judgment experiment indicated that subdivisional organization significantly affected the accuracy of second and fifth graders and adults. When decisions could not be based on subdivisional membership then evidence was obtained of the availability of metric information about locations within a subdivision of a route. The availability of such information varied with age and developmental level.

The literature on spatial cognition also distinguishes between route knowledge and survey knowledge (Shemyakin, 1962). Theoretically and empirically there is support for the notion that knowledge progresses from landmarks to specific paths to an integrated frame of reference system for structuring spatial information. As knowledge accumulates and as distance information becomes more precise, notions of angularity and direction emerge. The exact nature of how the transition from route knowledge to survey knowledge occurs is largely unknown. There is, however, a volume of work that argues for the importance of attention processes in selecting spatial sensory information for further processing (Gärling and Böök, 1981). Of critical importance appears to be information such as whether or not a reference point is visible from the origin, destination or some intermediate decision point. These authors hypothesize that the acquisition of information about non-visible reference points requires central information processing or deliberate attention processes. Interfering with deliberate attention processes reduces the probability that information about reference points can be integrated in an overall frame of reference system.

The progression from route knowledge to survey knowledge, a progression from one dimensional to two-dimensional knowledge implies a qualitative difference in the nature of spatial information represented in long-term memory (Chase and Chi, 1980). An integral part of route knowledge, survey knowledge, and the transition between them is a hierarchical ordering of reference nodes or anchor points. Thus anchor points represent the organizational nodes of the representation and together with the path connections between them, define the skeletal structure of a representation (Golledge, 1978). A definition of anchor points also facilitates the grouping or clustering of information thus allowing greater efficiency in coding, storage and recall than would otherwise be possible under the hypothesis of procedural descriptions of route knowledge.

Development and acquisition of spatial knowledge
Both developmental and learning theories of the acquisition of spatial knowledge emphasize the increasing complexity of an individual's knowledge of the environment. There is a general progression from landmark to route to configurational knowledge representing the successive coordination of knowledge units within an external frame of reference. Similarly, the relational information contained in the individual's representation of the environment progresses from topological to projective to metric properties. The emergence of these properties also depends upon the availability of referential frameworks that are nonegocentric. A child's knowledge of his environment is both a function of cognitive capacities (i.e. general cognitive developmental level) and amount of experience in the environment. The knowledge acquisition of an adult mirrors the general developmental sequence observed for children.

Piaget and Inhelder (1967) have described three stages in the development of spatial cognition that are in correspondence with Piaget's theory of the general stages of cognitive development. During the sensorimotor stage of development children develop landmark knowledge and this is associated with places of activity or emotion. Landmarks serve as organizers and guides for the child's visual and motor exploration of an area. The child has a primarily egocentric frame of reference and landmarks remain uncoordinated with one another. Topological information thus becomes the first type of spatial relational knowledge acquired and this occurs during the transition from sensorimotor to preoperational thought. Landmarks are then linked together through the establishment of path or route knowledge. Environmental cues are encoded in a form that retains decision-relevant features and the child's motoric responses to such cues. Path knowledge emerges from the existence of such representations. Projective spatial relations develop next during the concrete operational stage of development. During this stage, there is a shift from an egocentric to allocentric frame of reference. Finally, spatial relations develop metric properties and are represented in a system of axes and coordinates and the concepts of distance and angles are understood. This level of spatial knowledge emerges during the formal operational stage of development. The development stages postulated by Piaget and Inhelder have been discussed in terms of different frames of reference by Hart and Moore (1973).

A second developmental model has been proposed by Siegel and White (1975). In this model the first stage in the acquisition of spatial knowledge involves landmark information. After landmarks become known (i.e. recognizable) paths or routes are established between landmarks. Spatial knowledge along routes passes from topological properties to metric properties. Landmark and path sets are organized into clusters where there is a high level of coordination within cluster but only topological information about cluster relations. Finally, an overall coordinated frame of reference develops such that Euclidean properties are available within and across clusters. This is often referred to as survey knowledge. Siegel and White (1975) and Kuipers (1978) assume that a memory representation of a complex external environment such as a town contains hierarchically organized knowledge including landmarks or salient locations, routes, and configurations representing the integration of information about the relative location of landmarks. In other words, in a cognitive representation of an environment, there are different types of knowledge represented—places and paths (environmental descriptions), travel instructions (routes), and relative locations (local maps).

The Hart and Moore (1973) and Siegel and White (1975) developmental theories closely relate to the anchor point theory suggested by Golledge (1978), in which a hierarchical ordering of places within the spatial environment is based on the place's significance to the individual. The cognitive hierarchical ordering can be described by a system of primary, secondary and tertiary nodes and the paths that link them (Briggs, 1972). A person's home, place of work and shopping areas serve as primary nodes and are the anchor points from which the rest of the hierarchy develops. As interactions occur along the paths between primary nodes there is a spillover effect and neighborhoods surrounding the primary node set become known. Other areas of frequent use and importance to the individual become the secondary and tertiary nodes in the hierarchy, e.g. recreational and entertainment areas. Continued interactions along the node-path network strengthen the image of the environment for the individual.

The empirical literature on the acquisition of spatial knowledge in children has typically examined what is known about a familiar environment such as the child's neighborhood (e.g. Cousins *et al.*, 1983). Inferences about knowledge components and the acquisition process have been based upon comparisons of individuals differing in age or experience within a given area. Studies that have tried to examine the acquisition of route knowledge for a new environment have focused on tasks such as the accuracy of distance judgments for a pictorially presented route (e.g. Allen *et al.*, 1979). Tests of wayfinding and route-reversal in a new environment have been restricted to relatively small scale spaces and short routes, such as through a building (Hazen *et al.*, 1978).

Important issues for theory and research

The literature on spatial cognition and its development presents a broad picture of the types of knowledge that are presumed to exist and the way in which such knowledge is accumulated. Our research program focuses on two areas in need of detailed theoretical and empirical analysis. First, although there are competing theories of the representation and acquisition of spatial knowledge, these theories represent general positions rather than formal models dealing with the specifics of cognitive representations and processing. Second, empirical studies of the acquisition of spatial knowledge are limited in scope and not explicitly tied to a formal model. In the next two sections we describe our initial attempt to delineate such a formal model and empirically to validate assumptions and hypotheses embodied in the model.

Before considering the details of the conceptual and computational model, it is useful to summarize four major issues derived from the literature on spatial cognition that need to be addressed by any adequate computational model. A first set of issues concerns the nature of knowledge acquisition in spatial learning. Both environmental learning theories and developmental theories assume that knowledge of the environment is a function of accumulated and organized perceptual experiences representing interactions between the organism and the environment. Thus, from a series of episodes, a more generalized and context-independent knowledge structure emerges that is capable of sustaining a wide range of spatial behaviors (Kuipers, 1982).

A second set of issues concerns the representation of the several different components of spatial knowledge presumed to exist. Landmark, route and configurational knowledge are assumed to represent a mixture of declarative and procedural structures. A declarative representation provides for efficient storage and organization and permits the derivation of relationships that are only implicit in the structure. Knowledge of specific routes is more appropriately conceptualized as procedural in nature since it involves specifying goals, recognizing environmental features, and performing actions at specific locations.

A third set of issues concerns the 'inaccurate' or 'incomplete' nature of the knowledge base as a consequence of the learning process. Much of the literature on cognitive mapping has revealed that spatial knowledge is incomplete, partial and disconnected. Some portions of cognitive maps have fully metric properties whereas other portions are distorted by holes, folds, tears and cracks, and can be represented only in topological terms (Golledge and Hubert, 1982). The ability to produce such 'flaws' in the knowledge structure is an essential feature of a model of learning by experience.

A fourth and related set of issues concern the overt behavioral phenomena to which the knowledge structure directly give rise. For example, the information contained in an individual's knowledge structure often leads to systematic errors in distance and direction judgments. This is true for distance judgments within a single route (Allen, 1981) and within a larger configuration (Stevens and Coupe, 1978). The specific types of errors that occur have been presumed to reflect a hierarchically organized knowledge structure as the basis of both route and configurational knowledge (Golledge, 1978; Chase and Chi, 1980).

In summary, a model of the acquisition and representation of spatial knowledge must adequately account for (1) acquisition and representation based on episodic experience and subsequent generalization, (2) different types of knowledge and forms of representation, (3) systematic inaccuracies and distortions in the cognitive representation, and (4) behavioral errors associated with inaccurate and hierarchically organized knowledge. Each of these will be addressed in the following descriptions of our conceptual model and its empirical verification.

The Conceptual Model

Our conceptual model of route learning specifically addresses the four issues raised in the previous section. We present the model at three levels of abstraction: in terms of real-world entities, in terms of psychological and artificial intelligence constructs, and in terms of computer programming constructs. Primary attention is given to the first two levels.

The task environment

The task modeled here is how an individual learns to navigate between two fixed points in a specific environment. There are two aspects of the model: a representation of the environment and a representation of the cognitive structures and processes of an individual performing tasks in the environment. The conceptual model of the environment represents those features of the task environment that are salient for spatial navigation. The simulated environment is a typical suburban neighborhood, consisting of a network of intersecting streets with houses along both sides of each street.

First-level concepts. The real-world entities of the model correspond to major objects occurring in suburban environments, namely plots and choice points. A plot is defined as a unit of land containing residential structures, landscaping, and sidewalks. A choice point is any location at which a navigational decision must be made. Each plot is automatically linked to a choice point in a wayfinding task because at any location, an individual has the alternatives of continuing forward, reversing the direction or crossing the street. Some choice points such as four-way intersections offer the additional alternatives of right or left turns.

Plots and choice points form the basis of the first level of abstraction. Also at this level are the features and properties that describe these entities. Typical features include house, roof, garage, window, mailbox; typical properties are color, size and shape of the features.

Second-level concepts. At the second level of description are the data structures commonly called frames. Frames are the structures used here to organize knowledge about the features and properties of environmental objects. The concept was first introduced by Minsky (1975) to represent complex stereotyped situations in memory. Frames typically contain two types of information: (1) a general description or list of features that are common to all instances of the frame and (2) an organized collection of 'slots' which hold information concerning the current example of the frame. Hence, we represent a particular plot by a collection of features and properties organized into a frame. Data structures called feature–property–value triples (FPVs) are used to fill various slots in the frames.

Each FPV triple comprises one member of a class of features (e.g. houses, mailboxes, trees, etc.), one member of a class of properties (e.g. size, color, etc.) and one member of a class of values that can be taken on by the property (e.g. large, blue, etc.). For example, one FPV triple is house–color–blue. We also represent as FPV triples a large number of spatial and other relationships between features in terms of 'relational properties', as in mailbox–beside–tree. More complex FPV triples may be constructed through the logical connectives 'and', 'or', and 'not'. Such connectives may occur between features, values and properties.

Third-level concepts. The third level of concepts are constructions in the programming language, LISP, used to implement the conceptual model. At this level are programming features such as property lists. These concepts are necessary for implementation of the model as a computer simulation of human learning, but they have little psychological analog.

The decision-making individual

The second component of the conceptual model is a representation of the cognitive structures and processes of a decision-maker involved in a wayfinding task. The decision-maker is assumed to interact with the environment (or more strictly, with the 'objective' model of the environment) and to construct its own representation of it.

The first level of concepts relating to the individual corresponds to general cognitive structures and procedures that for the most part have been identified and used by previous researchers and long-term memories and mechanisms for sensory input. The second level of concepts involve data structures and procedures that have been developed by artificial intelligence (AI) researchers and cognitive scientists for modeling cognitive processing. These include semantic networks, production systems, and mechanisms of activation. As before, the third level of concepts relate to an implementation of the model in LISP.

Related research

Before describing the main features of our model of the individual, we briefly review relevant concepts from previous literature. A more detailed discussion may be found in Smith *et al.* (1982). Two major concepts at the first level are long-term memory (LTM), which represents a structure for storing 'permanent' data and procedures, and short-term memory (STM), which represents a structure for storing transient information that is either lost or transferred to LTM. These concepts are in widespread use in modeling cognitive phenomena (see, for example, Newell and Simon, 1972; Anderson, 1982, 1983).

Other important first-level concepts refer to aspects of LTM and STM. For example, within LTM a lexicon of 'concept definitions' is often assumed (Schank, 1975), while other data structures include sets of heuristic rules (Smith and Lundberg, 1984) or 'episodes' (Kintsch, 1977). There are also various concepts relating to STM, including the process of 'activation spread' (Anderson, 1982, 1983; Smith *et al.*, 1982), by which various data structures and procedures within LTM are brought to an excited state.

As indicated previously, the level two data structures and procedures have been adapted from the results of AI research, and they are specific structures and mechanisms by which the level-one concepts may be implemented. These concepts include semantic networks (Quillian, 1968), frames or scripts (Minsky, 1975), and production systems (Newell and Simon, 1972; Anderson, 1982). Semantic networks are node-link data structures that were originally designed to model human associative memory. A production system (PS) refers to an architecture for organizing computations. The heart of a PS is a control mechanism by which the most appropriate IF–THEN rule (or heuristic) from a set of such rules is applied during some computation. The appropriateness of different rules is assessed in terms of the degree to which the IF sides of the rules and the state of the computation match each other.

While there have been many applications of the previous level-one and level-two concepts in the modeling of cognitive processing in various contexts (see Smith *et al.*, 1982, for examples), there have been relatively few applications to the computational modeling of human spatial processing in large scale spaces. A notable exception has been the work of Kuipers (1978). The main contribution of Kuipers was his TOUR model which employed the second level concept of frames to model how individuals store and retrieve information in simple spatial decision-making situations. Kuipers has more recently examined the possibility of employing other representational techniques (Kuipers, 1983). Most recently, McDermott and Davis (1984) have constructed a computational model that reasons about route planning in uncertain territory. However, this model is not directly concerned with explaining human performance.

A model of wayfinding
In our model of wayfinding, we make three assumptions concerning human cognitive processing in spatial tasks. These assumptions are derived from the empirical and theoretical literature discussed above:

(1) During wayfinding, cognitive processes relating to perception, storage, retrieval and reorganization interact with memory structures and construct a symbolic representation of the environment (cognitive map) that is accessed during future wayfinding tasks.

(2) Specific cognitive processes, common to all individuals include automatic processes of feature extraction, concatenation, pattern matching and generalization, which respectively extract, store, access and reorganize information.

(3) There is variation among individuals in terms of the environmental features represented in a symbolic representation of the environment (cognitive map) that depend upon the innate saliency of different features for different individuals.

These assumptions address directly the first two issues defined previously, namely the acquisition and representation of episodic knowledge and the different types of knowledge structures required for the representation. The assumptions are embodied in four major components of the conceptual model: (1) a set of actions that the individual may take during navigation; (2) a set of knowledge structures that encode knowledge concerning the task environment; (3) a set of cognitive processes relating to perception, storage, retrieval and reorganization of environmental knowledge, that operate on the knowledge structures; and (4) a set of control processes that determine the interaction of the decision-maker with the environment. We now consider each of these components in turn.

(1) *The feasible actions.* At any plot, an individual has three possible courses of action: continue forward, reverse direction, or cross the street and continue the navigation task. At a choice point involving two or more alternate directions, the individual must choose one of the possible directions.

(2) *The knowledge structures.* The knowledge structures that incorporate a decision-maker's procedural and declarative knowledge of the task environment may be categorized as either permanent memory structures or temporary memory structures (see Figure 1). The permanent memory structures include a lexicon and a LTM. The temporary structures are a perceptual buffer and a STM.

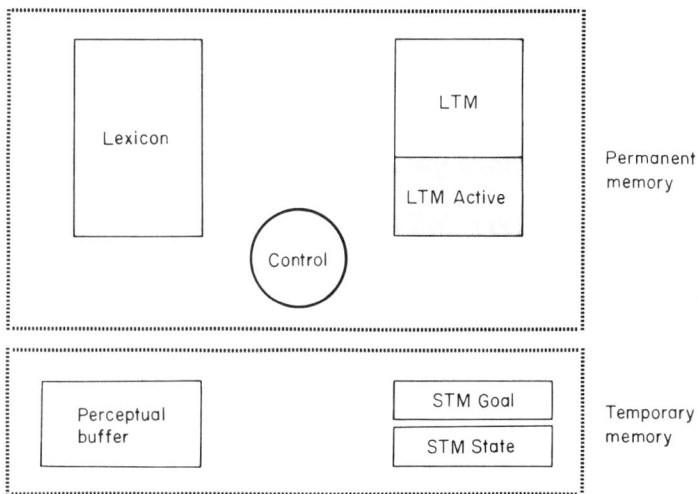

FIGURE 1. Diagrammatic representation of knowledge structure in the cognitive model. LTM = long-term memory. STM = short-term memory.

The perceptual buffer is a temporary structure that stores the results of perceiving a single environmental object (e.g. a plot or choice point). Not every feature of a plot or choice point enters the perceptual buffer. Only those with high saliencies for a particular individual will be noticed. (The saliencies are contained in a permanent structure, the lexicon, discussed below.) For example, at a plot containing house–color–white, car–type–porsche and fence–type–chainlink, an individual who typically notices vehicles might encode only car–type–Porsche while an individual with a propensity to notice houses might encode house–color–white.

Short-term memory is also a temporary structure that has two components. The first stores a record of the individual's perceptions of a limited number of environmental objects at the individual's most recent location. These objects are represented in the sequence in which they were perceived. Representations of objects enter this record only if they have first been held in the perceptual buffer. Thus, the buffer passes the representation of the objects to STM. This part of STM (or working memory as it is sometimes called) is finite, and there are known restrictions to its capacity (Miller, 1956). Therefore, its composition changes as an individual moves from one location to another, with some representations decaying or fading to make room for others.

The second part of STM contains goals that the individual attempts to achieve during a wayfinding task. Simple examples of such goals include plots or choice points along the route. These also change as an individual progresses along the route: existing goals are satisfied and new goals are defined.

As indicated above, there are also two knowledge structures characterizing permanent memory, the lexicon and the LTM. The lexicon is a static structure containing fixed lists of features, properties, and values. These lists can be used to describe choice points and plots in the task environment. The lexicon also implicitly contains real-valued saliencies for each FPV triple, indicating how 'noticeable' each is to an individual. These saliences are assumed to vary among individuals. The information stored in the lexicon determines which environmental features will be perceived by an individual and thus determines which features enter the perceptual buffer.

The LTM component of permanent memory is a record of objects perceived and actions taken during previous navigation in the environment. Elements in LTM are stored as three-part 'episodes': an initial state of the individual (including perceived environment, goals and subgoals), the navigational act taken and the resulting state (containing new perceptions and new goals). Long-term memory has two parts, an activated portion and a non-activated portion. As its name implies, the activated LTM is temporarily available for immediate processing. At any given time, activated LTM will contain representations of various environmental objects. Not every object perceived by an individual will enter activated memory. Only those with high saliency become part of it.

The objects represented in LTM may be addressed in two different ways: first, in terms of their recorded features (i.e. FPV triples) and, second, in terms of their spatial relationships to other objects that were encoded during wayfinding tasks.

(3) *The cognitive processes.* Four sets of cognitive processes are assumed to interact with the knowledge structures during wayfinding. These processes relate to perception, storage, retrieval and reorganization of environmental knowledge. The perception processes access the lexicon and construct a representation of the plot or choice point at which the individual is located. The representation is expressed in terms of FPV triples and is stored in the perceptual buffer. Only triples possessing a sufficient degree of salience are stored. The critical saliency level is a dynamic variable that depends upon several parameters, including, for plots, the relative proximity of major choice points.

The storage processes involve both temporary and permanent storage. As each object is perceived, the contents of the perceptual buffer are placed into STM, joining a list of previously perceived objects. The objects represented in STM are periodically 'aged' and removed in order to maintain constraints on the size of STM. The most recent object in STM is transferred to the activated portion of LTM if a real-valued function of the saliency of all its FPV triples exceeds a critical threshold.

In activated LTM, linkages describing spatial proximity are established between the newly added object and objects currently represented. Each object is gradually incorporated into a structure of linear links connecting salient plots and choice points. This process is termed 'concatenation'. The objects represented in activated LTM are periodically 'aged' and removed from the activated memory in order to satisfy capacity constraints. The pattern of connections established here varies among objects, depending upon their initial level of activation. When objects are removed from activated LTM, they are placed in the non-activated portion of LTM.

(4) The control processes. The control processes determine the order in which the processes of perception, storage, access and generalization occur. They are also responsible for carrying out navigation acts when an appropriate action has been determined from activated LTM and the goals of STM. In this connection, a plot or choice point is identified as a goal. Since each plot or choice point is stored in LTM with the links created previously in activated LTM, each one has associated with it other plots, choice points and actions. Given a goal, it is possible to determine for any plot whether the result of an hypothesized action matches the goal by tracing the links of the object to the action taken in some previous wayfinding task and to the result of that action. If the result matches the goal, taking the particular action will obtain the goal.

A return to basic issues
Our model can be evaluated by the success with which it addresses the four issues of acquisition of knowledge, types of knowledge, incomplete perception with inaccurate representation, and errors in performance. First, our model explicitly incorporates episodic knowledge as a basis of most route-finding knowledge. In particular, we assume that salient information about locations is stored as episodic knowledge in LTM. These basic episodes are processed and transformed during all wayfinding activities.

Second, our model is concerned with how different components of wayfinding knowledge are represented. Route knowledge is represented as essentially procedural in form, forming links between episodes in LTM. Other aspects of knowledge are represented declaratively, as in the representation of plots as frames.

Third, we model the incompleteness of human knowledge and perception in our model by varying the salience of different environmental objects. Some objects may have high salience and thus be easily activated or retrieved in LTM, while others will not attain sufficient salience to be stored permanently. At present, we do not incorporate genuine errors of perception, although our mechanisms of pattern matching may lead to inappropriate storage locations for perceived environmental cues.

Finally, the model allows errors in task performance. One long-range goal of the model is the simulation of incorrect performance and the changes in knowledge structures that occasion errors. By varying the depth at which our model of the individual perceives environmental objects, we can control the completeness with which the model stores details from the environment. We will then be able to compare the errors made by the model with those made by human decision makers operating in a similar environment.

A Case Study of the Acquisition of Route Knowledge

Earlier we indicated that empirical studies of the acquisition of route knowledge have been limited in scope. Typically, they employ either cross-sectional designs or limited learning trials with simple routes. There is little detailed data on the specifics of the acquisition process over successive learning trials. In addition, no study has been done to test or validate assumptions derived from an explicit conceptual model of the acquisition and representation of route knowledge. The study that we report was designed to provide detailed information about the acquisition of route knowledge by a single child. It also provides a test of the following hypotheses derived from our conceptual model of wayfinding.

(1) The overall saliency of a plot decreases as the distance of the plot from major choice points increases (i.e. more information is coded near choice points).

(2) The more alternative actions that can be taken at a choice point, the more information is coded at that point.

(3) The choice points of a route, at which navigation decisions are made, provide a natural segmentation of the route as represented in LTM.

(4) With repeated exposure, individual route segments in LTM become concatenated into longer segments and comprise a specific route.

(5) More information is encoded for objects at which errors in decision-making have occurred (i.e. more FPV triples are encoded).

(6) As the types of decisions made at choice points become more complex, errors are more likely to occur. Errors at less complex choice points persist longer over time.

(7) The generalization process leads to a hierarchical representation of the route elements in long term memory.

The empirical study that provided a test of the preceding hypotheses was conducted in a moderate density neighborhood in Goleta, California. The area was purely residential in character with environmental cues related to street, yard and dwelling features. There were no visibly dominant cues in the neighborhood to aid in orientation, and spatial knowledge had to be acquired in a sequential manner as the specific route was followed through the neighborhood.

The route is shown in Figure 2. It incorporates a variety of characteristics. The origin and destination points were not simultaneously visible and were identifiable only from locations on the first or final segments (respectively) of the route. The route consisted of left and right hand 90 degree turns, street crossings, long straightaways with good visibility, inconsequential sidestreets and cul-de-sacs, and curved sections. The route length was 0·8 miles but the origin and destination were only 0·4 miles apart by the shortest direct route through the neighborhood. General background and orientation information—mountains to the north and a noisy freeway to the south—were available if required, i.e. if the child became lost.

The basic cue structure of the general area in which the route was embedded was obtained from cue lists and sketch maps compiled from neighborhood children who were contacted both through local schools and scout groups. They were asked to list things and places in the environment that they considered important, giving names for places and things identified. A number of idiosyncratic 'popular' place names were uncovered for an otherwise featureless area. One of these, a park overlook,

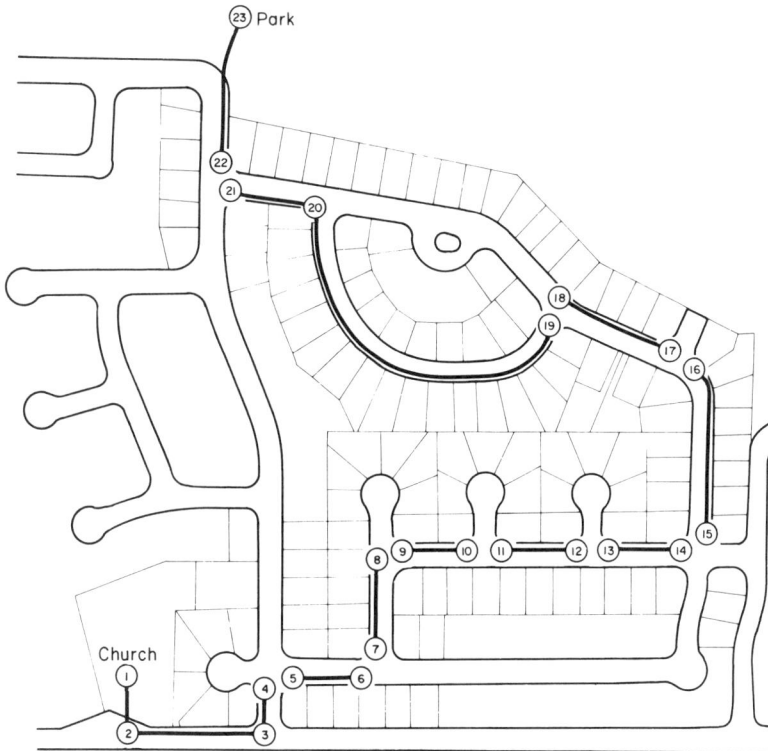

FIGURE 2. The test neighborhood and route.

was chosen to anchor one end of the route. The other end point was another frequently mentioned cue, a church, no city-wide well-known cues were located in the area of the route.

Initial testing was done with children of ages nine to 11 years to determine if the route was of sufficient difficulty and complexity that mastery would not occur within a single trial. The children walked the route with an experimenter. Upon completing the route they were then asked to either retrace the route in a forward or reverse direction. The experimenter recorded actions, errors, hesitations, etc., while the children were navigating the route. The pilot data obtained from these children indicated that the task was difficult and that knowledge of the route was still fragmentary after one or two exposures.

Extensive multiple trial testing was then done with an 11-year-old male. On five successive days, the boy was tested on route navigation and components of route knowledge. He was first taken over the route in a forward direction and then asked to navigate the route on his own in a forward direction. Upon reaching the end of the route, he was taken back to a mobile laboratory at the beginning of the route and underwent extensive probing. First, he was asked to describe his strategy for navigation and what salient aspects of the route he recalled. Next, he was asked to generate a set of verbal directions that would allow another child to get from the beginning to the end of the route. This was followed by a map-drawing task which required him to draw the route. The map was drawn on a blank piece of paper with the church and park indicated in their appropriate relative and absolute positions. Finally, the child was shown a videotape representing traversal of the route in the forward direction. He was asked to identify any cues, locations, etc., that he remembered, actions that needed to be taken at specific loci, and his expectations about what would appear next in the sequence. Following this detailed debriefing, he was taken to the end of the route and asked to navigate the route in reverse by returning to the start. Upon completion of the reversal task, he was given the same extensive debriefing except that the videotape presentation represented route reversal. On each of the five successive days, he completed one forward and one reverse navigation trial with debriefing following each trial. Total testing time on each day averaged five hours. All behaviors in route navigation were coded by the experimenter who provided no assistance to the child in finding his way.

This multi-trial testing provided several data sources relative to his behavior while navigating the route as well as his knowledge of cues and loci along the route and their acquisition. By the tenth trial the route appeared well known, was segmented consistently, traveled with confidence; little new information was produced either during the navigation tasks or in the debriefing period. Sketch maps included most of the information summarized in the verbal reports and reflected accuracy in reconstruction of cue and route information during the videotaped replays of the route.

The preceding is a very general overview of the final outcomes of multiple exposures and testing on the route. What follows is a more detailed description of the subject's knowledge of the route relative to the hypotheses previously stated.

Place and feature knowledge
From the debriefing procedures, a composite representation of cue and feature knowledge for loci along the entire route was constructed. Figure 3 shows the cumulative frequency for cues and features for the entire route. The x axis is a linear scaling of the route with choice points indicated and the y axis represents a cumulative composite of features mentioned and their frequency. The data are plotted separately for forward and reverse testing trials as well as combined over direction. Hypothesis (1) predicts that individuals will code more information at plots near choice points than at other plots in a wayfinding task. Figure 3 clearly illustrates that knowledge was concentrated in the vicinity of choice points for both forward and reverse navigation trials as hypothesized. The only difference between forward and reverse trials is the lower absolute level in the cumulative frequency. When the data from both forward and reverse trials are combined, the cumulative frequency graph over locations is even more dramatic, as shown in Figure 3.

Thus far we have shown that knowledge of route cues and features is concentrated at loci where real or potential actions occur. These loci can be further differentiated into four types: (1) a choice point where a change in direction *and* a street crossing occurs; (2) a choice point where only a change in direction occurs, i.e. turning a corner; (3) a choice point where a significant action occurs but no change in direction occurs, i.e. crossing a street but continuing straight ahead; and (4) loci where a choice does not occur, i.e. loci along a route segment between two choice points. Hypothesis (2) predicts that the amount and variety of information extracted at or near choice points varies according to the type of choice point. If we condense

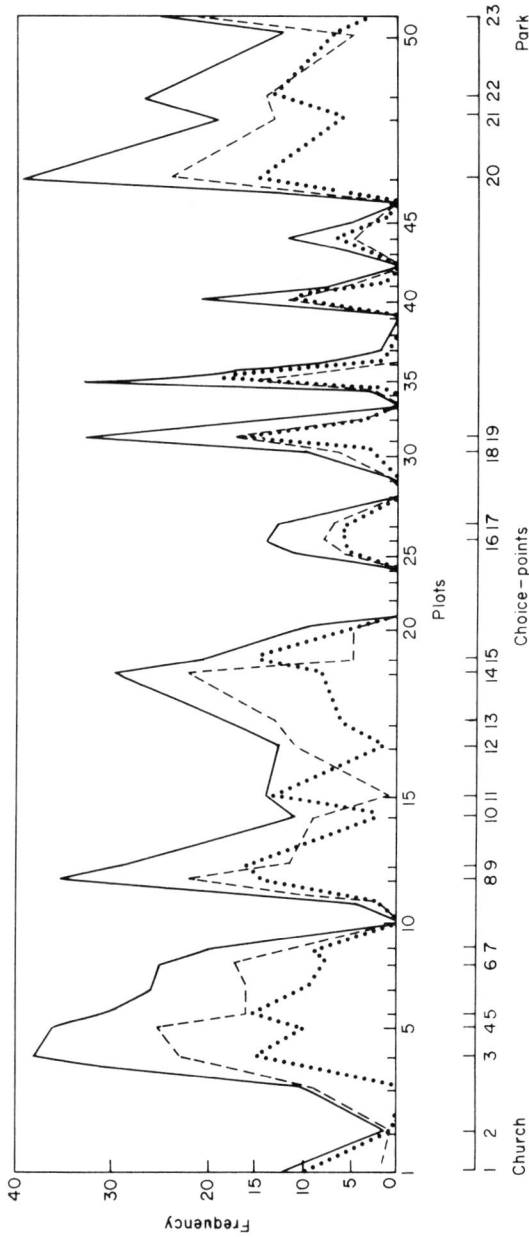

FIGURE 3. Cumulative frequency of cue recognition for the entire route for all trials and forward and reverse navigation trials by plot and choice point. ——, Total; ———, forward; ·····, reverse.

TABLE 1

*Cumulative totals of cue and feature knowledge as a function of
type of location*

| | Type of route location | | | |
Direction	Change direction and cross street	Change direction	Cross street	Other locations
Forward	16·8	10·25	11·0	7·64
Reverse	15·5	8·5	8·5	5·09
Totals	32·3	18·75	19·5	12·73

the data in our cumulative frequency graph in terms of these four types of loci, the results are as shown in Table 1.

Consistent with Hypothesis (2), knowledge of features and cues is a systematic function of the actions and decisions that must be made at specific loci. Coding and storage of information appears to be directly related to the potential importance of such information in correctly navigating a specific route. A choice point involving both a change in direction and street crossing has greater cue and feature coding than a point where only one such action occurs. Locations where no major action or choice is mandated are least well known.

The preceding data are cumulative for loci along the route. Of major interest is the course of acquiring such place and feature knowledge. Figure 4 shows cumulative frequency functions for successive days. The data clearly indicate a progressive differentiation of loci along the route with initial information primarily tied to the origin and destination. Even on initial trials, information is primarily known about the loci with actions or decisions. Over subsequent exposures, second and third order loci or nodes emerge, representing loci associated with specific localized actions. Loci without any specific action remain consistently low with respect to cue and feature knowledge.

Summary

The knowledge acquired about loci along the route and used for wayfinding appeared to consist of four types of nodes: (1) origin and destination nodes anchoring and defining the task environment; (2) interstitial second and third order nodes identifying key choice points where single or multiple actions were mandated; (3) lower order nodes that help define expectations with respect to the location of choice points; and (4) miscellaneous cues that were trial or episode specific. As the trials proceeded, it became obvious that cue and feature knowledge was greatest in the vicinity of higher order nodes. This phenomenon is well known in geography as the 'distance decay' effect, and in psychology as the 'spread' effect. In general geographic theory, it accounts for the development of 'neighborhoods' in the vicinity of major nodes and provides an important component of an individual's knowledge of environment.

At various points along the route, critical decisions have to be made. An incorrect decision involves the choice of an 'off-route' segment and the possibility of becoming lost. Critical second order nodes involving direction changes and street crossings quickly emerged in the subject's knowledge structure. Third order nodes, signaling

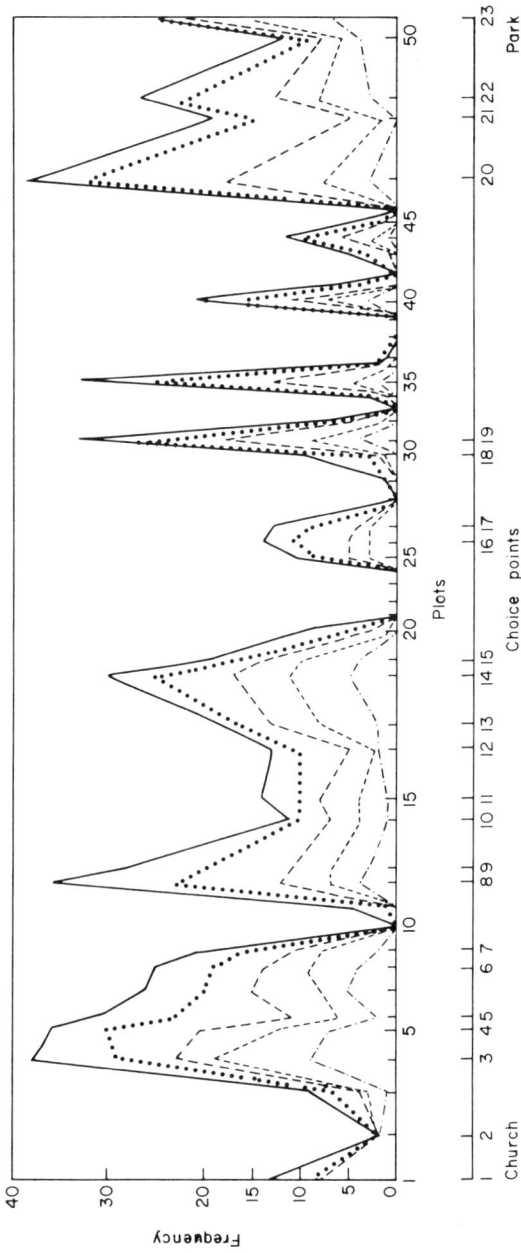

FIGURE 4. Cumulative frequency of cue recognition over successive days by plot and choice point. ——, Day 1; ----, day 2; —·—, day 3; ······, day 4; ———, day 5.

a direction change or street crossing also emerged. In the early stages of the task, these were not confidently recognized and used, resulting in some movement errors or in the development of incorrect expectations about the next segment of the route and its expected behaviors. Associated with each of the higher order nodes were a variety of lower order cues or features which helped identify and clarify these locations and arouse expectations regarding decisions and behaviors.

On various trials and at various places, trial-specific cues would emerge. Sometimes such cues would endure even when they were not permanent features, e.g. barking dogs in particular yards, cars parked in specific places and unique features not typical of the environment such as electrical wiring crossing the sidewalk and a fallen tree that served as a minor barrier to the normal path of locomotion. These cues did not signal decision points but impinged on the senses with sufficient impact to guarantee recall. Locationally they could be precise or imprecise and they need not signal any action (except 'continue') but could act as a general referent. They became attached to route segments and provided 'security' information during wayfinding.

Segmentation and sequencing

Hypothesis (3) predicts that routes naturally become segmented with respect to choice points. That is, the organization and acquisition of information with respect to choice points leads to a natural segmentation of the route. For the segments to be con-catenated into a unified structure representing a 'route', the individual must have sufficient knowledge of the major choice points along the route. We have shown that such knowledge emerges over trials. A natural consequence of this is an initial inability to generate the appropriate sequence of route segments. This was shown in two ways in the data. First, the verbal directions for the route were incomplete and gradually achieved precision on trials. Second, the route maps showed this same type of gradual progression. Figure 5 shows the route maps produced by the subject at different stages of testing. The maps on the left represent the first, third and fifth attempts at forward map drawing following forward navigation trials. The maps on the right represent the first, third, and fifth attempts at map drawing following reverse navigation trials.

Hypothesis (4) argues that, with repeated exposure to the task situation, individual route segments become concatenated and eventually form a complete representation of the whole route. These sketch maps show well identified route segments in the vicinity of the primary nodes on initial trials with incomplete segments and missing areas in the remainder of the total structure. There is a gradual elaboration with additional segments included but with some distortion, reversal and lack of segment differentiation. The final maps are complete with respect to segments, include additional detail about cues and features at specific choice points, and have a high degree of distance and directional accuracy relative to the actual route map. As cue and feature knowledge develops, the ability to coordinate and sequence this information into a coherent structure also develops.

An interesting phenomenon related to the appropriateness of sketch mapping as a representational technique was the marked difference in the trial sequence at which information about a choice point and segment first appeared. For many second and third order nodes, the subject was almost halfway through the trials before he was able to add detailed information to his maps and appropriately sequence all segments. However, cue and segment information regularly appeared in the verbal

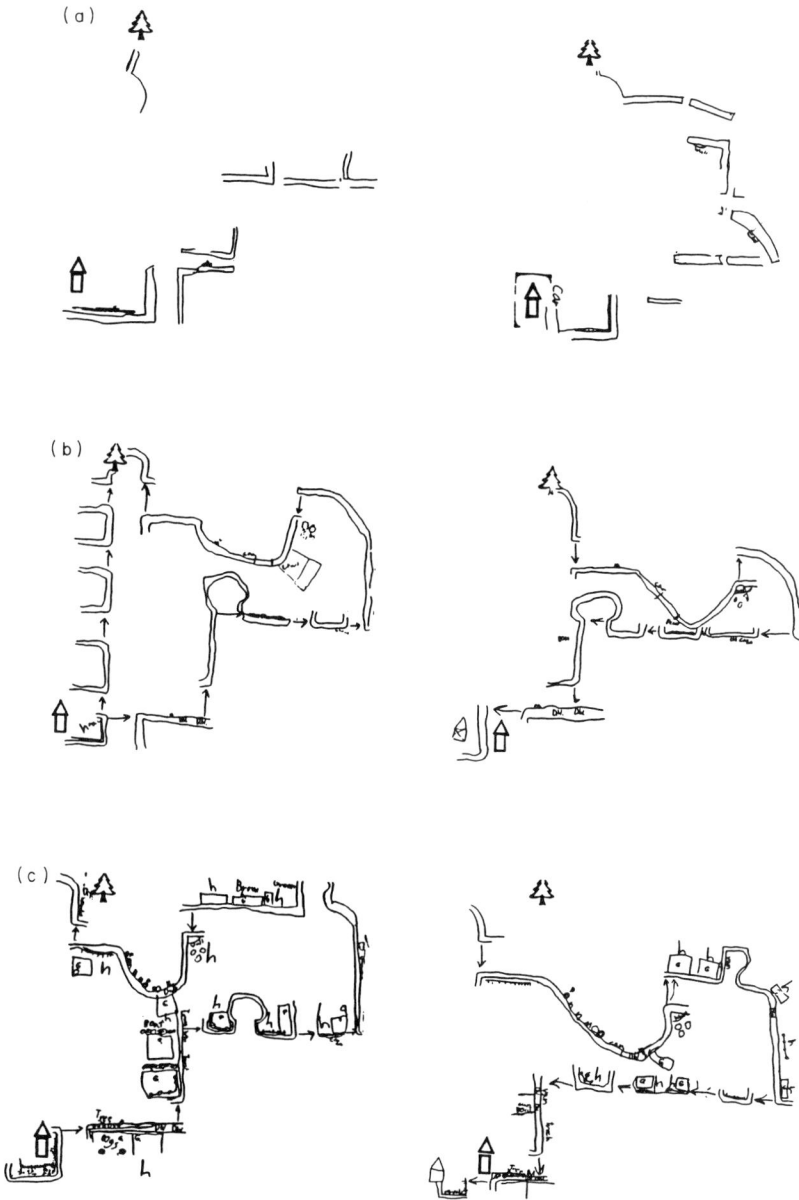

FIGURE 5. Sketch maps drawn after forward (left) and reverse (right) navigation trials. (a) Day 1. (b) Day 3. (c) Day 5.

descriptions and pictures of places along the route could be identified well before such information appeared on the maps. This is shown by data presented in Table 2. For segments representing the initial, middle, and final portions of the route, a number of specific features were mentioned in both the video and map tasks. Table 2 shows the proportion of such features mentioned on the same trial or mentioned first in either the video or map task. When a feature was in either task first, the

TABLE 2
*Proportion of features occurring on same trial and first in video or
map task and trial lag for different route segments*

Choice point boundaries for segments		Same trial	Video first	Map first
1–3	Proportion	0·50	0·33	0·17
	Trial Lag	—	8·00	4·00
7–8	Proportion	0·20	0·60	0·20
	Trial Lag	—	5·00	1·00
19–20	Proportion	0·23	0·69	0·08
	Trial Lag	—	2·80	1·00
20–23	Proportion	0·50	0·50	0·00
	Trial Lag	—	3·70	—

trial lag before it appeared in the other task was calculated. As can be seen in Table 2, for beginning and ending segments of the route, 50% of the features mentioned occurred on the same trials. When a feature was mentioned first in one task it typically occurred first in the video with a large trial lag before being included in the map task. For middle segments of the route, the largest proportion of features occurred first in the video task with a substantial trial lag before appearance in the map task. Thus it appears that individual bits of information were available in the knowledge structure before such information was organized into a serial structure where nodes were sequentially related. Further evidence for this progression was an initial inability to anticipate what would appear next in the videotape. Errors of prediction and anticipation were substantial on early trials and gradually diminished to zero errors on the final trials.

Errors
An extensive analysis and coding of errors was also conducted. We classified errors into two types: action and expectation errors. Errors in action consisted of navigational/procedural mistakes which caused the child to deviate from the established route. These errors appeared in three data sources: the behavioral protocol of performance during wayfinding, verbal directions for actions and actions mentioned during the viewing of the videotape recreation of the route. Errors in expectation consisted of a number of subcategories of confusions which did not directly result in a mistake action but indicated incomplete knowledge at a choice point location. Confusions that were categorized as errors in expectation included verbalized but nonexistent cues at choice point locations, substitution of cues from one route segment to another route segment, verbalized nonrecognition of a sequence of scenes, and incomplete coupling of cues and notions as indicated by vacillation about actions to be taken given a particular cue. Figure 6 represents the decreasing number of errors in action and expectation as well as the total errors over the ten trials. Errors diminished rapidly over the first two days and there were virtually no errors at choice points after the third day.

Of particular interest are the error data for individual choice points along the route. Figure 7 represents the frequency of errors at 23 potential choice points along

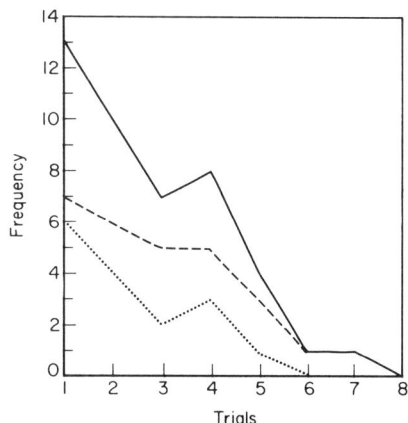

FIGURE 6. Error frequencies over successive trials. ——, Total errors at choice points; ----, actions; ·····, expectations.

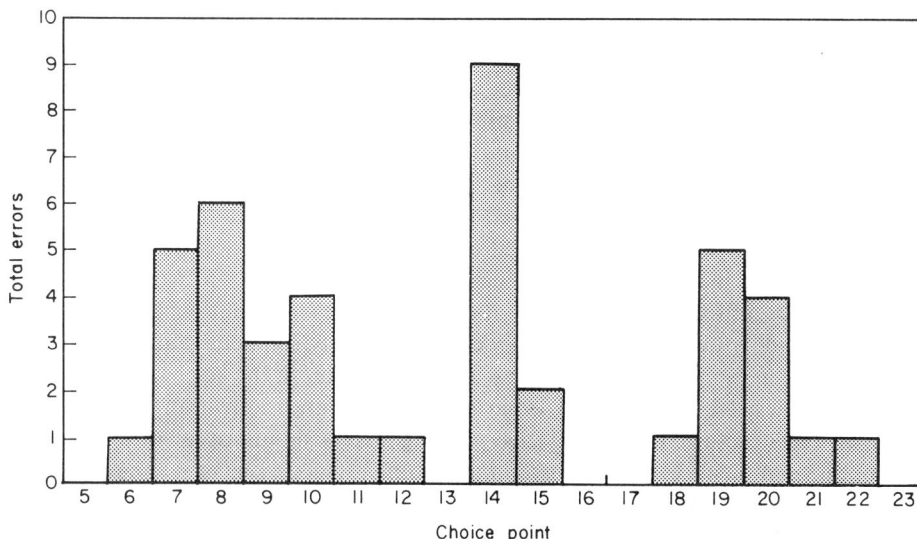

FIGURE 7. Error frequencies for individual choice points.

the route. The peak at choice point 14 corresponds to a location in the middle of the route where two actions, a street crossing and heading change, must be taken. Relatively few errors occur at loci near the beginning and end of the route and at loci such as 11 and 12 which involve a street crossing but with simplified decision making because this is an intersection involving a cul-de-sac. Errors that occur at these two loci are on the first trial. In contrast, errors at 14 persist until trial 7 and are the last recorded errors. Thus, the cue and feature knowledge data and the error data indicate that more is known about those loci which have more errors and more potential for errors in wayfinding. This is consistent with predictions based on Hypotheses (5) and (6). The total effect of differential weightings of choice points and route segments and the variations of error occurrence with choice point and segment complexity provides support for our final Hypothesis (7) relating to the hierarchical representation of route elements in LTM.

Additional empirical evidence

There is cause to be skeptical about the power of testing hypotheses derived from the model by a single child's data, no matter how detailed they might be. There are, however, other data which support the generality of the findings obtained in the case study. If knowledge of a route and, by extension, multiple routes through a neighborhood, is concentrated at choice points, then one would expect to find similar differences in knowledge of loci for children living in a given neighborhood. Doherty (1984) examined the latency and accuracy of scene recognition for specific loci in a suburban neighborhood, contrasting scenes at intersections (choice points) with scenes along routes. The children tested were from seven to 15 years of age and had been residents of the neighborhood for at least one year. As expected, scenes representing choice points were recognized more accurately than scenes along routes. This differentiation of loci in recognition accuracy only occurred in children of nine years old or above. As expected, older children had greater knowledge of their neighborhood and higher levels of recognition accuracy. Finally, Doherty (1984) showed that scenes depicting single plots were better recognized than scenes depicting views. Plots at intersections were particularly well recognized supporting the position that in a relatively undifferentiated suburban neighborhood, these plots serve as functional landmarks for navigation and anchor knowledge of route segments.

Other aspects of the conceptual model have been tested in a series of studies conducted by Shute (1984). She showed that recognition of familiar and unfamiliar loci in an urban setting is a systematic function of the degree of knowledge or familiarity of the individual with the loci and the amount of information contained in a slide. Decontextualization of a location reduces recognition accuracy, particularly for individuals with less knowledge of the environment. Shute (1974) also tested a basic assumption about the spread of activation in the knowledge structure. If, as hypothesized, spatial knowledge is organized and activation flows in such a knowledge structure, then facilitative priming effects should occur in scene recognition. She demonstrated that substantial priming effects occur in scene recognition when a given location is primed by slides depicting the adjacent physical locations. Such priming was demonstrated for locations of high and low familiarity and for individuals with high and low familiarity with the environment.

Discussion

In the first section we considered extant research and theory on spatial cognition, and delineated four issues that need to be addressed by any conceptual and computational model of the acquisition and representation of spatial knowledge. Our research program has attempted to deal directly with these issues by developing a conceptual model and testing some of its implications. We have sketched out our assumptions about how spatial knowledge is acquired, elements of our formal model embodying these assumptions, and the initial empirical work used to test both the hypotheses and the model. While the case study focuses on a single child learning a single route, the data are strongly in accord with the hypotheses derived from the conceptual model. Knowledge of a route is focused on key loci representing choice points. There is a natural segmentation of the route, with the segments becoming increasingly differentiated and appropriately sequenced. Routes appear to be hierarchically organized both with respect to the choice points and the segments they anchor. Errors occur more frequently and linger longer at points or in segments that occupy lower levels of the hierarchy. The child's knowledge of the route and its development are manifest in several aspects of behavior. These include the development of an ability to navigate the route without error and assistance, verbalize directions, draw sketches of the route, and outline verbal protocols about route details while viewing a videotape.

The results obtained from studies by Doherty (1984) and Shute (1984) provide strong additional support for basic assumptions about representation and processing. Children who are residents of a neighborhood have differential knowledge of locations consistent with the assumption that choice points constitute major 'nodes' in the route and survey knowledge of an environment. The differentiation of the location nodes is a function of accumulated episodic experience. Younger children who have limited range within their neighborhood fail to show much differentiation of locations. Locations that are more familiar to individuals are recognized faster and more accurately with more detailed information stored in memory. The greater detail permits more accurate recognition even when many of the FPV triples are not presented for recognition decisions. Evidence for the organization of locations is provided by their ability to prime each other and facilitate scene recognition.

While it appears that our initial assumptions serve our purpose at this stage of empirical testing, it is also apparent that a number of additional assumptions will be required as we make the task environment more complex and as we alter the scope of the model. For example, wayfinding tasks in environments that include dominant off-route cues within the task neighborhood may require additional assumptions relating to non-route orientations and frames of reference. While our theoretical conceptualization covers procedural and configurational (survey) knowledge, the model and task situations defined in this paper relate specifically to route-finding situations. Thus additional assumptions may be required regarding the general hierarchical organization of spatial knowledge, the process of generalizing to route finding activities in other unknown environments, and the mechanism by which the transfer from route to configurational knowledge structures take place.

With respect to the hypotheses outlined earlier in the paper, additional statistical work is needed to fully test some of them (e.g. those relating to cue richness (quality) by segment, and error occurrence). The computer simulations will relate more to testing concatenation, generalization and expectation procedures. Obviously the operational model needs to be tested by replicating actual subject moves through chosen task environments and then predicting such moves prior to exposing subjects to new task environments.

Our empirical studies have also pointed to other hypotheses whose testing will contribute to further development of the model. Examples of such hypotheses are that the amount and variety of information recalled about route segments varies according to the *type* of *object* anchoring the segment; and that characteristic *predictable errors* of recognition, action and expectation are a result of the activation of knowledge and the processes of perception, concatenation, pattern matching and generalization.

The essence of our conceptualization so far is that action or potential action is a critical component in deciding what environmental knowledge is accumulated and how it is used. The validity of this conceptual base will, in the long run, have to be tested in a non-wayfinding task or setting. This, along with the more extensive hypothesis testing proposed earlier, will provide a framework for continued research in the immediate future.

An obvious next step in the current program of research is to fully implement the model and determine its adequacy for simulating the data already obtained. The mechanisms of the model are one possible formalization of our general assumptions. However, formal implementation and testing against data will provide the necessary test to determine its initial sufficiency. Numerous additional predictions about performance can be derived and then tested in further empirical studies with children of different ages learning different routes. The sufficiency and generality of the model can then be further evaluated.

References

Allen, G. L. (1981). A developmental perspective on the effects of subdividing macrospatial experience. *Journal of Experimental Psychology: Human Learning and Memory*, 7, 120–132.

Allen, G. L., Siegel, G. W. and Rosinski, R. R. (1978). The role of perceptual context in structuring spatial knowledge. *Journal of Experimental Psychology: Human Learning and Memory*, 4, 617–630.

Allen, G. L., Kirasic, K. C., Siegel, A. W. and Herman, J. F. (1979). Developmental issues in cognitive mapping: The selection and utilization of environmental landmarks. *Child Development*, 50, 1062–1070.

Anderson, J. R. (1982). Acquisition of cognitive skill. *Psychological Review*, 89, 369–406.

Anderson, J. R. (1983). *The Architecture of Cognition*. Cambridge: Harvard University Press.

Anderson, J. R. and Bower, G. F. (1973). *Human Associative Memory*. New York: V. H. Winston & Sons.

Appleyard, D. (1970). Styles and methods of structuring a city. *Environment and Behavior*, 2, 100–118.

Briggs, R. (1972). *Cognitive distance in urban space*. Unpublished doctoral dissertation, The Ohio State University, Columbus, U.S.A.

Burrows, W. J. and Sadalla, E. K. (1979). Asymmetries in distance cognition. *Geographical Analysis*, 11, 414–421.

Cadwallader, M. (1979). Problems in cognitive distance and their implications for cognitive mapping. *Environment and Behavior*, 11, 559–576.

Chase, W. G. and Chi, M. T. H. (1980). Cognitive skill: Implications for spatial skill in large-scale environments. In J. Harvey (ed.), *Cognition, Social Behavior, and the Environment*. Potomac, Maryland: Lawrence Erlbaum Associates.

Cousins, J. H., Siegel, A. W. and Maxwell, S. E. (1983). Wayfinding and cognitive mapping in large-scale environments: A test for a development model. *Journal of Experimental Child Psychology*, 35, 1–20.

Doherty, S. E. (1984). *Developmental differences in cue recognition at spatial decision points*. Unpublished doctoral dissertation, University of California, Santa Barbara.

Downs, R. M. (1981). Maps and mappings as metaphors for spatial representation. In L. Liben, A. Patterson and N. Newcombe (eds), *Spatial Representation and Behavior Across the Life Span: Theory and Applications*. New York: Academic Press.

Downs, R. and Stea, D. (1973). *Image and Environment: Cognitive Mapping and Spatial Behavior*. Chicago: Aldine.

Ericksen, R. H. (1975). *The effects of perceived place attributes on cognition of urban residents* (Discussion Paper No. 23). Iowa City: University of Iowa, Department of Geography.

Gärling, T. and Böök, A. (1981). *The spatiotemporal sequencing of everyday activities: How people manage to find the shortest route to travel between places in their hometown.* Unpublished manuscript, University of Umeå, Department of Psychology, Sweden.

Golledge, R. G. (1978). Learning about urban environments. In T. Carlstein, D. Parkes and N. Thrift (eds), *Timing Space and Spacing Time.* London: Edward Arnold.

Golledge, R. G. and Hubert, L. J. (1982). Some comments on noneuclidean mental maps. *Environment and Planning A*, **14**, 107–118.

Golledge, R. G. and Rayner, J. N. (1976). *Cognitive Configurations of a City*, Vol. 2. Columbus, Ohio: The Ohio State University Research Foundation.

Golledge, R. G. and Spector, A. N. (1978). Comprehending the urban environment: Theory and practice. *Geographical Analysis*, **10**, 403–426.

Golledge, R. G., Briggs, R. and Demko, D. (1969). The configuration of distances in intra-urban space. *Proceedings of the Association of American Geographers*, **1**, 60–65.

Golledge, R. G., Parnicky, J. J. and Rayner, J. N. (1979). *The Spatial Competence of Selected Populations*, Vols 1 and 2. Columbus, OH: The Ohio State University Research Foundation.

Golledge, R. G., Rayner, J. N. and Rivizzigno, V. L. (1982). Comparing objective and cognitive representations of environmental cues. In R. G. Golledge and J. N. Rayner (eds), *Proximity and Preference: Problems in the Multidimensional Analysis of Large Data Sets.* Minneapolis: University of Minnesota Press, pp. 233–266.

Hart, R. A. and Moore, G. T. (1973). The development of spatial cognition: A review. In R. Downs and D. Stea (eds), *Image and Environment.* New York: Aldine, pp. 246–288.

Hazen, N. L., Lockman, J. J. and Pick, H. L., Jr. (1978). The development of children's representations of large-scale environments. *Child Development*, **4**, 71–115.

Herman, J. F. (1980). Children's cognitive maps of large-scale spaces: Effects of exploration, direction, and repeated experience. *Journal of Experimental Child Psychology*, **29**, 126–143.

Herman, J. F. and Siegel, A. W. (1978). The development of spatial representations of large-scale environments. *Journal of Experimental Child Psychology*, **26**, 389–406.

Kintsch, W. (1977). *Memory and Cognition.* New York: Wiley.

Kosslyn, S. M. (1981). The medium and the message in mental imagery: A theory. *Psychological Review*, **88**, 46–66.

Kosslyn, S. M. and Pomerantz, J. R. (1977). Imagery, propositions and the form of internal representations. *Cognitive Psychology*, **9**, 52–76.

Kosslyn, S. M., Pick, H. L. and Fariello, G. R. (1974). Cognitive maps in children and men. *Child Development*, **45**, 707–716.

Kuipers, B. (1978). Modeling spatial knowledge. *Cognitive Science*, **2**, 129–153.

Kuipers, B. (1982). The map in the head metaphor. *Environment and Behavior*, **4**, 202–220.

Kuipers, B. (1983). The cognitive map: Could it have been any other way? In H. L. Pick, Jr. and L. P. Acredolo (eds), *Spatial Orientation, Theory, Research and Application.* New York: Plenum Press, pp. 345–360.

Lloyd, R. (1982). A look at images. *Annals of the Association of American Geographers*, **72**, 532–548.

Lynch, K. (1960). *Image of the City.* Cambridge, Massachusetts: M.I.T. Press.

MacEachren, A. M. (1980). Travel time as the basis of cognitive distance. *The Professional Geographer*, **32**, 30–36.

MacKay, D. B., Olshavsky, R. W. and Sentell, G. (1975). Cognitive maps and spatial behavior of consumers. *Geographical Analysis*, **7**, 19–34.

McDermott, D. and Davis, E. (1984). Planning routes through uncertain territory. *Artificial Intelligence*, **22**, 107–156.

Miller, G. (1956). The magical number seven, plus or minus two: Some limits on our capacity for processing information. *Psychological Review*, **63**, 81–97.

Minsky, M. A. (1975). A framework for representing knowledge. In P. H. Winston (ed.), *The Psychology of Computer Vision.* New York: McGraw-Hill.

Moore, G. T. and Golledge, R. G. (1976). Environmental knowing: Concepts and theories. In G. T. Moore and R. G. Golledge (eds), *Environmental Knowing.* Stroudsburg, Pennsylvania: Dowden, Hutchinson & Ross.

Newell, A. and Simon, H. A. (1972). *Human Problem Solving*. Englewood Cliffs, NJ: Prentice Hall.

Paivio, A. (1969). Mental imagery in associative learning and memory. *Psychological Review*, **76**, 241–263.

Pezdek, K. and Evans, G. W. (1979). Visual and verbal memory for objects and their spatial locations. *Journal of Experimental Psychology: Human Learning and Memory*, **5**, 360–373.

Piaget, J. and Inhelder, B. (1967). *The Child's Conception of Space*. New York: W. W. Norton.

Quillian, M. R. (1968). Semantic memory. In M. Minsky (ed.), *Semantic Information Processing*. Cambridge, Massachusetts: M.I.T. Press.

Richardson, G. D. (1981). The appropriateness of using various Minkowskian metrics for representing cognitive configurations. *Environment and Planning A*, **13**, 475–485.

Sadalla, E. K. and Staplin, L. J. (1980*a*). The perception of traversed distance: Intersections. *Environment and Behavior*, **12**, 167–182.

Sadalla, E. K. and Staplin, L. J. (1980*b*). An information storage model for distance cognition. *Environment and Behavior*, **12**, 183–193.

Schank, R. (1975). The structure of episodes in memory. In D. Bobrow and A. Collins (eds), *Representation and Understanding*. New York: Academic Press.

Shemyakin, F. N. (1962). General problems of orientation in space and space representations. In B. G. Ananyev (ed.), *Psychological Science in the USSR*. Washington, D.C.: U.S. Office of Technical Reports (NTIS No. TT62-11083).

Shute, V. (1984). *Characteristics of Cognitive Cartography*. Unpublished doctoral dissertation, University of California, Santa Barbara.

Siegel, A. W. and White, S. H. (1975). The development of spatial representations of large-scale environments. In H. W. Reese (ed.), *Advances in Child Development and Behavior*, Vol. 10. New York: Academic Press.

Siegel, A. W., Allen, G. L. and Kirasic, K. C. (1979). Children's ability to make bi-directional distance comparisons: The advantage of thinking ahead. *Developmental Psychology*, **15**, 656–665.

Smith, T. R. and Lundberg, G. (1984). Psychological foundations of individual choice behavior and a new class of decision making models. In G. Bahrenberg, M. Fischer and P. Nijkamp (eds), *Recent Developments in Spatial Data Analysis*. Aldershot, England: Gower Publishing.

Smith, T. R., Pellegrino, J. W. and Golledge, R. G. (1982). Computational process modeling of spatial cognition and behavior. *Geographical Analysis*, **14**, 305–325.

Stevens, A. and Coupe, P. (1978). Distortions in judged spatial relations. *Cognitive Psychology*, **10**, 422–437.

Thorndyke, P. W. (1981). Distance estimation from cognitive maps. *Cognitive Psychology*, **13**, 526–550.

Tobler, W. R. (1976). The geometry of mental maps. In G. Rushton and R. G. Golledge (eds), *Spatial Choice and Spatial Behavior*. Columbus, Ohio: The Ohio State University Press.

Tolman, E. C. (1948). Cognitive maps in rats and men. *Psychological Review*, **55**, 189–208.

Trowbridge, C. C. (1913). Fundamental methods of orientation and imaginary maps. *Science*, **38**, 888–897.

THE ACQUISITION AND INTEGRATION OF ROUTE KNOWLEDGE IN AN UNFAMILIAR NEIGHBORHOOD

NATHAN GALE,* REGINALD G. GOLLEDGE,†‡
JAMES W. PELLEGRINO§ and SALLY DOHERTY**

*Robert D. Niehaus, Inc., Santa Barbara, California, U.S.A., †Department of Geography at the University of California, Santa Barbara 93106, U.S.A., §Vanderbilt University, and **Salk Institute

Abstract

This paper investigates the acquisition of neighborhood route knowledge by children of ages 9–12. Subjects were exposed to two routes through an unfamiliar suburban neighborhood, with five learning trials undertaken on each route. One route was acquired by actual field experience, while the other was acquired by viewing a video tape. Route knowledge acquisition was tested by performance on navigation, sketch mapping, and scene recognition tasks. Mode of experience had little effect on recognition performance; however, navigation performance following five video trials was inferior and approximated that of children with only one trial of field experience. These data support the differentiation of knowledge types and the need to engage in route navigation to proceduralize such knowledge.

Introduction

Spatial knowledge is often studied and discussed by decomposing it into distinguishable elements or components. These elements include zero-dimensional points, nodes, landmarks, places, or referent loci; one-dimensional lines, paths, or routes; and two-dimensional areas, configurations, or survey representations (e.g. Lynch, 1960; Shemyakin, 1962; Siegel & White, 1975; Golledge, 1978; Kuipers, 1978; Sadalla et al., 1980; Gärling et al., 1985). Although the parts have been delineated, there is less understanding of how the different knowledge components are integrated in the process of spatial knowledge acquisition. It is this problem that is the focus of our current research program, which involves studying both residents' knowledge of a familiar home neighborhood (e.g. Doherty et al., 1989) as well as the processes of knowledge acquisition in an unfamiliar setting (e.g. Gale, 1985). This paper concentrates on the latter problem.

The acquisition of spatial knowledge is based, for the most part, upon direct environmental experience. Indirect environmental experience may occur through a variety of media, such as maps, graphic or visual representations, verbal or written descriptions, and so on, but direct experience with the large-scale environment usually comes from physically moving through it (Siegel, 1981), though some attempts at building small-scale simulated environments have produced interesting results

Partial support for the research was provided by NSF Grant No. SES87-20597.

‡ Reprints requests and other correspondence should be directed to Reginald G. Golledge, Department of Geography, University of California, Santa Barbara 93106, U.S.A.

(Appleyard, 1977, 1981; Evans *et al.*, 1984*a*). The role of travel, and by necessity route navigation, is therefore considered fundamental in the process of the development and acquisition of a spatial knowledge base (Gärling *et al.*, 1982, 1986; Gale *et al.*, 1985; Teske & Balser, 1986). Route knowledge is the key to the integration of spatial knowledge components because it lies in an intermediary position between landmark and area or survey level knowledge (Allen, 1985), and because it is most closely associated with direct spatial behavior.

Spatial knowledge is acquired by individuals as a function of their experience within a given environment. At the simplest level, an individual must have knowledge of important objects and/or places, generally referred to as 'landmark' knowledge (Heft, 1979). This knowledge includes the ability to state with certainty that an object or place exists and the ability to recognize it when it is within the perceptual field. Such knowledge is declarative in nature. Spatial knowledge also includes information about the relationships among objects. In its simplest form, an individual possesses information about topological relationships such as proximity. A more detailed knowledge of relationships among objects or places involves Euclidean or metric properties, such as distance and direction within a coordinate space. Again, this information may be contained within an elaborated declarative knowledge structure. Knowledge of an environment also includes procedures for moving from a given location to another location and an ability to identify the routes that facilitate such actions. Both declarative and procedural knowledge must be available to perform way-finding tasks.

Empirical studies of spatial knowledge have focused on several issues related to landmark, route and configurational knowledge. One issue concerns the features of objects and locations that cause them to be selected as landmarks. Appleyard (1969) provided evidence that the most easily recalled and recognized buildings are those with high use and important symbolic functions. Physical features such as sharp contours, bright surfaces and noticeable size contrasts relative to the immediate surroundings also influence memory. These attributes of buildings, as well as attributes such as location near an intersection, also have been shown to affect the ability to locate buildings in a small-scale model reconstruction task (Herman & Siegel, 1978; Pezdek & Evans, 1979; Herman, 1980).

The importance of landmarks as components of routes has received less attention. Previous studies of route knowledge tended to focus on issues associated with distance judgements. The general finding was that a power function captures the relationship between actual and judged distance (e.g. Briggs, 1972; Ericksen, 1975; Allen *et al.*, 1978; Thorndyke, 1981; Allen & Kirasic, 1985). There also appear to be asymmetries in distance judgments depending upon choice of the reference point (e.g. Siegel *et al.*, 1979).

Explanations of route knowledge assume that, as a minimum, it consists of a series of procedural descriptions involving a sequential record of the starting point or anchor point (e.g. one's home), subsequent landmarks, intermediate stopping points and a final destination. The procedural representation must contain productions associated with those decision points where a change in direction takes place, as well as the appropriate actions to perform. More detailed route knowledge includes information about secondary and tertiary landmarks along a route and distances between and sequencing of landmarks. Routes of equivalent physical distance may differ substantially in the amount of stored information necessary to complete the route, i.e. the number of landmarks, intersections and locations where changes in direction are required. Estimates of distance between two points along a route may be a function of the amount of information stored in memory that must be processed to 'mentally' traverse the route. According to this hypothesis, an individual's estimate of distance between two points is based upon information pertaining to the effort involved in traversing the route. Effort can refer to directional changes and travel time, which are correlated. Empirical evidence has been obtained to support such a conjecture (MacEachren, 1980; Sadalla & Staplin, 1980*a,b*).

Recent studies by Allen have examined several components of route knowledge including landmark selection, distance judgments, and subdivisional processing by different age groups (Allen, 1981, 1987). The general paradigm involved a simulated walk along a complex route via the use of a series of successive photographic slides. Allen *et al.* (1978) showed that distance judgments for high landmark potential scenes were more accurate than corresponding judgments for low landmark potential scenes and this was the case after one or two presentations of the original route (see also Allen *et al.*, 1979). Allen (1981) has also explored the importance of landmarks with respect to their role in subdividing routes into separate segments or chunks of information. The basis for his study was the assumption that route information may be organized as a series of subdivisions bounded by distinct environmental features rather than being organized strictly as a sequence of landmarks connected by pathways. Thus, route knowledge may involve organization in terms of higher level units such as subdivisions and discrete loci within units. The data from Allen's (1981) forced-choice distance judgement experiment indicated that subdivisional organization significantly affected the accuracy of second and fifth graders and adults. When decisions could not be based on subdivisional membership then evidence was obtained on the availability of metric information about locations within a subdivision of a route. The availability of such information varied with age and developmental level or position in the life span (see also Liben, 1981; Walsh *et al.*, 1981; Evans *et al.*, 1984*b*).

The literature on spatial cognition also distinguishes between route knowledge and survey knowledge (Shemyakin, 1962). Theoretically and empirically there is support for the notion that knowledge progresses from landmarks to specific paths to an integrated frame of reference system for structuring spatial information. As knowledge accumulates and as distance information becomes more precise, notions of angularity and direction emerge. The exact nature of how the transition from route knowledge to survey knowledge occurs is largely unknown. There is, however, a volume of work that argues for the importance of attention processes in selecting spatial sensory information for further processing (Gärling & Böök, 1981). Of critical importance appears to be information such as whether or not a reference point is visible from the origin, destination or some intermediate decision point. These authors hypothesize that the acquisition of information about non-visible reference points requires central information processing or deliberate attention processes. Interfering with deliberate attention processes reduces the probability that information about reference points can be integrated in an overall frame of reference system.

Studies of the development of spatial knowledge have generally supported the progression outlined above (e.g. Curtis *et al.*, 1981; Liben, 1981; Cousins *et al.*, 1983). Inferences about knowledge components and the acquisition process have frequently been based on comparisons of individuals differing in age and experience within a given area (see e.g. Gärling *et al.*, 1982*b*). However, with notable exceptions (Spencer &

Darvizeh, 1981*a,b*; Hart, 1984; Matthews, 1984*a*; Golledge *et al.*, 1985) little work has been done that focuses on the acquisition of knowledge by an individual as a function of direct or simulated experience with a given route or set of routes in an area. Many learning studies have focused on adults and, to a lesser extent, children's ability to learn certain components of route knowledge either through direct experience or via slide or video presentation (Goldin & Thorndyke, 1982; Biel, 1983; Poag *et al.*, 1983; Carpman *et al.*, 1985).

There are several reasons why 'data gathering has lagged far behind theory building in the study of the development of spatial representations of large-scale environments' (Allen *et al.*, 1979, p. 1062). At a pragmatic level, study of the acquisition of knowledge of a single route or multiple routes through a large-scale space is exceedingly time consuming, with a single 'acquisition' trial often requiring 30 to 60 minutes. Exceptions to this include abbreviation of the procedure through presentation of selected views and/or the use of short routes. Second, it is often difficult to test efficiently for the existence of several components of spatial knowledge, i.e. the testing procedures to assess landmark knowledge, distance and directional accuracy, and navigational performance are time consuming. Third, studies of the acquisition of one or more routes in large-scale space must also deal with constraints imposed by the nature of the environment itself and the inability to control it directly. Thus, some attempts have been made to study route acquisition in impoverished and 'artificial' environments or simulations of real environments so that route characteristics can be systematically varied (e.g. Gärling *et al.*, 1982*a*; Evans *et al.*, 1984*a*; Stern & Leiser, 1988). Given these problems in studying the acquisition of spatial knowledge, it is not surprising that cross-sectional designs have typically been used to evaluate the general theory that the emergence of spatial knowledge is a systematic function of accumulated and organized perceptual experiences representing interactions between the organism and the environment (Kuipers, 1982).

Our approach in the study presented here is to examine spatial knowledge acquisition over time through repeated trials of well-defined route learning experiences. We pursue the issue of how the medium of experience affects the acquisition of knowledge by examining the performance of subjects who learn routes under two conditions: actual field experience and video tape presentation.

A battery of experimental tasks were designed to test various components of the acquisition process (Bryant, 1984; Matthews, 1984*b*; Waller, 1986). First, we recorded actual navigation behavior over prescribed routes as descriptive measures of task-oriented environmental learning. Second, we created scene recognition tasks to evaluate sensitivity for different types of scenes and locations. Third, sketch maps were used as a less-structured means of testing a variety of knowledge components: the features indicated on the maps reflect subjects' recall for important landmarks, the representation of the route itself gives information regarding segmentation and subdivision units, and the directional orientation of parts of the sketches can be used as a measure of the development of a more general spatial framework.

Method

Study area

The environment used for this field study was a residential neighborhood in Goleta, California (Figure 1). This is the same area used previously for studies of residents'

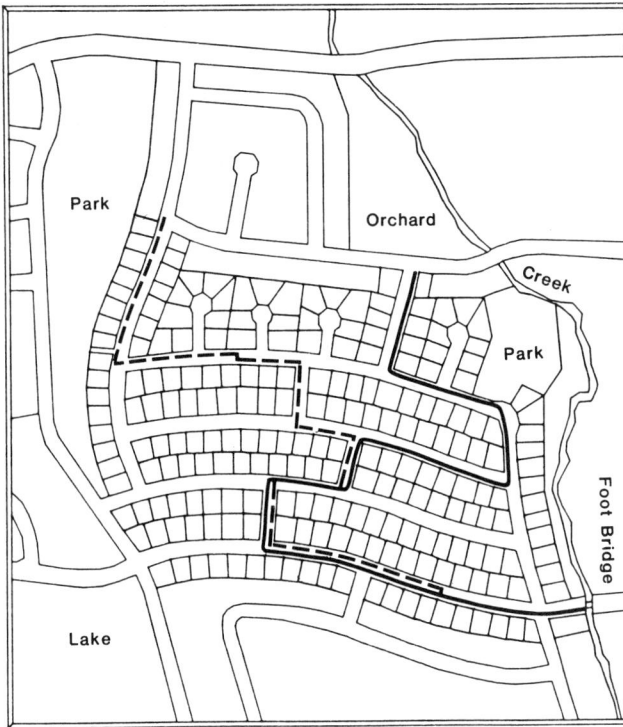

FIGURE 1. Study neighborhood map. —— = Route 1; ‐‐‐ = Route 2.

neighborhood knowledge (Doherty & Pellegrino, 1985; Doherty et al., 1989). This neighborhood was selected for route learning experiments, in part, because it provides a quite regular yet sufficiently complex configuration of streets and intersections. The streets include longer thoroughfares as well as much shorter connector streets and cul-de-sacs. All but three of the intersections in the study area are T-intersections, at which one street ends where it meets a street running perpendicular to it. There are two four-way intersections, one of which leads abruptly to a footbridge suitable only for bicycles and pedestrians. The remaining intersection is L-shaped, with two streets terminating at a right angle junction.

Two partially overlapping routes, both between 0·7 and 0·8 miles in length, were specified through the study area. The first route begins at the footbridge and ends at an orchard. It involves seven changes of direction, five of which are associated with crossing a street; additionally, there are three street crossings at intersections which involve no change of direction. The second route begins at the back entrance to the county park and ends at the footbridge. This route also necessitates seven major changes in direction, but only three of these are associated with street crossings; there are two street crossings at intersections with no change in direction. Further distinguishing the second route from the first are two additional street crossings that do not occur at natural intersections—these are included to examine the effect of arbitrary decision points.

Subjects
Sixteen children, eight females and eight males, were selected as subjects. There were
two females and two males from each of four ages—9, 10, 11 and 12 years. This range of
ages was targeted because it appears to be a pivotal period with regard to the
development of spatial representations (Doherty & Pellegrino, 1985). In previous
experiments, Doherty (1984) tested male and female children ranging in age from 6–14
years on scene (landmark) recognition, sketch mapping, and route knowledge of this
neighborhood. All children were long-term residents of the neighborhood. Her data
clearly differentiated performance on the basis of age, with significant differences
occurring between 6–8-year-olds and the older children (see also Doherty & Pellegrino,
1985). Children aged 9–12 years (regardless of sex) could be differentiated from both
younger and older groups on a variety of recognition, recall and movement tasks while
producing no significant within-group variation. Children in this age range have not
developed survey representation capabilities according to developmental theory
(Piaget & Inhelder, 1967; Siegel & White, 1975). Thus, their performance on route
learning tasks should be uncontaminated by survey procedures. Such subjects should
then provide the best insights into many facets of the route learning process.
    Subjects were selected for the study on the basis of their willingness to participate
and their consistent availability for daily testing over a two-week period. A further
requirement was that all subjects be completely unfamiliar with the study area. Upon
selection the children were alternately assigned to one of four experimental groups
which counterbalanced assignment of routes to both the first and second weeks of
testing and the field versus video learning conditions.

Materials
The materials developed for this study included a comprehensive set of slides and video
tapes of the routes. Depicted in the slides are two types of scenes designated as views
and plots. Slides of views were made to represent what would be seen by a subject
looking straight ahead while walking along the sidewalk of a given street. They were
taken from points on the sidewalk corresponding to the midpoint of each house plot
with the camera aimed parallel to the street. Two slides were taken at corner plots—
one from the midpoint to both sides of the plot. Approaching intersections, additional
scenes were photographed to represent the views of any potential change in direction
which could occur at that intersection. Slides of each of the 221 house plots in the
neighborhood were also made. Like the views, these were taken from a point at
midplot, but the camera was rotated 90 degrees to point perpendicularly to the street.
All slides, both views and plots, were taken with the same Nikon F2 camera with a
24 mm lens, yielding a viewing angle of 80 degrees. For slides of plots it was necessary to
ensure that the entire house could be depicted without including information from
adjacent plots. Slides of views and scenes from several other suburban neighborhoods
were also taken. These contrast neighborhoods were chosen on the basis of similarity of
character with the task environment and location; all were similar in terms of age and
type of housing, and all were at least three to four miles distant.
    In addition to the slides, both of the neighborhood routes were videotaped for the
laboratory presentation. Video taping was done with a Sony HVC 2200 video camera
fitted with a wide-angle lens to maximize the field of vision. The camera was operated
from a seated position by a person sitting in a wheelchair. This helped to obtain a
steady picture without requiring the use of expensive steady-cam equipment. The

camera was held pointing straight ahead in the direction of travel. Every effort was made during taping to maintain a constant, comfortable pace, in order to simulate as closely as possible an actual walk by the subject population. The camera was held at a height above the pavement similar to that of the normal viewing angle of a 9–12-year-old child. The completed video presentation was approximately 18 minutes in viewing time for each route.

Scene recognition tasks containing 80 photographs were constructed for both of the neighborhood routes. The two tasks were identical in design. Forty-eight (60%) of the photos in each task represented scenes on the route. Of the remaining 32 (40%), half were off-route but in the same neighborhood and half were from other neighborhoods. Within each of these scene groups, photos were divided evenly between those depicting views along the sidewalk and house plots. Moreover, the photos were also categorized by type of location, and again an equal split was made between scenes at intersections and non-intersections.

To construct the scene recognition tasks, both sets of 80 photographs were divided into four blocks of 20 scenes. The composition of each block was identical to that of the entire set divided by four. Photos in each scene category were randomly assigned to, and ordered within, the four blocks.

Procedures

Each child participated in the study for ten sessions—one per day, Monday through Friday, for a two-week period. Daily sessions typically lasted from 60–90 minutes depending on the combination of tasks for the day and the speed of the child. In all phases of the experiment the children were tested individually, and consequently the testing of 16 subjects required two experimenters working over a period of eight weeks.

Testing was carried out using a counterbalanced design. In each week a separate route was learned. The different conditions depended on whether Route 1 or Route 2 was learned in the first or second week, and whether learning was active (i.e. in the field) or passive (i.e. in the laboratory). Subjects were distributed equally between the conditions by age and sex. This schedule is described in Table 1.

To carry out the field testing, each child was first driven to the start of one of the routes via streets that were outside of the study area. Before beginning the walk the child was instructed to follow the experimenter and attend to the route. It was made clear that after the first day the roles would be reversed and the experimenter would follow the child. Furthermore, the child was told briefly about the scene recognition and map drawing tasks to be completed after traversing the route. During the walk, conversation was neither initiated nor encouraged by the experimenter so as not to aid nor distract the child in any way. A single trial consisted of one forward traversal of the route.

TABLE 1
Experimental design

Condition	Week 1	Week 2	Age and sex			
I	Field Rt 1	Video Rt 2	9 M	10 F	11 M	12 F
II	Field Rt 2	Video Rt 1	9 F	10 M	11 F	12 M
III	Video Rt 1	Field Rt 2	9 F	10 M	11 F	12 M
IV	Video Rt 2	Field Rt 1	9 M	10 F	11 M	12 F

On the second, third, fourth and fifth trials, the child was asked to navigate the same route with no help from the experimenter except an indication that a navigation mistake was made. On a map of the study area the experimenter recorded the child's behaviors by type and location. After making a mistake, the child was allowed to continue for a distance of one and a half to two house plots from the point of the mistake. This was done to give the child an opportunity to recognize the mistake and potentially correct it. If he or she did so, the behavior was recorded as a 'realized mistake', otherwise an 'unrealized mistake' was recorded, whereupon the child was stopped and returned to the location at which the mistake was made. In this manner the child would eventually make the right navigation decision, if only through the process of elimination.

After each subject completed the first, third and fifth trials of the experiment, testing of various aspects of spatial knowledge was carried out. The first test was a map drawing task. For this the child was given a piece of paper with the relative positions of the start and finish of the route marked with Xs. The instructions were to draw a map of the route as accurately as possible, including everything that could be remembered; no clues were given as to starting orientation.

Following the sketch mapping, the scene recognition task was administered. The child was shown the 80 slides, one at a time, and asked to respond 'yes' to those scenes that were judged to be on the route and 'no' to those not on the route, or about which they were uncertain. In each session the photographs were presented in different block order (e.g. 1, 2, 3, 4; 3, 1, 4, 2; 2, 4, 1, 3).

For the laboratory trials, instead of going to the neighborhood and actually navigating the route, the children viewed a simulated walk via video tape. An attempt was made to make the laboratory experience as close as possible to that of the field. Each child was told of the tasks to follow the presentation as was done in the case of the field experiment. The map drawing and slide recognition tasks were carried out on the first, third and fifth days after viewing the simulated walk, exactly as after the actual walk.

Finally, as a test of field navigation following laboratory learning, those eight subjects who saw the video presentations during the second week (i.e. Conditions I and II) were required to navigate the route. All other testing was completed before this task so that it would not in any way contaminate the results from the week of field testing. The task consisted of a single route navigation trial, with behaviors recorded for comparison to navigation performance following learning in the field.

Results and Discussion

Route learning in the field

For the field learning trials, behaviors associated with the route learning tasks were tabulated for all subjects (Conditions I, II, III and IV) using three categories: (1) Recorded Behaviors—including all mistakes, stops, hesitations, tentativeness and verbal expressions of disorientation; (2) Total Mistakes—including turns in the wrong direction, crossings to the wrong side of the street and failures to turn, whether realized by the subject without the aid of the experimenter or not; and (3) Unrealized Mistakes—including all mistakes except those recognized as such by the subject, i.e. only those mistakes which necessitated experimenter intervention.

As shown in Figure 2, route learning mistakes were almost eliminated by the end of the third trial, with little apparent change between the fourth and fifth trials. The lapsed time from start to finish of the routes also showed a similar leveling off by the fourth trial. Route 2, which included the two arbitrary crossings, appears to have been the more difficult of the two routes. With respect to the number of mistakes, however, by the fourth trial both routes had been learned almost equally well.

The data on each of the three behavioral measures and Time were analysed, in turn, using a 2×4 (route $\# \times$ trial) ANOVA design. No main effects were found for the group factor (route $\#$), indicating that the observed differences between the two routes were not statistically significant. In contrast, the within-subject factor (trial) was significant in all cases: Recorded Behaviors, $F(3, 42) = 84\cdot39$, $p < 0\cdot01$; Total Mistakes, $F(3, 42) = 35\cdot90$, $p < 0\cdot01$; Unrealized Mistakes, $F = 37\cdot14$, $p < 0\cdot01$; Time, $F(3, 42) = 17\cdot81$, $p < 0\cdot01$. No route $\# \times$ trial interactions were significant.

With respect to the variables collected from the field navigation task, there was strong evidence of relatively rapid and systematic learning across all measures. By the end of the third trial, learning was virtually complete. Two subjects, both 12-year-old girls, actually experienced one-trial learning, not once making an unrealized mistake, and only one subject made an unrealized mistake on the final day. These results emphasize the orderly manner in which routes were learned, and imply that, even in a completely unfamiliar environment, successful strategies for route learning can be developed readily by children in this age range.

Field navigation after video
The performance of those eight subjects who carried out the one-trial route navigation task in the field after viewing five video presentations (Conditions I and II) was

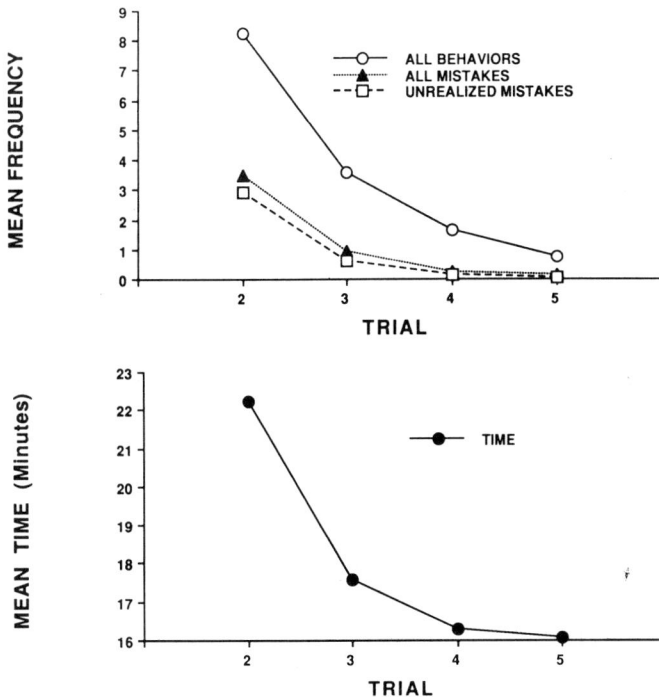

FIGURE 2. Navigation behavior as a function of trial and type of learning experience.

TABLE 2
*Navigation performance during field learning (first and last navigation trials)
and after video learning (only navigation trial)*

Performance measure	Field learning (n = 16)		Video learning (n = 8)
	First	Last	Only
All behaviors	8·25	0·75	7·25
All mistakes	3·50	0·13	3·63
Unrealized mistakes	2·94	0·06	2·71
Time (minutes)	22·22	16·09	21·07

compared to the performance of all 16 subjects who had five field learning trials. This comparison reveals major differences between the two learning experiences. The relevant data are shown in Table 2. For all four measures of performance there is a marked similarity between the video results and those obtained on the first free navigation trial in the field.

Statistical tests confirm these impressions. Differences between the video and field experiences on Trial 5 (last navigation trial) were found to be significant with respect to all of the variables: Recorded Behaviors, $t = 4.39$, $p < 0.01$; Total Mistakes, $t = 3.52$, $p < 0.05$; Unrealized Mistakes, $t = 2.73$, $p < 0.05$; Time, $t = 2.50$, $p < 0.05$. No significant differences were found when comparing the video data with the Trial 2 (first navigation trial) field results. Seeing the route five times in the laboratory had essentially the same effect on navigation as one presentation of the route in the field. This observed difference in navigation performance is an important finding, especially in light of the other experimental results, for it suggests that different types of knowledge were acquired in the two learning experiences.

Sketch mapping
Individual differences in graphic abilities and styles led to considerable variation in the sketch maps drawn following the first, third and fifth learning trials. It was, nonetheless, possible to define appropriate quantitative measures describing key elements of the sketches in terms of orientation, segmentation and features. The data for each of these map elements were then analysed, in turn, using a $2 \times 2 \times 3$ (learning experience × route # × trial) ANOVA design.

Orientation. A simple binary classification (correct/incorrect) was used to code the beginning and ending route segments on each of the sketch maps with respect to direction. Considering both segments together, a single map could thus be scored 0·0, 0·5 or 1·0. The results, in terms of percent correct representations, are shown in Figure 3. Interestingly, there was essentially no change over successive trials in the sketch maps following the video presentations. On the other hand, the field experience apparently helped the subjects gain a better sense of orientation over time, as the number of correct directional representations increased by 25 percentage points from the first to the fifth trial.

Analysis was done by testing the data against chance. Considering the four cardinal directions as possible alternatives, the chance of a correct representation for the beginning and ending route segments is 0·25. However, since the start and finish of the

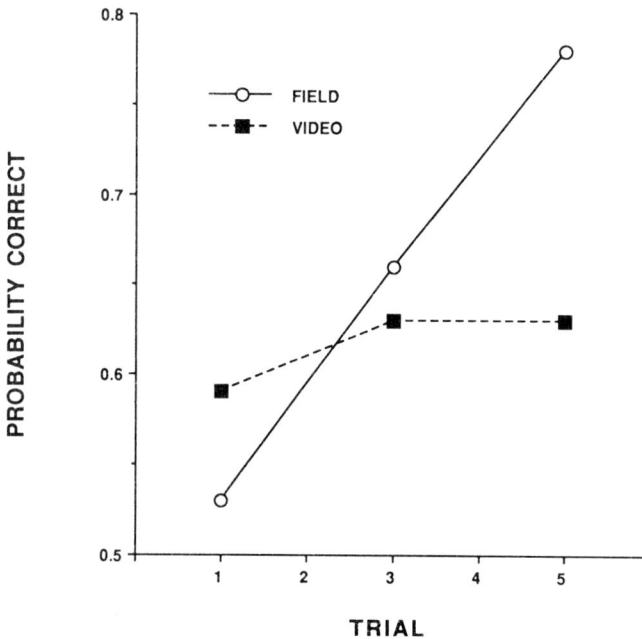

FIGURE 3. Sketch map orientation accuracy as a function of trial and type of learning experience.

route were given to the subjects, a more realistic and conservative assumption is that the probability of being correct is 0·50. On this basis, and assuming independence among events, the value of 0·53 obtained on Trial 1 of the field experience, and the value of 0·59 found on Trial 1 of the video experience, are not significant at the 0·05 level. The score of 0·63 found on both Trial 3 and Trial 5 of the video experience is marginally significant ($p = 0.052$). But the results on Trial 3 (0·66) and Trial 5 (0·78) of the field experience are quite significantly different from chance ($p < 0.05$ and $p < 0.01$, respectively). Hence it appears that although orientation accuracy improved only marginally with time following video presentations, actual experience in the field was of considerably more help in acquiring a general frame of reference and relative sense of direction.

Segmentation. To see how well the subjects were able to remember changes in direction, the sketch maps were scored according to the number of segments represented in each. The data were transformed into absolute deviations from the actual number and analysed using a $4 \times 2 \times 3$ (condition × type of learning experience × trial) design. The main effect of type of learning experience was not significant, $F(1, 12) = 0.41, p < 0.10$. There was, however, a marginally significant main effect of trial, $F(2, 24) = 2.91, p < 0.07$. Overall, the average number of mistakes by trial was: Trial 1, 1·72; Trial 3, 1·16; and Trial 5, 1·29, indicating a decrease between the first and third trials but no significant difference between the third and fifth trials. In general, the subjects were very accurate in estimating the number of turns involved on each of the routes. It seems that they were able to acquire a basic knowledge of the degree of complexity of the routes quite easily, as no improvement was evident after the third trial. Characterized in terms of the mistakes made in representing the number of route segments, the two learning experiences led to maps of essentially equivalent accuracy.

Features. Finally, the sketch maps were analysed with regard to the type and number of features represented along each route. Once again a $4 \times 2 \times 3$ (condition × type of learning experience × trial) ANOVA design was used to analyse the feature frequency data. The only significant result of this analysis was the main effect of trial, $F(2, 24) = 10.92$, $p < 0.001$. Over time the pattern of increase in the mean number of features represented was: Trial 1 = 1.16; Trial 3 = 2.22; and Trial 5 = 2.66.

With increased experience, more about the routes was remembered, but in most cases the number of features represented on the maps was quite small—an average of about two features per map. Many of the subjects, in fact, only indicated features on their maps after a prompt to include anything else that they remembered. It appears, therefore, that the recall and graphic representation of features is a relatively difficult task and may not be a good indicator of route learning activities.

Although the overall numbers for the field and video experiences were remarkably similar, some important differences were observable. First, a greater number of different features were represented after the video presentations; more commonality among a smaller number of features was evident following the field experience. Secondly, the predominance of street names was considerably stronger for the field versus the video. And thirdly, the video results reflect a more sidewalk oriented experience, with higher frequencies for such features as mailboxes and bus stops.

Clearly, the nature of the video presentation was largely responsible for the latter two results; some of the street signs were simply not legible in the video tapes, and obviously the constant straight ahead viewing angle limited the field of vision such that many peripheral objects (e.g. a boat parked at the far end of a driveway) could not be seen. But given the fact that the video presentations constituted a more restricted, controlled and consistent image of the environment, it seems surprising that this experience led to a greater variety of different features being recalled. Perhaps because street names were more important to the navigation task involved in the field experience, there was a trade-off between these and other features. On the other hand, in the video experience, either because they were not visible or since no navigation decisions had to be made, less attention was given to street names and more attention could hence be focused on a larger variety of other objects.

Scene recognition

Primary analysis. A nonparametric signal detection measure for sensitivity, calculated using the formulae given in detail in Grier (1971), was used to analyse the scene recognition data. The sensitivity index (ranging from 0.5 to 1.0) is derived from an adjustment, or 'correction', of the 'hits' (correct recognitions of on-route scenes) by the number of 'false alarms' (incorrect positive responses to off-route scenes). Analysis was carried out using a $4 \times 2 \times 3 \times 2 \times 2$ (condition × type of learning experience × trial × location × scene) mixed design.

Significant main effects were found for trial, $F(2, 24) = 24.78$, $p < 0.001$, and scene, $F(1, 12) = 25.38$, $p < 0.001$; significant two-factor interactions were limited to scene × location, $F(1, 12) = 9.56$, $p < 0.01$, and scene × condition, $F(3, 12) = 4.90$, $p < 0.05$.

The main effect of trial reflects an increase in sensitivity over time. From the first to the third trial the largest increase in sensitivity occurred, from an initial score of 0.60 to a value of 0.69. A more modest increase between the third and fifth trials brought the final overall sensitivity to 0.72. Combining trials, the overall mean sensitivity was 0.67. With respect to scene, subjects showed a higher sensitivity for views (0.71) than for plots

(0·63). The pattern of increase in sensitivity over time is quite consistent across all factors, yielding no significant two-factor interactions involving trial. Thus, the increase in sensitivity cannot be attributed to any particular kind of scene, but rather must be considered a broadly based general effect.

The fact that sensitivity for views was consistently higher than sensitivity for plots is not at all surprising considering that the slides of views more closely represented the task environment as perceived in both the field and laboratory settings. Since the video tapes were filmed with the camera pointed straight down the sidewalk, the images presented to the subject were very similar to those seen in the slides of views. Sensitivity for plots, under these circumstances, implies the ability to recognize scenes from a novel angle, rotated 90° from the angle of initial presentation. In the field, the navigation task also directed attention in the line of travel, but the subjects could easily turn their heads, and thus directly perceive a house from essentially the same viewpoint as represented in the plot slides. One would therefore expect higher sensitivities for plots following the field experience than after viewing the video tapes. This was indeed the case, with plot sensitivity scores of 0·65 and 0·61 found for field and video, respectively. The interaction between experience and scene, however, was not significant.

Perhaps as important as the significant main effects is the fact that no significant effects were found for type of learning experience or location. In terms of scene recognition, response patterns were virtually identical for learning experiences in both real and simulated environments. And in addition to the absence of a main effect, no significant interactions involving experience were found. Thus, it appears that the differences between active (actual navigation) and passive (simulated navigation) were not reflected in differential recognition abilities. Apparently very similar amounts and kinds of propositional information were acquired in both experiences.

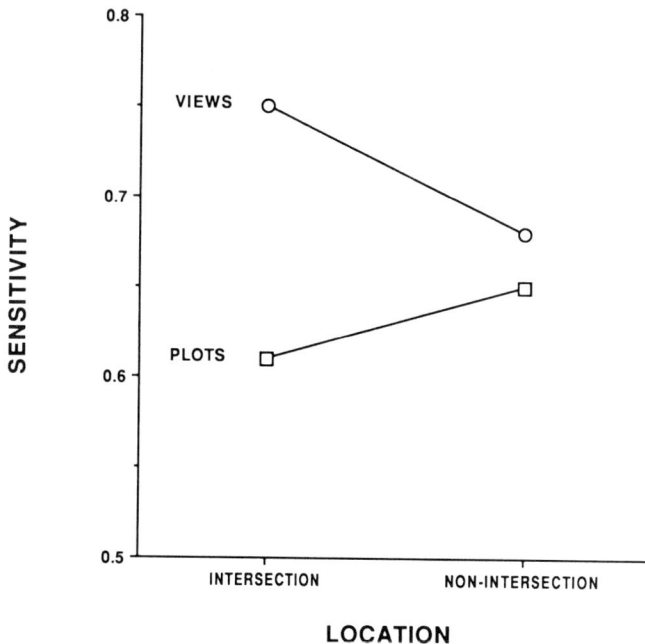

FIGURE 4. Scene recognition sensitivity as a function of scene location and type of scene.

The lack of a main effect of location appears at first to contradict both theory and previous empirical findings (Doherty & Pellegrino, 1985). Many theoretical conceptualizations suggest that a major characteristic of spatial knowledge is its concentration on landmarks or important nodes. Further, it is argued that choice points in a route act as such landmarks for the organization of knowledge; and, indeed, the studies undertaken to date support this notion.

With respect to the present study, an explanation for the absence of a main effect of location may be found in light of the significant interaction between location and scene. As depicted in Figure 4, the highest sensitivity was found for views at intersections, with considerably lower scores for views not at intersections. Bearing in mind that views are the type of scene which best represent the task environment, this result does, in fact, support the hypothesis of the concentration of knowledge at choice points. The reason for no main effect associated with location is that plots at intersections had the lowest overall sensitivity. It appears, then, that there may have been a trade-off at choice points between views and plots. The perceptual patterns encoded at intersections, where navigational decisions are made, are quite possibly more general 'wide angle' views rather than narrowly focused images of particular objects or house plots.

One additional factor concerning the difference between views and plots at intersections is the fact that in the slide scenes of plots there are generally no clues to indicate the location of the plot. A slide depicting a plot at an intersection is virtually indistinguishable from a slide of a plot along the way. Thus, there is no information given in the stimulus to associate the scene with a choice point. On the other hand, by their very nature views at intersections can be directly associated with actual or potential navigation decisions, possibly facilitating sensitivity for these scenes.

The scene × location interaction is also of some interest in the context of comparing the two learning experiences. As shown in Figure 5, the effect of location appeared to be qualitatively very similar in both the real and simulated environments, emphasized by the fact that the three-way experience × scene × location interaction was not significant. There is some evidence, however, of a quantitative difference between experiences as the video experience tended to facilitate sensitivity for views, and adversely affect sensitivity for plots.

The same distinction is evident, in part, in the interaction between scene and condition. The subjects in Conditions III and IV, who saw the video first before going into the field, had higher sensitivities for views than for plots.

Neighborhood effects. To examine the effect of neighborhood on scene recognition, the off-route slides included scenes from inside the study area and scenes from other similar neighborhoods. For the primary analysis these were all classified together as off-route scenes; in this secondary analysis, 'in' and 'out' of neighborhood scenes were treated separately. Thus, neighborhood (in versus out of the study area) was introduced as a within-subject factor in an analysis of response accuracy for the off-route scenes. The design used for the ANOVA was therefore $4 \times 2 \times 3 \times 2 \times 2 \times 2$ (condition × type of learning experience × trial × location × scene × neighborhood) mixed design.

A significant main effect was found for neighborhood, $F(1, 12) = 23.51, p < 0.001$, and several significant interactions also involved this factor. These results suggest that in the process of learning the specific routes through the study area, the subjects also learned some basic neighborhood characteristics. Correct rejections were higher for out of neighborhood scenes (0.88) than for inside neighborhood scenes (0.79), indicating that sensitivities for on-route scenes were increased by the subjects' ability to

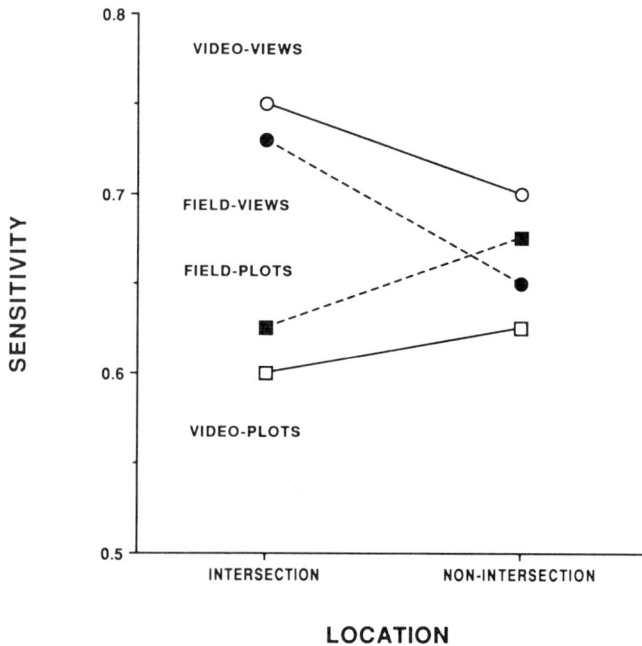

FIGURE 5.   Scene recognition sensitivity as a function of scene location, type of scene and type of learning experience.

identify those scenes which did not belong in the neighborhood, and hence could not be on the route.

General Discussion

Learning experience, tasks and outcomes
There appears to be no adequate substitute for field work, as it is the closest we can come to replicating actual learning and behavior in real world environments. Experimental work in natural settings will always be fraught with problems stemming from the inability to control the environment and to ensure that conditions remain consistent over the course of an experiment. The clear advantage of using a simulated representation of the environment, such as videotaped walks, is that the exposure to the environment, if not totally controlled, is at least the same each time. One can thus eliminate many of the inconsistencies in environmental experience that inevitably are a part of field research. On the other hand, without field testing one can never really know how behavior in the laboratory relates to behavior in actual large-scale environments.

Interpretations of the results from experimental testing are based on assumptions concerning the type of information provided by the different tasks. In this study it is assumed that the slide recognition task is essentially a test of declarative knowledge about neighborhood scenes. This type of spatial knowledge is assumed to consist of sets, or patterns, of propositional facts about environmental features and their properties. Recognition of a scene depicted on a slide is assumed to imply the ability to perform a successful pattern match between the visual stimulus and stored information. The recognition task is perhaps the least ambiguous of those employed in the study because it attempts to test what is considered to be the simplest initial stage in spatial knowledge acquisition, namely place identification.

The map drawing task is more complex. It attempts not only to elicit information about specific place or feature characteristics but also information about spatial properties. In an admittedly less than perfect way, the mapping task is used to gather simultaneously partial data on all three components of spatial knowledge—landmark, route and configuration. The features represented on the maps presumably are tied to the identification of particular places or landmarks; the number of direction changes that are indicated is assumed to be an approximate measure of route length and spatial complexity; and the orientations of the concatenated route segments, together with the location of any off-route or marginal information, are taken to be a reflection of the development of survey knowledge.

As with the slide recognition task, it is assumed that the information obtained from the sketch mapping task is primarily indicative of the declarative knowledge structure of the subjects. What is elicited in this manner is some measure of how much is known about the salient features and spatial properties of a route. In addition, the route representations possibly contain information about the procedures necessary for navigation, but map drawing is not a test of these procedures as such. Although it could be argued that producing a map of a route requires procedural knowledge, it is clear that the procedures required for graphic representation are not necessarily related to those required for route navigation. The lack of ability to create a symbolic externalized reproduction of a route does not mean that the procedures for navigation are absent and, likewise, being able to navigate a route does not imply that one can draw a map of it.

The quality of the sketch map data is thus tainted by reliance on the recall and graphic abilities of the subjects. Indeed, an explanation for the relatively small number of features included on the maps may well be related to the difficulty in recalling stored information which, given a visual stimulus, could be accessed quite easily but otherwise lies beyond retrieval. Moreover, even if such information is recalled the subject may be reluctant or unable to provide a corresponding graphic representation. Such problems with the use of sketch maps, however, are most critical in making between-subject comparisons. They are less important in the present study since, in regard to the mapping data, the major concern is with the within-subject factors of experience and trial.

The experimental procedures followed in the field testing were designed to obtain descriptive measures of navigation performance. It is assumed here that these measures reflect the procedural route knowledge of the subject. To navigate a route one must acquire a declarative database, a set of rules for locomotion, and the appropriate associations between place identification and motor response. In this context, therefore, a mistake in navigation implies either a failure in recognition or a missing or incorrect rule of action; conversely, successful navigation implies the embedding of factual knowledge into productions which, if followed in sequence, lead to achievement of the goal.

Specific findings deriving from the sets of tasks previously described are as follows:

(1) The conceptual differentiation between declarative and procedural knowledge types can be verified empirically with respect to the acquisition of route knowledge.

Strong support for this hypothesis is derived from the fact that results of the laboratory tests involving recognition and recall of specific cues showed virtually no differences between the field and video learning experiences, while navigation performance differed greatly. If spatial knowledge with respect to routes were primarily of one basic type, then the similarities in knowledge acquisition evident in the laboratory test results should have been reflected in actual navigation performance. The differences in ability to navigate after five trials in the field and five trials of the video presentation cannot be attributed to quantitative differences in the amount of factual knowledge acquired. In this respect, on practically every measure investigated, no significant differences between the learning experiences were found. Thus, it must be assumed that very similar declarative knowledge about the task environment was acquired, stored and could be accessed, whether learning took place in the real or simulated environments.

That such similar knowledge could lead to quite different levels of navigation performance can best be explained by the differentiation of knowledge types. Learning in the field involved not only the acquisition of declarative knowledge about locations but also the integration of this data into the procedural knowledge, or productions, necessary for the successful completion of the navigation task. Even though the same basic spatial information was available after learning from the video presentations, this information was apparently not coupled with the appropriate action directives to the same degree.

(2) Active engagement in navigation facilitates the proceduralization of route knowledge.

Although it seems quite clear that both field and video learning experiences involved the acquisition of declarative spatial knowledge, and also that the field experience led to the development of a much greater amount of procedural knowledge than did the video experience, the question remains as to why this was so. One might argue that because the field experience was characterized by active navigation in a real environment, while the video experience involved passive navigation in a simulated environment, the two factors of type of environment and type of activity were inextricably confounded. Under these circumstances it is difficult to ascertain whether poor navigation performance after laboratory learning was due to the lack of active engagement in decision making or the experience of an impoverished environment. However, since no major differences were found between experiences with respect to recognition, we may assume that the video tapes were a reasonable representation of the real route and not significantly impoverished in terms of the portrayal of essential spatial information. Therefore we must conclude that it was the difference in type of engagement, not type of environment, that facilitated the development of procedural knowledge. In other words, learning 'about' may come by seeing, but apparently learning 'how to' comes only by doing (cf. Anderson, 1982).

(3) During route learning, more information is coded at intersections where choices are made than between intersections.

Support for this hypothesis must be qualified but is nonetheless quite strong. In all of the analyses of the recognition data no main effects of location were found. However, the highest sensitivity scores were obtained for views at intersection. Thus, the hypothesis holds for the type of scene most representative of the route learning task. Moreover, several plausible explanations have been outlined for the low sensitivities found for plots at intersections which account for the absence of a significant main effect of location.

Since the effect of location was found to be similar for both field and video experiences, it cannot be assumed that encoding of information at intersections is dependent on actually making a navigation decision. The potential for making a decision, or the simulated representation of a change in direction, may be equally as important in invoking environmental learning. Perhaps it is the combination of actual or potential navigation choices, coupled with the greater complexity and richness of the scenes, that make intersections the foci of route knowledge.

(4) Knowledge acquisition of a specific route involves concurrent learning of the environmental context at a more general level.

In the process of route learning, information about particular places was acquired and stored; and apparently, on another level, some knowledge of the basic commonalities found in this information was also developed, providing a set of general neighborhood descriptions. This synthesized knowledge facilitated the correct rejection of those off-route scenes that were not from the study area. In addition to a more general knowledge of neighborhood characteristics, there was some evidence that concurrent with route learning was at least partial development of a survey representation of the environment. The results in this regard were not conclusive, but based on the sketch map data it appears that the field experience led increasingly to a better overall sense of orientation and a knowledge of directional relationships. The same degree of survey knowledge acquisition was not evident in the video experience.

Theoretical implications

Although primarily empirical in focus, this study has produced several results of importance to the theory and modeling of spatial knowledge acquisition. First, it has shown that some of the concepts developed in work on non-spatial knowledge acquisition can usefully be applied in a spatial context. Specifically, the concepts related to declarative and procedural types, or stages, of knowledge appear to be crucial. Any valid theory or model of the acquisition of spatial knowledge must deal with the development of both of these types of knowledge structures. Of particular interest to route learning are the mechanisms and processes by which productions are created, compiled and proceduralized such that they appropriately integrate information from the declarative spatial database with the action directives necessary for the completion of navigation goals.

In this regard, the research results suggest that having a certain amount of declarative knowledge does not imply the procedures to operate on that knowledge have also been developed—they may or may not have been, depending on the type of learning experience. From the results of the video learning experience it seems clear that one can know a good deal about a route, and even perhaps have some understanding of the procedures involved in traversing the route, without having acquired the procedural skills necessary for navigation. The term 'route knowledge', therefore, becomes quite ambiguous if not carefully defined. A clear distinction must be made between knowledge (propositions, facts) about a route and knowledge (ability, skill) of how to navigate the route.

Theories of spatial knowledge acquisition must also address questions relating to the effect of different kinds of interaction with the environment, particularly with respect to the degree of active engagement in goal oriented tasks. The evidence of this study indicates that the kind of interaction is a more important factor than the medium of environmental presentation in the differential acquisition of procedural (route) knowledge. If the encoding of environmental information is conceptualized in terms of a process such as the 'Now Print!' mechanism proposed by Livingston (1967a,b), then whenever an individual experiences a biologically important event a general order is sent out for all recently active patterns to be printed. Such patterns, therefore, would include, in addition to the sensory information from the environment, the associated motor patterns of bodily locomotion. Thus, passive exposure to routes, such as the video presentations, may allow for the printing and memory of visual patterns, but without any further engagement of the subject the appropriate actions to be taken at any given place may not be learned. On the other hand, during active involvement in route learning, whenever the nervous system 'takes a picture', as it were, it records both what it sees and what it is doing; in this way actions can be associated with places to create the productions necessary for navigation.

Another finding is that route knowledge is quite parsimonious. Successful navigation apparently does not require extensive knowledge about all the scenes along a route. On the last field trial almost all the subjects made no mistakes in actual navigation, and yet less than half of the on-route scenes in the slide task were correctly recognized. This degree of sensitivity is well below that found in the study of children living in the neighborhood (Doherty & Pellegrino, 1985). Knowledge acquisition during route learning, therefore, is a highly selective process, and much environmental information can be ignored without adverse effects on navigation. Efficiency can have its drawbacks, however, in that a parsimonious route knowledge system—selectively created for a given task—might consequently lead to errors in behavioral performance if the goal of the task changes.

To understand the mechanisms of route learning more completely the criteria for selectivity must be known. In this regard, several questions come to mind. What is it that induces attention to be given to certain scenes? Why is it that significance is attached to some places and not to others? Is it the properties of the environment, the type of interaction, the goals of the decision-maker or all of these factors which determine what places and/or actions are 'printed' and integrated into spatial knowledge structures?

The scene type that was selected and stored with the highest frequency in both real and simulated exposures to routes was views at intersections. Several factors may contribute to the explanation of this result. First, in a specific route learning task what is learned may be determined by the goal of the task—one learns first what one needs to learn. Strictly speaking, in order to accomplish the requirement of reaching a particular location by a prescribed route, all that is necessary is to be able to recognize the choice points and know the correct action to take at each of these locations. Thus, in light of the goal, the places where action must be taken are the places that must be known. They become salient by definition of the task.

Views at intersections may also accrue salience from their physical characteristics. Of all neighborhood scenes, they are the most complex, affording a much more expanded spatial context than scenes along streets with houses on both sides. Also, the number and variety of features that can be perceived, as well as the variation in distances and viewing angles, is greatest at intersections.

With respect to an individual's interaction with the environment, intersections represent the points at which new views are first seen (Heft, 1983). The novelty of fresh visual stimulus may have a significant impact on focusing attention and encoding information. In the video presentations the impact of each new view is perhaps accentuated by the panning of the camera that takes place as the corner is turned. In the field, the experience of a new view at an intersection is usually accompanied by more extensive visual panning in several directions, motivated by fear of traffic, if not curiosity.

The fact that scenes of views were recognized more often than scenes of individual house plots gives further evidence of the importance to spatial learning of the kind of interaction with the environment. An important comparison with previous research is that, for children living in the neighborhood, sensitivities were higher for plots than for views (Doherty & Pellegrino, 1985). Taken together these results imply that extensive and varied types of interaction with the environment tend to facilitate the acquisition and storage of information about specific features, while more limited and controlled environmental exposure through a route learning task tends to lead to knowledge of more general scenes. With respect to modeling route knowledge acquisition, it appears crucial that representations of the environment be constructed not solely of particular objects or features but also of composite views which include many features and the spatial relations among them. It may be not so much the features themselves but rather integrated patterns of features that are the most salient components of the knowledge acquired while learning a route.

A number of questions are raised by this research that warrant further investigation. We now know something more about the relations between recognition and navigation. Without further research on properties such as the ability to make correct interpoint distance judgements between points on the route, we find that we know little about the ability to make spatial judgements. To investigate such questions, further experiments must be carried out that incorporate tasks requiring subjective estimations of the inherent spatial properties of routes involving sequence, distance and directional relations. In terms of the components of spatial knowledge, the present research has concentrated on the integration of landmarks and routes. How both the declarative and procedural knowledge of individual routes contribute to the development of relational information and its eventual integration into survey knowledge remains a largely unexplored question, a question that is now the focus of the authors' ongoing research program.

References

Allen, G. (1981). A developmental perspective on the effects of 'subdividing' macrospatial experience. *Journal of Experimental Psychology*, 7, 120–132.
Allen, G. (1985). Strengthening weak links in the study of the development of macrospatial cognition. In R. Cohen, Ed., *The Development of Spatial Cognition*. Hillsdale, NJ: Lawrence Erlbaum Associates.
Allen, G. (1987). Cognitive influences on the acquisition of route knowledge in children and adults. In P. Ellen & C. Thinus-Blanc, Eds., *Cognitive Processes and Spatial Orientation in Animal and Man*. Dordrecht: Martinus Nijhoff, vol. II. Neurophysiology and developmental aspects. NATO ASI Series, Series D: Behavioural and social sciences—no. 37, pp. 274–283.
Allen, G. & Kirasic, K. (1985). Effects of the cognitive organization of route knowledge on judgments of macrospatial distance. *Memory and Cognition*, 13, 218–227.

Allen, G., Kirasic, K., Siegel, G. & Herman, J. (1979). Developmental issues in cognitive mapping: the selection and utilization of environmental landmarks. *Child Development*, **50**, 1062–1070.

Allen, G., Siegel, G. & Rosinski, R. (1978). The role of perceptual context in structuring spatial knowledge. *Journal of Experimental Psychology: Human Learning and Memory*, **4**, 617–630.

Anderson, J. R. (1982). Acquisition of cognitive skill. *Psychological Review*, **89**, 369–406.

Appleyard, D. (1969). Why buildings are known. *Environment and Behavior*, **1**, 131–159.

Appleyard, D. (1977). Understanding professional media: issues, theory and a research agenda. In I. Altman & J. F. Wohlwill, Eds., *Human Behavior and Environment*. New York: Plenum Press, vol. 2.

Appleyard, D. (1981). *Livable Streets*. Berkeley, CA: University of California Press.

Biel, A. (1983). Children's spatial representation of their neighborhood: a step towards a general spatial competence. *Journal of Environmental Psychology*, **2**, 193–200.

Briggs, R. (1972). *Cognitive Distance in Urban Space*. Ph.D. dissertation. Columbus, OH: Ohio State University.

Bryant, K. (1984). Methodological convergence as an issue in environmental cognition research. *Journal of Environmental Psychology*, **4**, 43–60.

Carpman, J., Grant, M. & Simmons, D. (1985). Hospital design and wayfinding: a video simulation study. *Environment and Behavior*, **17**(3), 296–314.

Cousins, J., Siegel, A. & Maxwell, S. (1983). Wayfinding and cognitive mapping in large-scale environments: a test of a developmental model. *Journal of Experimental Child Psychology*, **35**, 1–20.

Curtis, L., Siegel, A. & Furlong, N. (1981). Developmental differences in cognitive mapping: configurational knowledge of familiar large-scale environments. *Journal of Experimental Child Psychology*, **31**, 456–469.

Doherty, S. (1984). *Developmental Differences in Cue Recognition and Spatial Decision Making*. Ph.D. dissertation, Graduate School of Education, University of California, Santa Barbara.

Doherty, S. & Pellegrino, J. W. (1985). Developmental changes in neighborhood scene recognition. *Children's Environments Quarterly*, **2**(3), 38–43.

Doherty, S., Gale, N., Pellegrino, J. & Golledge, R. (1989). Children's vs adults' knowledge of places and distances in a familiar neighborhood environment. *Children's Environments Quarterly* **6**, 65–71.

Ericksen, R. (1975). *The Affects of Perceived Place Attributes in Cognition of Urban Residents*. Discussion paper no. 23, Department of Geography, University of Iowa.

Evans, G., Skorpanich, M., Gärling, T., Bruant, K. & Bresolin, B. (1984*a*). The effects of pathway configuration, landmarks, and stress on environmental cognition. *Journal of Environmental Psychology*, **4**, 323–335.

Evans, G., Brennan, P., Skorpanich, M. & Held, D. (1984*b*). Cognitive mapping and elderly adults: verbal and location memory for urban landmarks. *The Journal of Gerontology*, **39**, 452–457.

Gale, N. (1985). *Route Learning by Children in Real and Simulated Environments*. Ph.D. dissertation, University of California, Santa Barbara.

Gale, N., Doherty, S., Pellegrino, J. W. & Golledge, R. G. (1985). Toward reassembling the image. *Children's Environments Quarterly*, **2**(3), 10–18.

Gärling, T. & Böök, A. (1981). *The Spatio Temporal Sequencing of Everyday Activities: How People Manage to Find the Shortest Route to Travel Between Places in Their Home Town*. Unpublished manuscript. Department of Psychology, University of Umeå, Sweden.

Gärling, T., Böök, A. & Ergezen, N. (1982*a*). Memory for the spatial layout of the everyday physical environment: differential rates of acquisition of different types of information. *Scandinavian Journal of Psychology*, **23**, 23–35.

Gärling, T., Böök, A. & Lindberg, E. (1982*b*). Adults' memory representations of the spatial properties of their everyday physical environment. In R. Cohen, Ed., *The Development of Spatial Cognition*. Hillsdale, NJ: Lawrence Erlbaum Associates, pp. 141–148.

Gärling, T., Böök, A. & Lindberg, E. (1985). Adults' memory representations of the spatial properties of the everyday physical environment. In R. Cohen, Ed., *The Development of Spatial Cognition*. Hillsdale, NJ: Lawrence Erlbaum Associates, pp. 141–148.

Goldin, S. & Thorndyke, P. (1982). Simulating navigation for spatial knowledge acquisition. *Human Factors*, **24**, 457–471.

Golledge, R. G. (1978). Learning about urban environments. In T. Carlstein, D. Parkes & N. Thrift, Eds., *Making Sense of Time*. London: Edward Arnold, vol. 1.

Golledge, R. G., Smith, T. R., Pellegrino, J. W., Doherty, S. & Marshall, S. P. (1985). A conceptual model and empirical analysis of children's acquisition of spatial knowledge. *Journal of Environmental Psychology*, **5**, 125–152.

Grier, J. B. (1971). Nonparametric indexes for sensitivity and bias: computing formulas. *Psychological Bulletin*, **75**, 424–429.

Hart, R. (1984). The geography of children and children's geographies. In T. Saarinen, D. Seamon & J. Sell, Eds., *Environmental Perception and Behavior: An Inventory and Prospect*. University of Chicago, Department of Geography Research Paper No. 209.

Heft, H. (1979). The role of environmental features in route learning: two exploratory studies of wayfinding. *Environmental Psychology and Non-Verbal Behavior*, **3**, 172–185.

Heft, H. (1983). Way-finding as the perception of information over time. *Population and Environment: Behavioral and Social Issues*, **6**, 133–150.

Herman, J. (1980). Children's cognitive maps of large-scale spaces: effects of exploration, direction, and repeated experience. *Journal of Experimental Child Psychology*, **29**, 126–143.

Herman, J. & Siegel, A. (1978). The development of spatial representation of large-scale environments. *Journal of Experimental Child Psychology*, **26**, 389–406.

Kuipers, B. (1978). Modeling spatial knowledge. *Cognitive Science*, **2**, 129–153.

Kuipers, B. (1982). The map in the head metaphor. *Environment and Behavior*, **4**, 202–220.

Liben, L. (1981). Spatial representation and behavior: multiple perspectives. In L. Liben, A. Patterson & N. Newcombe, Eds., *Spatial Representation and Behavior Across the Lifespan*. New York: Academic Press, pp. 3–32.

Livingston, R. B. (1967a). Brain circuitry relating to complex behavior. In G. C. Quarton, T. Melnechuk & F. O. Schmitt, Eds., *The Neuro-Sciences: A Study Program*. New York: Rockefeller University Press.

Livingston, R. B. (1967b). Reinforcement. In G. C. Quarton, T. Melnechuk & F. O. Schmitt, Eds., *The Neuro-Sciences: A Study Program*. New York: Rockefeller University Press.

Lynch, K. (1960). *The Image of the City*. Cambridge, MA: M.I.T. Press.

MacEachren, A. (1980). Travel time as the basis of cognitive distance. *The Professional Geographer*, **32**, 30–36.

Matthews, M. (1984a). Environmental cognition of young children: images of journey to school and home area. *Transactions of the Institute of British Geographers*, NS, **9**, 89–105.

Matthews, M. (1984b). Cognitive maps: a comparison of graphic and iconic techniques. *Area*, **16**, 33–40.

Pezdek, K. & Evans, G. (1979). Visual and variable memory for objects and their spatial locations. *Journal of Experimental Psychology: Human Learning and Memory*, **5**, 360–373.

Piaget, J. & Inhelder, B. (1967). *The Child's Conception of Space*. London: Routledge and Kegan Paul; New York: Humanities Press.

Poag, C., Cohen, R. & Weatherford, D. (1983). Spatial representations of young children: the role of self- versus adult-directed movement and viewing. *Journal of Experimental Child Psychology*, **35**, 172–179.

Sadalla, E. & Staplin, L. (1980a). An information storage model for distance cognition. *Environment and Behavior*, **12**, 183–193.

Sadalla, E. & Staplin, L. (1980b). The perception of traversed distance: intersections. *Environment and Behavior*, **12**, 167–182.

Sadalla, E., Burroughs, W. & Staplin, L. (1980). Reference points in spatial cognition. *Journal of Experimental Psychology: Human Learning and Memory*, **6**, 516–528.

Shemyakin, F. N. (1962). General problems of orientation in space and space representations. In B. G. Ananyev, Ed., *Psychological Science in the USSR*. Arlington, VA: U.S. Office of Technical Reports (NTIS No. TT62-11083), vol. 1.

Siegel, A. W. (1981). The externalization of cognitive maps by children and adults: in search of ways to ask better questions. In L. Liben, A. Patterson & N. Newcombe, Eds., *Spatial Representation and Behavior Across the Life Span: Theory and Application*. New York: Academic Press.

Siegel, A. W. & White, S. H. (1975). The development of spatial representations of large-scale environments. In H. W. Reese, Ed., *Advances in Child Development and Behavior*. New York: Academic Press, vol. 10.

Siegel, A. W., Allen, G. & Kirasic, K. (1979). The development of cognitive maps in large and small scale spaces. *Child Development*, **50**, 582–585.

Spencer, C. & Darvizeh, Z. (1981a). The case for developing a cognitive environmental psychology that does not underestimate the abilities of young children. *Journal of Environmental Psychology*, **1**, 21–31.

Spencer, C. & Darvizeh, Z. (1981b). Young children's descriptions of their local environment: a comparison of information elicited by recall, recognition and performance techniques of investigation. *Environmental Education and Information*, **1**, 275–284.

Stern, E. & Leiser, D. (1988). Levels of spatial knowledge and urban travel modeling. *Geographical Analysis*, **20**, 140–156.

Teske, J. & Balser, D. (1986). Levels of organization in urban navigation. *Journal of Environmental Psychology*, **6**, 305–327.

Thorndyke, P. (1981). Distance estimations from cognitive maps. *Cognitive Psychology*, **13**, 526–550.

Waller, G. (1986). The development of route knowledge: multiple dimensions? *Journal of Environmental Psychology*, **6**, 109–120.

Walsh, D., Krauss, I. & Regnier, V. (1981). Spatial ability, environmental knowledge and environmental use: the elderly. In L. Liben, A. Patterson & N. Newcombe, Eds., *Spatial Representation and Behavior Across the Lifespan*. New York: Academic Press, pp. 321–357.

THE DEVELOPMENT OF ROUTE KNOWLEDGE: MULTIPLE DIMENSIONS?

GLENN WALLER

Department of Experimental Psychology, University of Oxford, South Parks Road, Oxford OX1 3UD, U.K.

Abstract

Children are better at way-finding than at externalizing their spatial knowledge in abstract settings (e.g. in tests of route recall). It is suggested that this is due to experimenters' tendencies to use single criteria for development, such as the acquisition of landmark knowledge. Children of different ages may be using different types of information in their route representations. However, if only one aspect of the child's knowledge is examined at a time, then developmental changes of style may go unnoticed. Previous experiments have tested children's spatial knowledge as if it were qualitatively similar to that of adults, which is not always appropriate.

A study is reported in which young schoolchildren were asked simply to describe routes in a familiar area. There was a developmental shift from a 'Directions—End Information' format to one of 'Landmark—End Information'. From this it is concluded that children may be capable of using different information earlier in life, which is adequate to allow way-finding but which has not been recorded by the traditional unidimensional measures of spatial knowledge. A case is made for the use of combined measures when investigating the many skills involved in environmental cognition.

Introduction

Theories of children's spatial knowledge have generally concentrated on the young child's deficiencies relative to older subjects. In particular, young children's route knowledge has been characterized as being very poor. They appear to lack skills of landmark selection (Allen *et al.*, 1980), sequential ordering of landmarks (Cousins *et al.*, 1983), and 'vector' mapping (Conning and Byrne, 1984). However, such poor route knowledge seems to be at odds with the very obvious way-finding skills which children actually have in realistic settings (Cousins *et al.*, 1983; Spencer and Darvizeh, 1983; Conning and Byrne, 1984). This study looks at whether it is fair to describe children as having deficient spatial representations, when most studies only test one aspect of their knowledge. In particular, the child's ability to use directional indicators has been ignored, and the use of landmarks has been emphasized. These components are both tested in this study, to examine whether children's spatial knowledge is really deficient, or whether it simply differs from the adult form.

The most obvious explanation for the discrepancy between children's way-finding skills and their apparent route knowledge is that children do not have the same type of cognitive representation of routes as adults. Both Piaget and Inhelder (1956) and Siegel and White (1975) have proposed that young children are unable to represent linear sequences, until they have reached some qualitative developmental 'stage' of representational skill. These theoretical positions can still accomodate the child's way-finding skills. Newcombe (1981) points out that children could lack any kind of integrated route representation, yet could be recognizing the landmarks as they proceed.

In the case of verbal descriptions of routes, it is possible that children's apparent route knowledge is restricted more by their poor spatial reference skills. Young children respond to instructions containing locatives in ways which show that they do not understand the spatial relationship intended by an adult. In particular, the child responds according to conceptual preferences (Clark, 1973) and according to the specific task employed (Durkin, 1981), rather than to any 'objective' criteria. Children's poor comprehension of the terms involved in spatial reference offers a further explanation of why they will be unable to describe routes adequately, yet still be able to find their way using recognition of landmarks.

A further possibility, to be taken up in this study, is that children are able to represent and describe routes, but that their skills are being under-estimated by the use of inappropriate measures (cf. Spencer and Darvizeh, 1981), which are more suited to adult subjects. There is evidence that young children can represent landmarks and routes. Acredolo et al. (1975) have shown that even three- to four-year-olds must be aware of landmarks, as they depend upon adjacent landmarks when trying to recall where a bunch of keys was dropped. Possibly then, the poor landmark selection and utilization shown by seven-year-olds (Allen et al., 1979) and the improvement in landmark selection between seven and 13 years (Cousins et al., 1983) are products of the tasks used, irrelevant to younger children's real-life performance. Similarly, Spencer and Darvizeh (1983) have shown that three- to five-year-olds can describe routes, but that British children prefer to use directional indicators, rather than landmarks (in contrast to Iranian children, who tend to use landmarks more often). From this it seems that any study of route representation which measures only the child's use of landmarks will fail to give a full picture of the child's skills.

This study will attempt to show that there is a development of spatial knowledge in middle childhood, but that this development is not dependent simply upon the acquisition of a specific form of information. It is expected that young children will be able to express some form of knowledge, given the results of previous studies (e.g. Spencer and Darvizeh, 1983), and that the form of this knowledge will change with age. In this way it is hoped to marry the findings, among western children, that three- to five-year-olds use directional indicators (Spencer and Darvizeh, 1983) and that older children use landmarks (Allen et al., 1979) in route representations.

In addition, the effect of the age of addressees will be considered, as other researchers have found that even young children are able to adjust their communications according to the age and presumed skills of the listener. Shatz and Gelman (1973) have shown that children of four years of age can use different message structures in speech, according to the age of the listener. They use shorter utterances, more attention-getting devices and different syntactic constructions to two-year-olds than to same-age peers or adults. Similar results are shown by Sachs and Devin (1976), with three-year-olds addressing 'babies' in a role-play task. More recent work (Greenberg et al., 1983; Waller, in press a) suggests that it is only in early school years that the child's spontaneous message adjustments begin to have any pragmatic informational benefits for the listener. This body of research suggests that the older children in the present study may respond to the age of listener, possibly by reducing the use of directional indicators for some listeners, as such forms are not useful in instructions to children aged up to six years (Waller, in press b).

A problem which is always present in attempting to elicit children's spatial representation is whether the product is a fair reflection of the spatial storage (Liben,

1981). All spatial production will probably involve some degree of dissociation from the stored representation. This study analyses the verbal descriptions given, and therefore will suffer from some drawbacks. For example, the language which is readily available to young children may be unable to express certain ideas. On the other hand, this method has the positive value of tapping a skill which the child will use in realistic situations. The results of the present study, as with other investigations of children's spatial skills, will not represent the full extent of the child's knowledge. However, they will be directly relevant to the child's use of spatial knowledge for an everyday task, involving social interaction.

Method

Subjects
Forty children from an Oxfordshire First School acted as subjects. They were divided into two groups of 20 children each, with mean ages of 5:8 years (range 5:1–6:1 years) and 8:8 years (range 8:0–9:3 years). All children were fluent English speakers.

Materials
'Prizes' (of pieces of fruit) were selected by the children as the objects to be hidden. There were eight possible hiding places, all in the school playground area. All messages were given whilst inside the school building, and were recorded for later transcription.

Procedure
Four groups of ten pairs of children were created, so that there were all possible age-pairings of speaker and listener (5–5, 5–8, 8–5, 8–8). Same-age pairs were matched to within two months, while different-age pairs were matched to 3:2 years ± 3 months. Each child participated twice (with an interval of at least four days), as speaker and listener, and completed four trials in each role. The order of roles was counterbalanced. No child ever communicated twice about the same hiding place, either as speaker or listener.

The children were told that they could help each other in a game of 'Hide-and-Seek'. They were then assigned to the roles of speaker and listener, and the game was described to them. The experimenter then took the speaker outside to hide the prize at a site decided by the experimenter. The experimenter and the speaker then returned to the building, where the listener was waiting. The speaker was asked to 'tell . . . where you hid the fruit, so that she/he can find it'. The following description was recorded (and the experimenter noted use of gesture) for later written transcription. The listener then searched for the prize, with the experimenter's 'assistance' to ensure that it was found on the few occasions when the listener was unsuccessful. This procedure was repeated a further three times before the session ended and the children were given their prizes.

This procedure yielded 160 written transcripts (four trials × four groups × ten pairs) for scoring as outlined below. Each transcript consisted of all the child's utterances, as well as the gestures (added in from the experimenter's notes, which were taken by hand during the description).

Scoring
The 160 transcripts were analysed—by a judge who knew the area but was blind to

the purpose of the study—for the following categories of information. The judge was asked to consider the context of the full transcript when deciding upon the categories.

(a) *Directions:* any comment in which the direction/bearing of the prize was given (includes Gestures—added from the experimenter's record). This category was defined in the same way as the category of 'directional indicators' used by Spencer and Darvizeh (1983) (i.e. instructions to turn left or right, or 'this way', accompanied by a gesture; and words such as *'coming* to', *'crossing* the road', 'going *up* the hill'). The use of previously established criteria allows for comparison between studies, even where the individual categories might be open to debate.

(b) *Landmarks:* reference to features of the environment which were used to specify the route to be taken (unless the feature could be classified as End Information— see below). This included places, objects and buildings which were not used as directional indicators (as used by Spencer and Darvizeh (1983), though they did not formalize their operational definition). So 'playground' could be a landmark ('you see a playground . . .') or a directional indicator ('you go across a playground'), depending on the context.

(c) *End Information:* reference to the general area or precise point where the prize was hidden. The criterion here was that the goal ('prize') should be mentioned directly (e.g. 'the bag', 'the fruit'), indirectly (e.g. 'it's there'), or through anaphoric reference (where the goal is understood to be the subject of a statement, even though it is not mentioned), in conjunction with some directly relevant landmark information.

Each was scored as present/absent, regardless of the actual quantity used in the transcripts. Any unclassifiable parts of messages (e.g. 'Don't tell Mrs Bennett') were ignored for this purpose. The transcripts were also divided into halves, to give some measures of the distribution of information categories within the message. The additional variables of Age of Speaker and Age of Listener were also used. Examples of the scoring system can be seen in Table 1.

Results

A number of examples of the types of message given are presented in Table 1, with the categorical analysis provided by the judge. These examples are chosen for their typical nature, not to fully illustrate the results of the analysis.

Log-linear (Logit) modelling was used to examine the pattern of associations between the dependent variables of Direction, Landmark and End Information use. The independent variables of Age of Speaker and Age of Listener were used to test for the development of the use of types of environmental information. Finally, a crude measure of the structure of the message—Half—was used as an independent variable, to indicate whether the pattern of use of environmental information changed within the message.

The Logit analysis (Knoke and Burke, 1980) is an analogue of the classical ANOVA, but is applicable to categorical data. This gives it a much wider flexibility than other methods of analysing contingency tables. The analysis finds the pattern of associations which are required to account for the variance within a multi-dimensional table of frequencies. In contrast with most statistical techniques, it requires the largest probability value which can be obtained while still introducing significant effects and interactions, in order to define the most parsimonious model of Goodness-of-Fit. This

TABLE 1

Examples of children's descriptions of hiding places and judge's analyses

Ch Th (5:11) to Ke Si (5:10)
    Description   'It's round the corner (points),/up a little bit (points),/under the garage/with the cars,/under a ... and under a bench./Under a bench.'
    Analysis   Direction/Direction/Landmark/Landmark/End Information/End Information
Ma Bi (5:9) to Le Gr (8:10)
    Description   'You go round the corner (points)/and then there's a fence,/and you walk down there a bit,/and you have to stop ... somewhere,/and it'll be in behind a gate./On that side (points)./And some leaves might be covering it.'
    Analysis   Direction/Landmark/Direction/Landmark/End Information/Direction/End Information
Br Ha (8:7) to Ry Gr (5:8)
    Description   'The bag is in the first playground./When you come out of the door/you see a playground/and there's one pair of gates in front of you,/and just off to one side (points)/is the bag with the apples in.'
    Analysis   End Information/Landmark/Landmark/Landmark/Direction/End Information
Ro Ca (8:0) to Ch Si (8:1)
    Description   'It's in the big boy's playground,/behind the long logs./Behind the second one along/... the one nearest the little logs/... and it's closest to the nursery.'
    Analysis   End Information/Landmark/Landmark/End Information/End Information

overall Goodness-of-Fit is best expressed in terms of the Likelihood Ratio (L^2), while the significance of individual effects and interactions is expressed in terms of standard scores (z-values; Nie, 1983). The analysis also gives the coefficient of association of significant main effects and interactions. From this coefficient, it is possible to calculate the comparative likelihood (the 'odds') that the particular variables will occur in each other's presence or absence. So, given a significant interaction between two variables, A and B, it is possible to say that A is x times more likely to occur if B occurs than if B is absent. Knoke and Burke (1980) give a fuller account of the uses and intricacies of the statistical method of log-linear modelling.

The most parsimonious model consisted of the effects of Landmarks and End Information, and the interactions of Directions × End Information, Landmarks × End Information, End Information × Half, Directions × Landmarks × Half, Landmarks × Age of Speaker, Directions × Age of Speaker, and Landmarks × End Information × Age of Speaker. This model yielded an impressively high L^2 of 38·55 (df = 47, $P = 0·805$), using only nine of the 56 possible effects and interactions. As an overall frequency table might be unintelligible, the separate effects, summarized in Table 2, will be discussed individually with reference to the frequency counts of those effects, presented in Tables 3(a)–(g). Table 2 contains the significant main effects and interactions, the z-scores (and the probability levels) for those effects, and the comparative likelihoods (the 'odds') of the variables involved. The z-scores and 'odds' are discussed in explaining the separate effects below.

The effects of Landmarks and End Information will not be discussed, as each figures in a number of higher order interactions. The interactions of variables will be presented in terms of (a) those involving only the use of the environmental information (the content of the message); (b) those involving the position of the environmental

TABLE 2
Summary statistics for separate effects in the Logit model

Effect[a]	z	P	Odds
Land.	−2·282	0·025	1·449
End.	−8·785	0·001	5·685
Land. × End.	−4·694	0·001	2·062
Dir. × End.	−6·722	0·001	2·276
End. × Half	5·459	0·001	2·404
Dir. × Land. × Half	2·168	0·05	1·280
Dir. × Age S.	−2·102	0·05	1·285
Land. × Age S.	6·194	0·001	2·752
Land. × End. × Age S.	2·580	0·025	1·545

[a]Land. = Landmark, End. = End Information, Dir. = Directions,
Age S. = Age of Speaker.

TABLE 3
Frequencies of association for significant interactions in the Logit model

(a)

End information	Landmarks N	Y
N	18	53
Y	145	104

(b)

End information	Directions N	Y
N	16	55
Y	162	87

(c)

End information	Half 1	2
N	57	14
Y	103	146

(d)

		Half 1		2	
		Direction		Direction	
Landmarks	N	Y	N	Y	
N	43	31	52	37	
Y	35	51	48	23	

(e)

Age of speaker	Directions N	Y
5	83	77
8	95	65

(f)

Age of speaker	Landmarks N	Y
5	110	50
8	53	107

(g)

	Age of speaker 5		8	
End information	Landmarks N	Y	N	Y
N	18	14	1	38
Y	92	36	52	69

information within the message (the structure of the message); and (c) those involving the ages of the children communicating. The frequency count relating to each of the significant interactions is presented in Tables 3(a)–3(g).

The use of environmental information
There are two relevant interactions, the associations of Landmarks × End Information and of Directions × End Information. Table 3(a) shows that there is a tendency for landmarks and end information to be negatively associated, particularly due to the high tendency for end information to be used without landmarks. This association was significant ($z = -4.694$, $P<0.001$), showing that, other things being equal, if end information is used then landmarks are 2·06 times more likely to be omitted. Table 3(b) shows a tendency for directional indicators and end information to be negatively associated, particularly due to the high tendency for end information to be used without direction information. This association was also highly significant ($z = -6.722$, $P<0.001$), showing that, other things being equal, if end information is used then directional indicators are 2·28 times more likely to be omitted.

To summarize, these interactions show that the children use end information in preference to both landmark and direction information. This shows that they are far more likely to omit landmarks and directional indicators than to omit end information.

The structuring of environmental information
Two interactions involve the position of types of information within the message, the End Information × Half and the Directions × Landmarks × Half associations. Table 3(c) shows that there is a tendency for end information and half to be positively associated, with a stronger use of end information in the second half. This association was significant ($z = 5.459$, $P<0.001$), showing that, other things being equal, end information is 2·40 times more likely to be used in the second half than in the first half of the message. Table 3(d) shows that directional indicators and landmarks are positively associated in the first half, but that they rarely coincide in the second. This effect was also significant ($z = 2.168$, $P<0.05$), showing that, other things being equal, when directional indicators are used in the first half, landmarks are 1·28 times more likely to be used than omitted.

To summarize, these interactions show that the first half of the message is likely to contain less end information, and will be likely to contain both landmarks and directional indicators or neither. In the second half, end information is highly likely to be used, while direction and landmark information are likely to be omitted. This suggests that the children are using directional indicators and landmarks in the early part of a message, and using end information in the later part. Incorporating the earlier results, the children seem to see end information as more important to the message than landmark or direction information, but will incorporate landmarks and directional indicators together in the first part of the message if at all.

The effect of the children's age upon the environmental information used
The final three significant interactions involve the age of speaker. Table 3(e) shows that the use of directional indicators tends to be negatively associated with the age of speaker, with eight-year-olds being more likely to omit this form of information. This association was significant ($z = -2.102$, $P<0.05$), showing that, other things being

equal, older children are 1·28 times more likely to omit direction information than to use it. Table 3(f) suggests that older children are far more likely to use landmark information, while young children omit it. This association was significant ($z = 6·194$, $P < 0·001$), showing that, other things being equal, older children are 2·75 times more likely to include landmark information than younger children. The final interaction to be presented, in Table 3(g), shows that five-year-old speakers are more likely to give only end information, while eight-year-olds tend to combine landmark and end information in the same message. This association was significant ($z = 2·580$, $P < 0·025$), showing that, other things being equal, younger speakers tend to use end information alone, while eight-year-olds are 1·55 times more likely to use both land-marks and end information in a message.

To summarize the effects of the age of speaker, younger speakers are more likely to use direction and end information, omitting landmarks, while older speakers prefer to use landmarks and end information, omitting directional indicators. This is of particular relevance to the (cited earlier) interactions of Landmarks × End Information and Directions × End Information [Tables 3(a) and 3(b)], suggesting that their interpretation cannot be continuous across ages.

Discussion

A simple schematic representation of the differences in message construction by five- and eight-year-olds is that five-year-olds tend to specify: *'Directions—End Information'*; while the older children tend to specify: *'Landmarks—End Information'*. There are also consistencies across this age range, with all children preferring to use end information as an absolute minimum. However, if these route descriptions are indeed valid reflections of the children's preferred use of particular forms, then it is not surprising that young children should appear to lack landmark knowledge (Allen *et al.*, 1979). Nor is it surprising that children should succeed on way-finding tasks (Spencer and Darvizeh, 1983). These two findings are easily reconciled by recognizing that development is in more than one dimension of the child's knowledge, and that a link between landmark knowledge and way-finding in late childhood and adulthood does not require that the same link will have existed earlier in life. These results show that the theories and research of Piaget and Inhelder (1956) and Siegel and White (1975; Cousins *et al.*, 1983) are inadequate to fully describe and explain the child's developing route knowledge and route descriptions. They assume that the child will be developing a unitary form of route representation, and do not allow for alternative strategies at different ages.

There are two developmental factors which might explain the difference found between the five- and eight-year-old children. Firstly, it was suggested in the Introduc-tion that there might be an increase in speaker's spatial reference skills, and the ability to use landmarks and directions as verbal labels. However, Waller (in press *b*) found that five- to seven-year-olds were capable of producing both landmarks and direc-tional labels accurately in spatial reference. This finding suggests that both landmarks and directional indicators should be available linguistic forms for the children in the present study.

A second, more likely explanation for the greater use of directional indicators by five-year-old speakers may lie in their level of experience and their learning skills. Directional indicators ('left', 'right', pointing, etc.) are relatively independent of context, while the use of landmarks depends upon context-specific knowledge (such as shared labels). As the five-year-olds all had less experience of the school than the eight-year-olds, they would have had less opportunity to learn those labels with confidence or in order to retrieve them consistently. This explanation does not require young children to be unable to use landmarks, particularly in a familiar environment such as the home. However, it does suggest that young children—due to a generally lower level of experience than older children—will be more likely to use the context-independent form.

This second explanation for the greater use of directional indicators by younger speakers can be related to the findings of Golledge *et al.* (1985), who tested a model of the acquisition of spatial knowledge by close scrutiny of an 11-year-old who was learning a route. They found that, over the first five days, learning was concentrated upon those 'loci' where some form of choice of behaviour was required (e.g. change of direction, crossing the street). Such 'loci' will tend to be those categorized in this study (and that of Spencer and Darvizeh, 1983) as 'directional indicators', rather than 'landmarks'. Thus, the younger children should be expected to produce more directional indicators than landmarks in spatial reference, given their lower level of experience of that specific environment.

The lack of any effect of age of listener suggests that the children were not varying the content or structure of their messages according to their audience's perceived status. This may be because the children see the landmark and directional information as equivalent in value, and so cannot make any informational or structural adjustments to the message. This is not the conclusion reached by Waller (in press *b*), who showed that seven-year-old speakers can use directional indicators flexibly, according to the age of the listener. However, that study used a much simpler situation, where a single piece of information was sufficient to distinguish the target. The ability to alter the content of spatial reference is probably not adequate to deal with the more complex task demands in this study, though eight-year-olds can make effective adjustments to the structure of spatial messages, allowing for the age of listener (Waller, in press *a*). It is probably fair to conclude that young school-children are aware of the relevance of the listener's age (e.g. Shatz and Gelman, 1973), but that more complex settings may present too large an informational load for the child to use that awareness to alter the content of spatial reference.

Summary

The apparently contradictory findings of research into children's route knowledge and way-finding abilities may be partly due to the view that development takes place in only one dimension of the child's knowledge. In fact, there is not a simple development of landmark knowledge in middle childhood, but a gradual change in preferred representational style. This change in preference may be explained in terms of experience of the environment, rather than any more complex development.

This study has only investigated the related development of landmark and direction knowledge, but there are many other aspects of spatial knowledge which are developing concurrently. It is clear that these different aspects of spatial knowledge should be studied as if they were multiple dimensions within the same individual, so that their interactive effects will not be overlooked. There is a need for more research employing combined measures of environmental cognition, both in children and adults, if these multiple dimensions are to be understood.

Acknowledgements

The author would like to thank the staff and pupils of West Oxford First School, for their very kind cooperation, and Paul Harris, for his helpful comments on earlier drafts of this paper. This research was carried out with the assistance of a grant from the Science and Engineering Research Council of Great Britain.

References

Acredolo, L. P., Pick, H. L. and Olsen, M. G. (1975). Environmental differentation and familarity as determinants of children's memory for spatial location. *Developmental Psychology*, **11**, 495–501.

Allen, G. L., Kirasic, K. C., Siegel, A. W. and Herman, J. F. (1979). Developmental issues in cognitive mapping; the selection and utilisation of environmental landmarks. *Child Development*, **50**, 1062–1070.

Clark, E. V. (1973). Non-linguistic strategies and the acquisition of word meanings. *Cognition*, **2**, 161–182.

Conning, A. M. and Byrne, R. W. (1984). Pointing to preschool children's spatial competence: a study in natural settings. *Journal of Environmental Psychology*, **4**, 165–175.

Cousins, J. H., Siegel, A. W. and Maxwell, S. E. (1983). Way-finding and cognitive mapping in large-scale environments: a test of a developmental model. *Journal of Experimental Child Psychology*, **35**, 1–20.

Durkin, K. (1981). Aspects of late language acquisition: school children's use and comprehension of prepositions. *First Language*, **2**, 47–59.

Golledge, R. G., Smith, T. R., Pellegrino, J. W., Doherty, S. and Marshall, S. P. (1985). A conceptual model and empirical analysis of children's acquisition of spatial knowledge. *Journal of Environmental Psychology*, **5**, 125–152.

Greenberg, J., Kuczaj II, S. A. and Suppiger, A. E. (1983). An examination of adapted communication in young children. *First Language*, **4**, 31–40.

Knoke, D. and Burke, P. J. (1980). *Log-Linear Models*. Sage University Series on Quantitative Applications in the Social Sciences, 07-020. Beverly Hills: Sage Publications.

Liben, L. S. (1981). Spatial representation and behaviour: multiple perspectives. In L. S. Liben, A. H. Patterson and N. Newcombe (eds), *Spatial Representation and Behaviour Across the Life Span: Theory and Application*. London: Academic Press.

Newcombe, N. (1981). Spatial representation and behaviour: retrospect and prospect. In L. S. Liben, A. H. Patterson and N. Newcombe (eds.), *Spatial Representation and Behaviour Across the Life Span: Theory and Application*. London: Academic Press.

Nie, N. H. (ed.) (1983). *SPSS^X User's Guide*. New York: McGraw-Hill.

Piaget, J. and Inhelder, B. (1956). *The Child's Conception of Space*. London: Routledge & Kegan Paul.

Sachs, J. and Devin, J. (1976). Young children's use of age-appropriate speech styles in social interaction and role-playing. *Journal of Child Language*, **3**, 81–98.

Shatz, M. and Gelman, R. (1973). The development of communication skills: Modification in the speech of young children as a function of listener. *Monographs of the Society for Research in Child Development*, **38**, (5, Serial no. 138).

Siegel, A. W. and White, S. H. (1975). The development of spatial representations of large-scale environments. In H. W. Reese (ed.), *Advances in Child Development and Behaviour*, Vol. 10. New York: Academic Press.

Spencer, C. and Darvizeh, Z. (1981). The case for developing a cognitive environmental psychology which does not underestimate the abilities of young children. *Journal of Environmental Psychology*, **1**, 21–31.

Spencer, C. and Darvizeh, Z. (1983). Young children's place-descriptions, maps and route-finding: a comparison of nursery school children in Iran and Britain. *International Journal of Early Childhood*, **15**, 26–31.

Waller, G. (in press *a*). Linear organization of spatial instructions: development of comprehension and production. *First Language*.

Waller, G. (in press *b*). The use of 'left' and 'right' in speech: the development of listener-specific skills. *Journal of Child Language*.

YOUNG CHILDREN'S REPRESENTATIONS OF THE ENVIRONMENT: A COMPARISON OF TECHNIQUES

M. H. MATTHEWS

Geography Department, Coventry (Lanchester) Polytechnic, Priory Street, Coventry CV1 5FB, U.K.

Abstract

The way in which young children aged between six and 11 years are able to represent their journey to school and home area by means of free-recall mapping, verbal description, and the interpretation of large-scale plans and aerial photographs is examined. Children's place 'whereness' and spatial awareness are shown to be influenced by the stimulus techniques used to assess their environmental knowing. Generalization is difficult as children's performances fluctuate dependent upon the place description. When describing their home area children achieve the best results using structured stimuli. Conversely, when recounting their journey to school children are able to recall most detail by free-recall drawing. Verbal reporting appears to inhibit the young child severely, suppressing our understanding of their true environmental capability. What also emerges is that by the ages of six and seven years many children are able to demonstrate a grasp of intra- and inter-place relationships, revealing a sound appreciation of 'objective spatial thought'. These findings lend further support to those who suggest that by using inappropriate methods of assessment in the past the young child's capacity to structure environmental information has been previously underestimated.

Introduction

The way in which young children acquire spatial awareness and place 'whereness' has been of interest to psychologists and geographers alike (Hart and Moore, 1973; Moore and Golledge, 1976; Altman and Wohlwill, 1978; Matthews, 1984a,b). The cognitive skills involved in mentally mapping the environment have been much discussed and the manner of their emergence in children is a topic of some complexity. Recent research has renewed the debate between the protagonists of two different schools of thought. On the one hand, constructivists, strongly influenced by the work of Piaget (1926; 1937; Piaget and Inhelder, 1967), argue that cognitive structures are seen as resulting from the unfolding of different kinds of thought processes through successive stages, which although not age-specific are characterized by sequential patterning. On the other hand, the incrementalists suggest that children share an innate ability to comprehend spatial relationships, which simply opens up with experience (Pick *et al.*, 1973; Pick, 1976).

These alternative viewpoints correspond to different theoretical expectations with respect to children's environmental competence. According to the former school, the young child responds on the basis of an egocentric frame of reference, whereby spatial locations are encoded in relation to the subject, and only older children are likely to use an objective frame in which environmental elements are fully co-ordinated in space. Convincing evidence exists for Piaget's developmental sequences (Laurendeau and Pinard, 1970; Mark, 1972; Acredolo *et al.*, 1975; Acredolo, 1976, 1977), all suggesting that young children are not capable of Euclidean spatial relationships. However, a number of recent studies have revealed that very young children possess a range of mapping and spatial modelling skills, often demonstrating a level of environmental competence more advanced than credited by the Piagetian tradition (Stea and Blaut, 1973; Stea and Taphanel, 1974; Neisser, 1976; Siegel and Schadler, 1977). These findings support the incrementalist assertion that the young child's capacity to structure and organize environmental information may well have been underestimated.

A common problem when studying young children is to find a suitable medium with which to examine their knowledge and awareness of large-scale environments. Siegel *et al.* (1978) have laid particular emphasis on the methodological difficulties of working with such subjects. They point out that the process of revealing a child's cognitive map is a complicated affair and that any method of assessment should provide a view of cognitive competence which is unbiased by intervening performance factors. As Neisser (1976) suggests, young children may well have developed considerable environmental knowing long before they can recount where they have been or how they got there. Hart (1979) and Spencer and Darvizeh (1981) argue that much of the younger child's apparent difficulty in externalizing about place is occasioned by the test material. Considerable debate exists as to what techniques and approaches can be regarded as suitable. Constructivists are keen to dismiss actual respondent cartography, arguing that the skill of graphicacy intervenes between knowledge and its depiction (Murray and Spencer, 1979). In contrast, incrementalists suggest that this fear is not justified since graphic responses are closely related to cognitive abilities and reflect 'truly visible thinking' (Lemen, 1966; Goodnow, 1977). Conversely, Piaget (1968) used verbal techniques to assert that young children were incapable of reversible thought, adding further support to his notions of egocentrism. This method has been subsequently criticized by Brown (1976) who points out that verbal recall is a skill which the very young find most difficult and as such would inhibit their performance.

A further difficulty when investigating the spatial cognition of young children has been that much generalization about environmental abilities has been based upon laboratory and small-scale tests. These studies often request the participants to undertake some novel or unusual behaviour (Hardwick *et al.* 1976) after relatively complex verbal instructions. A number of recent studies (Huttenlocher and Presson, 1973; Bluestein and Acredolo, 1977; Spencer and Darvizeh, 1981) cast doubt upon the validity of such methodologies asserting that the ability to translate from large-scale to a small-scale model is a particularly onerous task for the young mind, requiring changes of perspective which inevitably suppress absolute competences and true understanding. Piché (1981) and Spencer and Darvizeh (1981) have found that by using altered methodologies demanding direct contact with large-scale places the very young reveal a much greater environmental potential than hitherto suggested. Some of the strongest claims for young children possessing complex spatial comprehension has come from geographers and psychologists employing aerial photographs to assess environmental cognition, unfortunately these studies have not been examined in a full age-related context (Blaut *et al.*, 1970; Blaut and Stea, 1971; Dale, 1971; Spencer *et al.*, 1980).

Such contradictory evidence supports the views of White and Siegel (1976) who argue that children's performances seem to vary according to situational contexts and that competence levels can be moved 'upward' or 'downward' according to the load or difficulty of the task. Accordingly Siegel *et al.* (1978, p. 228) suggest that if children fluctuate in their levels of ability it becomes 'critical to measure a child's performances in several conditions or contexts before reaching conclusions regarding that child's competence'. In a recent review Piché (1981) notes that a comparison of alternative techniques designed to reveal environmental competences has never received formal attention, especially with children spanning an age range from six to 11. In consequence, this paper attempts to address this deficiency by comparing the way in which young children are able to represent large-scale spaces using a variety of iconic, graphic and verbal techniques.

Method

The study was carried out in four junior and infant schools within Coventry. These schools were carefully selected in order to ensure comparability both amongst the sampled populations and between the local environmental settings. An earlier investigation into the socio-economic character of school catchments (Matthews et al. 1978) provided a means for distinguishing school neighbourhoods of broadly similar socio-economic structure, size, housing style and density, and local service provision. This information was supplemented by a field and map survey undertaken within potential locations recording the number of environmental elements in successive $\frac{1}{2}$ km zones around each school. Areas of similar element diversity and magnitude were sought in an attempt to control the range of environmental features that children were likely to encounter on their journey to school. The chosen schools were all situated in the outer suburbs serving adjacent and often overlapping areas.

Every child aged between six and 11 years in these four schools participated in the survey, amounting to 155, 172, 174 and 192 cases: the mean ages of the children for each year group are shown in Table 1. Sampling within this age range was avoided

TABLE 1
Mean ages of children

School	Age group					
	6	7	8	9	10	11
1	6·4	7·6	8·6	9·6	10·5	11·5
2	6·6	7·4	8·5	9·4	10·6	11·4
3	6·4	7·6	8·5	9·5	10·6	11·4
4	6·5	7·4	8·5	9·6	10·6	11·5

in order to minimize the risk of unnecessary bias. For example, it was presumed that within each school and age group there would be a mix of intellectual abilities which would be upset by any selection procedure. Also, a preliminary inspection of school registers revealed similarities in the sex composition of each year group with boys dominant throughout. This is an important consideration especially if consensus viewpoints are to be compared as a number of studies have suggested that the cognitive mapping abilities of young girls and boys are likely to be different (Nerlove et al., 1971; Saegart and Hart, 1978; Matthews, 1984c). A check of home addresses disclosed a scatter throughout the local areas, with no locational concentrations evident for any particular age. None of the schools was involved in any environmental education initiatives. Their locational proximity encouraged close and friendly contact amongst the teaching staff and there was a broad agreement over curricula activities.

The investigation focused upon children's images of two large-scale environments, journey to school and home area, both of which become familiar during early schooling and yet are likely to be different in terms of their spatial form and in the way in which children interact with them. Children were asked to represent these environments by means of four different techniques of stimulus presentation, of a structured and unstructured kind: (i) free-recall mapping (graphic unstructured); (ii) map interpretation (graphic structured); (iii) air photograph interpretation (iconic structured) and (iv) free verbal recall (verbal unstructured).

As no results were available on the intellectual capability of the children, it was considered unlikely that sampling would provide four equally endowed groups of children for all age ranges in four schools. The risk that this factor would contaminate valid comparison of the techniques was thought greater than the likelihood of pollution by the influence of environmental diversity around the school, especially as the schools were chosen to be as similar as possible in terms of their local settings. Furthermore, accounts about the home area and of journeys to school are most likely to be influenced by the place of residence of the individual rather than the siting of the school. In this sense the school emerges as a convenient data collection point to assess the influence of methods of stimulus presentation upon the environmental cognition of children. In consequence only one of these techniques was used within a school.

The project was introduced in general terms to the children, who were then seen individually. It was stressed that no assessment was involved. Each child was asked to respond to the following questions:

> Imagine that you were taking me with you on your journey from your home to this school. Please would you draw me a map (tell me) of the way we would go, showing (telling) me the things that we would pass on the way. Name any of these features that come to mind;

and

> Imagine I was staying at your home and you were going to show me the area around your home. Please would you draw me a map (tell me) of the area around your home, showing (telling) me some of the things I might see nearby. Name any of these features that come to mind.

It was thought important that the same questions should be asked on each occasion, and by demanding a drawn response wherever possible a more meaningful comparison of the techniques would be achieved. Therefore, if differences were found to exist, it could be inferred that cartographic skill was not the sole determining factor in terms of what was being represented, but rather variations were the result of stimulus presentation. For verbal recall the questions were slightly modified (see parentheses), but in a manner thought unlikely to interfere with the nature of the task. In addition children were asked to consider their description as if they were about to draw a map of the area. Both instructions started with the home as the initial cue, as this would be the environment most familiar, allowing responses to be developed from the well-known. For each task a 20 minute time-limit was set and this proved adequate for nearly every respondent. In each case a note was made of length of residence and mode of transport usually used to get to school, as both these factors have been shown to influence urban imagery (Carr and Schissler, 1969; Appleyard, 1970; Moore, 1975), but for over 90% of the entire population most had lived in their localities for more than one year and walked to school.

The sketch mapping exercise involved 'actual respondent cartography' upon plain sheets of A4 paper. If the term 'map' was not understood this was substituted by 'picture', an instruction which was only relevant for two six-year-olds.

The published maps were derived from the O.S. 1:1250 plans. This scale was chosen as it showed the environment in a form which corresponded most closely to a vertical snapshot, employing the minimum of graphic symbols and a recent revision ensured that it was fully up to date. Although most children had encountered maps in some guise, the scale and format of this presentation was new to every respondent. This meant that the nature of the plan had to be explained and this was achieved by reference to an extract from another area. In order that printed prompts should not influence children's recall and to ensure that all verbalization should represent spontaneous response from the child, all written descriptions were erased from the plans. The exercise was initiated by a sheet of tracing paper placed over the map and children's attention drawn to their home and the school.

The iconic stimulus was presented at a scale of 1:4087, which was the largest reproduction available from the Ordnance Survey. The aerial photograph was centred on the school and covered an area of over 10 km². Ideally, the scale of the two structured media should have been identical for the most meaningful comparison, although this was not feasible the aerial photograph was sufficiently clear to allow individual buildings and features to be readily identified. Interestingly, none of the children had previously scrutinized an aerial photograph. In order that they should familiarize themselves with its format each child was asked to comment upon what it represented. The interviewer then explained it was a photograph of the local area taken from an aeroplane. On completion of this routine children were then shown their home and the school and a sheet of tracing paper was superimposed on the photograph. Only three children, two six-year-olds and one eight-year-old, were not able to recognize at least one feature correctly. The final technique involved tape recording the verbal recollections of each child. Children were made fully aware that their conversations were to be registered in this way, although during the interview the microphone was concealed from the child's view.

The data analyses examine the influence that stimulus presentation has upon young children's ability to represent place and externalize about space. A cross-sectional strategy is adopted implying that 'the structure of images held by different age groups at one point in time provides some insight into the whole learning process' (Matthews, 1984a). The results are considered in two sections. The first focuses upon the quantitative accretion of environmental knowledge. Children's responses are examined to assess the extent of their awareness surfaces and the information on place which they contained. Second, qualitative differences in environmental cognition are discussed, including the types of features recalled and general spatial competence.

The Quantitative Accretion of Environmental Knowledge

Information on place

Children's responses were sorted according to the amount of information on place which they contained. For each technique and by age group the mean number of elements correctly represented per account were assessed. These results were examined to see whether the method of stimulus presentation significantly influenced performance by means of the Kruskal–Wallis H test, a non-parametric method of analysis of variance. Trend analysis employing correlation procedures considered the influence of age upon information levels. What emerges is that not only do the techniques have a considerable bearing upon the extent of recalled environmental information, but also the places themselves are recounted in different terms. Furthermore, even the youngest children demonstrate an awareness of locations outside their home. These ideas will now be explored in turn.

First, it would appear that the method of stimulus presentation greatly influences how young children are able to recall or represent familiar environments. The Kruskal–Wallis H test confirms that between-technique variation was statistically significant for both environments (journey to school, $H = 13.96$, df = 3, $P = <0.005$; home area, $H = 8.9$, df = 3, $P = <0.05$). However, this simple generalization must be qualified since the amount of detail correctly located varies according to the place description (Table 2). For example, when describing their home area children per-

TABLE 2

Information on place: mean number of elements correctly identified per map

	Age group								
	6	7	8	9	10	11	r	t	P
Journey to school									
Free-recall sketching	2·6	3·6	5.9	7·1	8·3	9·7	0·99	18·9	<0·001
Air photo interpretation	0·7	0·9	1·3	1·0	1·6	1·6	0·89	4·0	<0·001
Map interpretation	0·7	0·6	1·3	1·4	1·5	0·9	0·53	1·3	NSa
Verbal reporting	0·5	0·7	0·9	1·0	1·2	1·1	0·94	5·6	<0·01
Home area									
Free-recall sketching	3·1	5·0	6·3	10·2	8·9	9·2	0·89	3·9	<0·02
Air photo interpretation	3·5	5·4	11·2	13·1	15·1	20·4	0·98	11·8	<0·001
Map interpretation	4·5	7·9	13·9	13·7	17·8	20·2	0·97	9·2	<0·001
Verbal reporting	1·8	2·4	3·8	4·1	6·6	7·5	0·97	9·1	<0·001

For journey to school $H = 13.96$, df = 3, $P = <0.005$.
For home area $H = 8.9$, df = 3, $P = <0.05$.
aNS = Not significant.

formed much better when using structured stimuli. In particular, consistently good responses were achieved with the use of large-scale plans. Even the very young were able to remember more than four places around the home and by the age of 11 years, a wide amount of information was successfully recalled, with many scoring over 20 correct identifications. Children employing free-recall mapping were much less successful, their responses rarely exceeded ten sitings and altogether this technique produced a narrower more constrained view of the environment. Conversely, when recounting their journey to school free-recall mapping yielded the best performances. In this instance both air photograph and map interpretations were characterized by a paucity of information with little variation according to age. For both environments children's verbal accounts were mostly hazy and ambiguous in detail, although there was some improvement with age little information on place was gained by this technique.

It would seem that response levels are related to particular techniques although this observation is complicated by the pronounced variation in the way in which different environments are best represented. The instructional set may have some bearing upon the results, as children were requested to start their descriptions from their home. When presented with a plain sheet of paper and asked to recall a journey in this manner, the effect of drawing a route may be to jostle the memory in such a way that a structured image appears in the child's mind, building upon what has just been drawn. Similar, although less strong imagery may be conjured up by verbal accounts but the difficulty of this medium, particularly for the very young, severely impinges upon the overall description. However, the effects of visual stimuli in which the home and the school were always in sight may mean that thought becomes blurred by haphazard and unsystematic recollection. On the other hand, the higher level of accuracy achieved by children using structured media for their home area descriptions suggests that these visual prompts allow children to scan randomly over areas near to their residence enabling a greater amount of information to be pieced together than achieved by mental recall alone.

Second, children appear to remember different environments in different ways. Table 3 shows the highest amount of correct identifications made by each age group

TABLE 3

Environmental description: comparison of details recorded on journey to school and home area responses by highest mean number of elements correctly identified per map

	Age group					
	6	7	8	9	10	11
Home area (a)	4·5	7·9	13·9	13·7	17·8	20·2
Journey to school (b)	2·6	3·6	5·9	7·1	8·3	9·7
Ratio a/b	1·7	2·2	2·4	1·9	2·1	2·1

expressed as a ratio of journey to school to home area. All age groups were able to describe the area around their home in the richest terms, usually noting more than twice the amount of detail than on their journey to school responses. Clearly, children's spatial know-how and confidence in describing an environment is dependent not only upon the stimulus but also on the place under investigation. Considerable variations exist even when recalling everyday environments, such that if generalizations were based upon journeys to school the environmental skills and capabilities of young children would have been significantly under-estimated.

Third, these results emphasise that with age children show an increasing awareness of their everyday worlds. The data were recast to permit Pearson's product moment correlation analysis which highlights the strength of the relationship between mean age and mean information levels for each technique (Table 2). All the results were statistically significant with one anomaly: when undertaking map interpretation to describe their journey to school within-group variation was sufficient to obscure any age-related trend. It would seem that on this occasion that both the technique and environmental context combined to confound children's representational ability. Otherwise, a clear age-related progression is apparent in terms of the amount of environmental information portrayed, for all techniques and for both places. By the age of 11 years a complex mosaic of locations is becoming pieced together in the child's mind, such that a broad array of detail is readily recalled (Table 3). More significantly,

Table 2 suggests that environmental knowing does not suddenly develop in these middle years of schooling, children appear to be actively accumulating environmental knowledge from an early age. Even six-year-olds possess some grasp of places away from their home. When using large-scale plans the youngest children frequently remembered more than five locations in their neighbourhood.

Awareness of place

Children's awareness of place was assessed by examining the spatial spread of the information recounted when describing the area around their home. The standard distance statistic provides a useful means for comparing the cognitive maps of different age groups (Kellerman, 1981). Similar to the standard deviation of a numerical distribution, it measures the degree of dispersion or clustering about a spatial mean centre of gravity. In order to employ this method all the data recorded by each age group were transformed into point locations. This was relatively straightforward for most of the environmental detail such as buildings, shops and landmarks. If districts or streets were indicated the centre of these areal and linear features were taken. Attention was confined to the extent of the awareness surfaces and so points were recorded as places away from the home, only distance and direction being collectively preserved. The procedure consisted of marking a central point on a sheet of tracing paper which was re-oriented to correspond to the home of each respondent located on a 1:1250 O.S. base map. Features were identified from this site, regardless of their accuracy. A centimetre grid was superimposed upon the distribution for analyses. Additionally, information was gathered on the distance of the furthest point from the home noted by each child.

The varying sizes of the circles and the areas that they bound confirm a growing awareness of space with age. In all cases statistically significant positive correlations were derived by Pearson's analyses (Table 4). The information surface of 11-year-olds

TABLE 4

Awareness of place: information surfaces around the home. Figures are area bounded by standard distances circle (km²)

| | Age group | | | | | | | | |
	6	7	8	9	10	11	*r*	*t*	*P*
Free-recall sketching	0·1	0·2	0·1	0·8	0·8	2·4	0·85	3·2	<0·05
Air photo interpretation	0·9	0·9	1·1	2·1	2·4	3·3	0·90	6·4	<0·01
Map interpretation	0·7	1·4	1·8	2·2	2·9	3·4	0·90	22·9	<0·001
Verbal reporting	0·1	0·1	0·2	0·6	0·7	1·8	0·88	3·7	<0·02

$H = 10·44$, df $= 3$, $P = <0·05$.

not only contain more detail than their juniors but are much broader in extent. By this age children are able to recall places near and far from their residence, such that the mean furthest distance of places away from the home recounted by children using large-scale plans was more than 800 m (Table 5). However, even the youngest children were able to show a grasp of complex spaces a considerable distance from their home, ranging from 193 metres in the case of verbal reporting to 415 metres with the aid of an aerial photograph.

Of particular significance is that these results lend further support to the assertion

TABLE 5

Awareness of place: knowledge of furthest places away from the home. Figures give mean distance of furthest point from home (m)

| | Age group | | | | | | | | |
	6	7	8	9	10	11	r	t	P
Free-recall sketching	210	329	260	505	530	556	0·91	4·6	<0·01
Air photo interpretation	415	420	508	592	610	748	0·97	8·2	<0·001
Map interpretation	366	448	521	595	618	803	0·97	8·7	<0·001
Verbal reporting	193	227	271	411	508	540	0·97	9·5	<0·001

that the method of stimulus presentation has an important influence upon how young children represent familiar places. Children's performances by age group are very uneven, suggesting that levels of environmental knowing are intricately interwoven with the 'externalizing' technique. This observation is supported by the results of the Kruskal–Wallis H test which was used to examine the influence of the method of stimulus presentation upon the extent of children's information surfaces around the home and their knowledge of furthest places away from the home ($H = 10·44$, df $= 3$, $P = <0·05$; $H = 17·1$, df $= 3$, $P = <0·001$) (Tables 4 and 5). In this instance, children of all ages show a much broader conception of space when using structured media and this is especially so for those under nine-years-old. Young children's lack of verbal and graphical skill severely limits their attempts to describe places, such that they are only able to demonstrate a rudimentary and narrow understanding of the environment.

Furthermore, the implication of these findings has some bearing upon attempts to explain how young children acquire environmental and spatial skills. Performance not only varies according to the method of stimulus presentation but also the rates at which children develop an appreciation of the environment seems dependent on the technique (Figure 1). For example, when employing free-recall mapping a stage-like

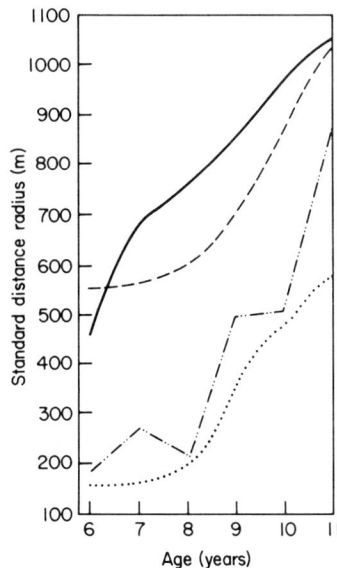

FIGURE 1. Awareness of places around the home: the influence of stimulus techniques upon environmental recall. – – –, Air photograph interpretation; —, map interpretation; –··–, free-recall sketching; ....., verbal reporting.

acquisition of spatial knowledge seems apparent. Strong similarities exist between the abilities and memories of six-, seven- and eight-year-olds; in turn these are differentiated from the maps drawn by nine- and ten-year-old children, which, themselves, are sharply different to those compiled by the oldest group. The results provided by air photograph interpretation confirm a strong correspondence in the images of children in their early school years, but emphasize that there is a continual growth of environmental knowledge with age, quickening between the ages of eight and ten years and slackening thereafter. The pattern produced by those using large-scale plans contradicts any similarities between age groups revealing a pronounced and almost consistent acquisition of place knowledge with age. Children's verbal accounts are often hazy, regardless of age. Locations are sometimes difficult to pin-point and directions are only given in the vaguest terms. Nevertheless, an age-related progression is discernible. The reports of the youngest children lack attention to detail and until the age of eight years there is little difference in how they are able to recall environmental information. From then on, there is a growing appreciation of place with age, such that many 10- and 11-year-olds are able to talk about places in a structured and reasonably coherent manner.

These observations are particularly relevant when attempting explanation about how young children gain an understanding of the world about them. Consider, for example, the Piagetian finding that young children develop through sequential stages which essentially are characterized by egocentric, topological and euclidean views of space. The results derived from free-recall mapping and, to a lesser extent, verbal reporting seem to confirm such developmental patterning. In fact, Piaget's (1968) assertion that pre-operational children are capable only of egocentricism and non-reversible thought was based upon verbal reasoning. However, these techniques seem restrictive in their applicability to young children. Young children's verbal and graphical skills are limited. When employing other methods children are able to represent space in broader and richer terms.

These results accord with the sentiments of Siegel et al. (1978, p. 228), in their debate on methodological issues they assert 'it is surely the case that young children's cognitive maps of the environment are more accurate than the products they draw to represent these environments'. Having a child draw a sketch map or talk about a place may obscure true cognitive mapping ability, which when employing alternative structured media becomes readily apparent. Such methodologies may be quite appropriate when attempting to understand adult's comprehension of space, but may lead to the environmental capabilities of young children being underestimated. Children may be categorized as pre-operational when using graphical and verbal media not because their appreciation of space is narrow and egocentric, but because the research techniques are measuring only these skills and not environmental knowing.

Qualitative Differences in Environmental Cognition

Element designation

In order to assess whether the method of stimulus presentation influenced what young children were able to recall about the environment, the features which they noted were classified into six groups: functional (aspects of the built environment), recreational (play and leisure spaces), natural (aspects of the natural environment) personal

TABLE 6
Environmental designation around the home: types of element (%)

		Age group					
		6	7	8	9	10	11
Functional	1	50	52	61	66	70	75
	2	76	75	81	93	87	89
	3	82	79	85	89	85	92
	4	58	51	68	74	72	80
Recreational	1	8	7	10	12	11	10
	2	4	5	6	3	5	6
	3	6	5	—	4	8	4
	4	2	11	9	10	15	10
Natural	1	25	22	19	19	18	15
	2	12	10	7	4	8	5
	3	8	13	6	7	7	4
	4	22	20	15	14	13	10
Transportational	1	5	2	3	3	—	—
	2	—	—	—	—	—	—
	3	—	—	—	—	—	—
	4	4	1	—	—	—	—
Personal	1	3	10	5	—	—	—
	2	4	6	6	—	—	—
	3	4	3	5	—	—	—
	4	10	12	8	2	—	—
Animal	1	10	7	6	—	—	—
	2	4	4	—	—	—	—
	3	—	—	1	—	—	—
	4	5	5	—	—	—	—

1 = Free-recall sketching, 2 = air photo interpretation, 3 = map interpretation, 4 = verbal reporting.

(people), transportational (modes of transport) and animal (Table 6). For ease of description this discussion will focus only upon the home area responses, especially as the journey to school results were similar in almost every respect.

Spencer and Lloyd (1974) in their 'draw me a map' exercizes noted that young children tended to see the environment in primarily human and natural terms. These results lend little support to this observation. In all cases a strong functional bias was evident with shops, buildings, and pathways characteristically represented. Nevertheless, the younger children employing free-recall mapping did personalize their accounts by drawing pictures of animals in nearby fields, moving cars, 'lollipop ladies' (road crossing supervisor) and friends and relatives. Features clearly insignificant to their elders were often given prominence: a favourite climbing tree, a path to some special site, or as many six- and seven-year-olds described, a tree known locally as 'the hawk-moth tree' because of its attraction to moths in summer.

Some variation was evident between the results produced by the different techniques. Young children using structured media were more inclined to identify aspects of the built environment at the expense of natural features. Their accounts lacked the idiosyncratic attention to detail commonly demonstrated by those drawing or talking about their home area. Nevertheless, by the age of nine years, formal environmental designation was accomplished by most, regardless of the methodology, but even here those using graphical and verbal means tended to give greater prominence to natural and recreational space.

Map accuracy

Successful externalization of cognitive imagery depends largely upon the capacity to arrange elements within a spatial structure. Research suggests that the young child's understanding and ability to depict place and space develops through successive levels, each associated with progressively better appreciations of euclidean relationships: these may be termed egocentric, objective and abstract stages of spatial understanding (Catling, 1979). These ideas were examined by applying Moore's (1973) three-fold classification to the data sets. Level I maps contain unorganized elements or features arranged topologically from an egocentric perspective. Level II maps are partially differentiated with some degree of intra-cluster accuracy. Level III maps demonstrate an awareness that all the parts are parts of a whole, such that intra-cluster accuracy and the location of elements is high throughout the map. Two independent judges sorted the maps into each category.

Each data set was examined statistically by means of a Kolmogorov–Smirnov (χ^2) two-sample one-tail test to see whether map accuracy varied with age (Tables 7 and 8). This test focused solely upon the occurrence of Level I and Level III maps. In all cases children's spatial comprehension improved with age, regardless of the technique. The highest incidence of Level I maps was drawn by six-year-olds and progressively as children became older fewer responses were classified in this way. By the age of 11 years

TABLE 7

Map accuracy: Moore's classification; journey to school responses (%)

	Age group					
	6	7	8	9	10	11
Free-recall sketching						
Level I	30	27	11	—	2	—
Level II	70	67	89	90	63	60
Level III	—	6	—	10	35	40
	$\chi^2 = 35.04$, df = 2, $P < 0.001$					
Air photograph interpretation						
Level I	48	42	42	32	10	6
Level II	52	51	58	60	82	80
Level III	—	7	—	8	8	14
	$\chi^2 = 38.7$, df = 2, $P < 0.001$					
Map interpretation						
Level I	45	33	29	18	5	—
Level II	55	67	70	71	72	80
Level III	—	—	1	11	23	20
	$\chi^2 = 25.8$, df = 2, $P < 0.001$					
Verbal reporting						
Level I	80	77	76	49	29	19
Level II	20	23	24	50	62	62
Level III	—	—	—	1	9	19
	$\chi^2 = 20.5$, df = 2, $P < 0.001$					

TABLE 8

Map accuracy: Moore's classification; home area responses (%)

	Age group					
	6	7	8	9	10	11
Free-recall sketching						
Level I	65	46	25	17	8	—
Level II	35	52	71	72	57	50
Level III	—	—	4	10	35	50
	$\chi^2 = 35{\cdot}7$, df = 2, $P < 0{\cdot}001$					
Air photograph interpretation						
Level I	51	22	4	—	—	—
Level II	49	74	89	88	61	52
Level III	—	6	7	12	39	48
	$\chi^2 = 39{\cdot}7$, df = 2, $P < 0{\cdot}001$					
Map interpretation						
Level I	48	15	8	—	—	—
Level II	52	75	75	81	58	49
Level III	—	10	17	19	42	51
	$\delta^2 = 43{\cdot}5$, df = 2, $P < 0{\cdot}001$					
Verbal reporting						
Level I	83	86	72	68	41	18
Level II	17	14	28	29	39	44
Level III	—	—	—	3	20	38
	$\chi^2 = 35{\cdot}9$, df = 2, $P < 0{\cdot}001$					

many children were able to represent space in a highly organized manner, demonstrating a sound awareness of intra- and inter-cluster relationships.

Two further points need to be made about these results. First, the ability to portray spatial relationships accurately is not entirely age-related, in that the same child can show varying appreciations of space when describing different environments. The task of depicting an undefined 'area' as opposed to a defined 'route' is likely to induce variations in terms of the ability to reveal euclidean relationships. For example, in this study there appears to be slight preference throughout for those using free-recall mapping to draw Level II maps when describing their journey to school, whereas when recalling the area around their home the same children were unable to demonstrate a grasp of the relative position of places. However, it may be that, as Beck and Wood (1976) found with adults, familiarity influences euclidean accuracy, so that children's representation of different environments will be dependent upon their interaction with those places. Further, Moore (1975) and Acredolo *et al.* (1975) suggest that within-individual developmental variation can result from the nature of the environment under investigation. In a very complex environment people are likely to resort to simpler frames of reference, such as that defined by egocentric perspectives, even though they may possess the ability to comprehend space in a more holistic manner. On this occasion the tendency for children to draw simpler maps when describing their home area may reflect the difficulty of summarizing, by memory alone, such heterogeneous places, whereas once prompted by visual stimuli order and cohesion are readily grasped.

Second, the stimulus–response techniques clearly exert a considerable influence upon accuracy levels. When describing their route from home to school many six-year-olds revealed a competent appreciation of within-place relationships by free-recall mapping (Table 7). This skill would be completely underestimated if children's verbal accounts were solely analysed. Not until the age of nine years do most children demonstrate an ability to articulate about their journey in a structured manner. Whereas 40% of 11-year-olds describing their route were able to free-draw Level III maps, less than 20% showed this skill when using the other techniques. On the other hand, when recalling their home area some seven-year-olds were able to show an euclidean grasp of space, especially those prompted by air photographs and large-scale plans. In particular, older children performed very well with the aid of these structured media and by the age of nine years none drew Level I maps, unlike those talking about or freely drawing their environment.

What also emerges is that apart from those called to make verbal reports, by the age of six years most children had developed sufficient grasp of the world about them to be classed within the stage of 'objective spatial understanding'. To some extent these findings support those derived from an experiment carried out by Acredolo (1976) who investigated the degree to which Moore's hypothesis could describe age-related progressions in very young children. She suggests that whereas three-year-olds rely largely on egocentric reference systems, many pre-school children were able to use a fixed frame of reference based only on objects. However, whilst young children may acquire knowledge about space, their young minds find it difficult to integrate this into a sense of wholeness, such that Level II maps provide the dominant mode of representation for all age groups within this survey.

These observations upon the qualitative structure of children's mental maps are further testimony to the complexity of environmental cognition. Children's place awareness and spatial comprehension are closely linked with the method of assessment and the environmental context.

Verbal reporting seems particularly inappropriate to young children and drawing skills show great variability dependent on the place to be described. When directed to their home area children's performances are consistently enhanced by the use of both structured media, suggesting that the young child's awareness of places and place relationships does not suddenly spring up in the middle years of childhood, but is often keenly developed at an earlier age.

Conclusion

The process of uncovering young children's cognitive maps of the environment is a difficult task. That young children's ability to understand and represent large-scale space improves as they mature is an obvious, yet important observation. This is not a simple and straightforward evolution as different skills seem to be acquired differentially. In particular, this study has attempted to consider the influence that methodology has upon cognitive competence. Clearly, children's place 'whereness' and spatial awareness are influenced by the stimulus techniques used to assess their environmental knowing. This has been demonstrated with respect both to the quantitative accretion of environmental knowledge as revealed by children's responses and in the way such accounts have been qualitatively organized.

For example, verbal reporting appears to inhibit the young child severely, suppressing our understanding of children's ability. Young children's verbal descriptions are never particularly accurate or reliable. Such reports are usually hazy with weak reference to directions and subsequently, locations are difficult to pin-point. When describing their home area distant places are not clearly remembered and responses are characterized by the paucity of environmental detail. The results derived by free-recall sketching reveal a wider appreciation of place although the environmental context is important. When recounting their journey to school by this method children of all ages were able to recall more information than by any other technique. However, home area descriptions were not as rich in detail as those derived from structured stimuli, nor were far places accurately drawn. In terms of spatial comprehension some seven-year-olds using free-recall mapping were able to reveal an appreciation of euclidean principles, suggesting that this method cannot be dismissed as too difficult for the young mind.

By far the best representations of complex macro-space were derived by the interpretation of large-scale plans and vertical aerial photographs. Some qualification to this observation is needed. Children using these media to describe a journey seemed little inclined to indicate features that they passed on their route, although intra-cluster accuracy was high, even amongst the youngest age groups. The usefulness of these techniques was more apparent when children's home area descriptions were analysed. Performance levels on these occasions, were much higher than by any other procedure, which may indicate that these methods allow even the very young to get near towards disclosing their true environmental understanding. Places, some considerable distance from the home, were often recounted and relationships between and within clusters were grasped from the age of seven years onwards.

Such findings need to be viewed cautiously as many more studies are needed with this age range and with younger children, using a variety of techniques, both representational and experimental, before suitable generalizations can be made. However some tentative conclusions can be offered, especially with regard to the psychological and educational implications of these results.

This paper supports the findings of those who have pointed out that a sense of place has already emerged by the time children have arrived at infant school. Many six- and seven-year-olds were able to trace complex routes and show a good appreciation of place relationships in the area around their homes. Clearly these were skills which had been developed at an earlier age and recent work by Spencer and Darvizeh (1983) suggests that such understanding and ability may be evident in the young child of three and four years.

Furthermore, the poor relative performance of young children using verbal reporting and the considerable influence that the method of stimulus presentation has upon response levels, especially awareness of space, qualifies Piaget's theoretical expectations of how young children are able to see the world. By the age of seven years some children have already acquired an euclidean grasp of space, revealing the beginnings of 'abstract spatial thought'. This observation needs to be viewed carefully, as one of the defining characteristics of Piaget's idea of abstract representation is the extent to which children's performances are independent of stimulus prompts. It could be argued that the two structured media would encourage high levels of performance at earlier ages simply because they provide more stimulus information. However, two findings demonstrate the complexity of this issue; first, some seven-year-olds describing their journey to school by free-recall methods demonstrate an incipient grasp of spatial relationships; second, children perform unevenly when using maps and aerial photographs in different environmental settings, such that their home area accounts reveal a better sense of relatedness than their route descriptions. It is surely the case that by using inappropriate methods of assessment in the past the young child's capacity to structure environmental information has been severely underestimated.

Educationally, the implications are that young children should be allowed to experience a range of spatial stimuli from their earliest school days. As Blaut and Stea (1971) have shown, even the youngest age groups were comfortable when using large-scale plans and air photographs. The ability to transform and rotate space may be too difficult for many six-year-olds when called to draw a map (Matthews, 1984a), but most were able to comprehend these visual forms, and this is at an age well before they are traditionally exposed to such media. The work of Spencer et al. (1980) with three-year-olds shows that many pre-school children have already acquired the ability to interpret aerial photographs, which suggests that the nursery class may not be too early to stimulate the young child in this way.

In summary, the method of assessing a young child's cognitive map of the environment exerts a considerable influence upon performance levels. Children's skills seem to fluctuate quite widely, dependent upon whether they have been asked to talk about, draw or rely upon structured stimuli, when describing familiar places. Ideally a situation is sought where true environmental potential can be revealed. It may be that by asking children to represent models of places, no matter by what technique, the instructional set automatically handicaps cognitive environmental capability. Children develop their awareness of places about them through repeated contact; if full environmental knowing is to be appreciated then research may need to focus upon such real world experiences at first hand.

Acknowledgements

I am especially grateful to Diane Rogers who assisted in the data collection, the head teachers and staff of all the schools who participated in the survey, and Coventry Education Authority for permission to undertake this research. Thanks too to Tony Airey who made valuable comments upon an earlier draft of this paper. The project was funded by The Nuffield Foundation.

References

Acredolo, L. P. (1976). Frames of reference used by children for orientation in unfamiliar spaces. In G. T. Moore and R. G. Golledge (eds), *Environmental Knowing*. Stroudsburg, Pennsylvania: Dowden, Hutchinson and Ross.

Acredolo, L. P. (1977). Developmental changes in the ability to co-ordinate perspectives of a large-scale space. *Developmental Psychology*, **13**, 1–8.

Acredolo, L. P., Pick, H. L. and Olsen, M. G. (1975). Environmental differentiation and familiarity as determinants of childrens' memory for spatial location. *Developmental Psychology*, **11**, 495–501.

Altman, I. and Wohlwill, J. F. (eds) (1978). *Human Behavior and Environment: Advances in Theory and Research, Vol. 3 Children and the Environment*. New York: Plenum Press.

Appleyard, D. (1970). Styles and methods of structuring a city. *Environment and Behavior*, **2**, 100–118.

Beck, R. and Wood, D. (1976). Cognitive transformation of information from urban geographic fields to mental maps. *Environment and Behavior*, **8**, 199–238.

Blaut, J. M. and Stea, D. (1971). Studies in geographical learning. *Annals of the Association of American Geographers*, **61**, 387–393.

Blaut, J. M., McCleary, G. S. and Blaut, A. S. (1970). Environmental mapping in young children. *Environment and Behavior*, **2**, 335–349.

Bluestein, N. and Acredolo, L. P. (1977). *Developmental Changes in Map-Reading Skills*. *E.R.I.C. Document No. ED 154902*.

Brown, A. L. (1976). The construction of temporal succession by pre-operational children. In A. D. Pick (ed.), *Minnesota Symposium on Child Psychology, Vol. 10*. Minneapolis: University of Minnesota Press.

Carr, S. and Schissler, D. (1969). The city as a trip: perceptual selection and memory in the view from the road. *Environment and Behavior*, **1**, 7–35.

Catling, S. (1979). Maps and cognitive maps: the young child's perception. *Geography*, **64**, 288–296.

Dale, P. F. (1971). Children's reactions to maps and aerial photographs. *Area*, **3**, 170–77.

Goodnow, J. (1977). *Children's Drawings*. London, Fontana.

Hardwick, D. A., McIntyre, C. W. and Pick, H. L. (1976). The content and manipulation of cognitive maps in children and adults. *Monographs of the Society for Research in Child Development*, **41**, Serial No. 166.

Hart, R. (1979). *Children's Experience of Place*. New York. Irvington.

Hart, R. and Moore, G. T. (1973). The development of spatial cognition: a review. In R. M. Downs and D. Stea (eds.), *Image and Environment*. London: Edward Arnold.

Huttenlocher, J. and Presson, C. C. (1973). Mental rotation and the perspective problem. *Cognitive Psychology*, **4**, 277–299.

Kellerman, A. (1981). Centrographic measures in geography. *Catmog*, **32**.

Laurendeau, M. and Pinard, A. (1970). *The Development of the Concept of Space in the Child*. New York: International Universities Press.

Lemen, J. (1966). *L'Espace Figuratif et les Structures de la Personnalité*. Paris: Presses Universitaires de France.

Mark, K. S. (1972). Mapping through toy-play: methodology for eliciting topographical representations in children. In W. J. Mitchell (ed.), *Environmental Design: Research and Practice*. Los Angeles. University of California Press.

Matthews, M. H. (1984*a*). Environmental cognition of young children: images of journey to school and home area. *Transactions of the Institute of British Geographers*, **9**, 89–106.

Matthews, M. H. (1984*b*). Cognitive maps: a comparison of graphic and iconic techniques. *Area*, **16**, 33–40.

Matthews, M. H. (1984*c*). Cognitive mapping abilities of young girls and boys. *Geography*, **69**, 327–336.

Matthews, M. H., Clark, D. and Parton, A. G. (1978). *Social Patterns in Coventry as Defined by School Catchment Area Profiles*. Coventry: Geography Department Research Monograph, Coventry (Lanchester) Polytechnic.

Moore, G. T. (1973). Developmental difference in environmental cognition. In W. Preisser (ed.), *Environmental Design and Research, Vol. 2*. Stroudsburg, Pennsylvania: Dowden, Hutchinson and Ross.

Moore, G. T. (1975). The development of environmental knowing: an overview of interactional constructivist theory and some data on within-individual developmental variations. In D. Canter and T. Lee (eds), *Psychology and the Built Environment*. London: Architectural Press.

Moore, G. T. and Golledge, R. G. (eds) (1976). *Environmental Knowing*. Stroudsburg, Pennsylvania: Dowden, Hutchinson and Ross.

Murray, D. and Spencer, C. P. (1979). Individual differences in the drawing of cognitive maps: the effects of geographical mobility, strength of mental imagery and basic graphic ability. *Transactions of the Institute of British Geographers*, **4**, 385–391.

Neisser, U. (1976). *Cognition and Reality: Principles and Implications of Cognitive Psychology*. San Francisco: W. H. Freeman and Co.

Nerlove, S. B., Munroe, R. H. and Munroe, R. L. (1971). Effect of environmental experience on spatial ability: a replication. *Journal of Social Psychology*, **84**, 3–10.

Piaget, J. (1926). *The Child's Conception of the World*. London: Paladin.

Piaget, J. (1937). *The Construction of Reality in the Child*. New York: Basic Books.

Piaget, J. (1968). *On the Development of Memory and Identity*. Worcester, Massachusetts: Clark University Press.

Piaget, J. and Inhelder, B. (1967). *The Child's Conception of Space*. New York: Norton.

Piché, D. (1981). The spontaneous geography of the urban child. In D. T. Herbert and R. J. Johnston (eds), *Geography and the Urban Environment, Vol IV*. Chichester: John Wiley.

Pick, H. L., Acredolo, L. P. and Gronseth, M. (1973). Children's knowledge of the spatial layout of their homes. Paper presented at the Society for Research in Child Development.

Pick, H. L. (1976). Transactional-constructivist approach to environmental knowing. In G. T. Moore and R. G. Golledge (eds), *Environmental Knowing*. Stroudsburg, Pennsylvania: Dowden, Hutchinson and Ross.

Saegart, S. and Hart, R. (1978). The development of environmental competence in boys and girls. In M. Salter (ed), *Play: Anthropological Perspectives*. New York: Leisure Press.

Siegel, A. W. and Schadler, M. (1977). Children's representations of their classrooms. *Child Development*, **48**, 388–394.

Siegel, A. W., Kirasic, K. C. and Kail, R. (1978). Stalking the elusive cognitive map. In I. Altman and J. F. Wohlwill (eds), *Children and Environment*. New York: Plenum.

Spencer, C. P. and Darvizeh, Z. (1981). The case for developing a cognitive environmental psychology that does not underestimate the abilities of young children. *Journal of Environmental Psychology*, **1**, 21–31.

Spencer, C. P. and Darvizeh, Z. (1983). Young children's place descriptions, maps and route finding: a comparison of nursery school children in Iran and Britain. *International Journal of Early Childhood*, **15**, 26–31.

Spencer, C. P., Harrison, N. and Darvizeh, Z. (1980). The development of iconic mapping ability in young children. *International Journal of Early Childhood*, **12**, 57–64.

Spencer, D. and Lloyd, J. (1974). *A Child's Eye View of Small Heath, Birmingham: Perception Studies for Environmental Education. Research Memorandum No. 34*. Birmingham: University of Birmingham.

Stea, D. and Blaut, J. M. (1973). Some preliminary observations on spatial learning in schoolchildren. In R. M. Downs and D. Stea (eds), *Image and Environment*. London: Edward Arnold.

Stea, D. and Taphanel, S. (1974). Theory and experiment on the relation between environmental cognition. In D. Canter and T. Lee (eds), *Psychology and the Built Environment*. London: Architectural Press.

White, S. H. and Siegel, A. W. (1976). Cognitive development: the new inquiry. *Young Children*, **31**, 425–435.

DEVELOPMENTAL DIFFERENCES IN THE ABILITY TO GIVE ROUTE DIRECTIONS FROM A MAP

MARK BLADES AND LOUISE MEDLICOTT

Department of Psychology, University of Sheffield, Sheffield S10 2TN, U.K.

Abstract

The ability to give accurate route directions is an important way of expressing environmental information, but the development of this ability has received little attention, therefore an experiment was designed to assess how children and adults gave route directions from a map. In an extension of a previous, small scale, experiment (Brewster & Blades, 1989 *Journal of Environmental Education and Information*, 8, 141–156). Four groups of children (aged 6, 8, 10, and 12 years) and one group of adults described two routes from maps and the route descriptions were assessed for both their accuracy and content. The six and eight year olds were unable to give correct route directions, but a few of the ten year olds and many of the 12 year old children were able to provide directions for most of the route, and all the adults gave accurate route descriptions from the maps. Analyses were carried out on the content of the descriptions, the effect of the presence of landmarks at turns on the route, and the effects of direction of travel when approaching turns. The main finding was a major developmental contrast in the style and content of the descriptions: the younger children relied predominantly on landmarks and vague indications of direction; the older children and adults included information about the type of road junction at the turn and about the road sequence (e.g. 'first left', 'second right'). The implications of the results for further research are discussed.

Introduction

An individual's environmental cognition may be expressed in different ways. Researchers have used several methods to externalize individuals' knowledge of the environment—for example, subjects have been asked to give verbal or written descriptions, draw sketch maps, estimate directions, judge distances and recognize photographs (Golledge, 1987; Heft & Wohlwill, 1987; Spencer *et al.*, 1989). But outside the laboratory the most common situation which involves the externalization of a cognitive representation is when one person asks another for information about how to find the way through the environment.

Successful wayfinding in an unfamiliar area may depend on information which is part of the environment (e.g. sign posts) or on information which is additional to the environment (e.g. maps)—see Blades (1989). But most frequently it depends on asking for directions from other people who are assumed to be more familiar with the environment (Petchenik, 1985; Mark & McGranaghan, 1988). Several surveys and experiments have shown that adults often rely on verbal or written directions (e.g. Gordon & Wood, 1970; Lunn, 1978) and that adults can use directions very effectively (e.g. Streeter *et al.*, 1985; Kovach *et al.*, 1988; Kirasic & Mathes, 1990). Adults are also very proficient at giving accurate directions (Hill, 1987).

Although there have been several studies of how adults give directions most of these studies have concentrated on direction giving as a form of communication (Psathas & Henslin, 1967; Psathas & Kozloff, 1976; Klein, 1982; Wunderlich & Reinelt, 1982; Psathas, 1987). These studies have described the pattern or sequence of communication which takes place when people ask for directions with the emphasis on the structure of the interaction, rather than on the content or accuracy of the route description itself, and there are only a couple of studies which have considered the content of adults' route descriptions (e.g. Vanetti & Allen, 1988; Ward et al., 1986). For example, in the study by Ward et al. (1986) college students were asked to describe a route between two places on an invented map, and they did so mainly by reference to landmarks and the use of the terms left and right, but their descriptions were only assessed by scoring the number of times that landmarks, 'relational terms' (left and right), cardinal directions and distance estimates were mentioned, and Ward et al. did not provide any more detailed analysis.

Children's route directions have been studied in more detail than adults— children's descriptions were first investigated by Piaget who asked children to describe a route from memory (Piaget et al., 1960). He argued that children pass through several stages of ability—in Stage I children's knowledge of a route is very limited and does not extend beyond a general sense of direction. Piaget suggested that children in Stage II base their route descriptions, not on the environment as such, but on their memory of movement and travel through the environment, and he quoted one child (aged six years) who gave a description without referring to any features or landmarks along the route:

I go straight along, I turn there, I go straight along again, I turn there, I keep going straight and I turn once more (Piaget et al., 1960, p. 11).

This type of description led Piaget to suggest that children's earliest appreciation of routes is a 'sensorimotor' understanding, and that it is only after this under- standing is achieved that children can 'attach' landmarks and other information to their memory for the route. It is not until Stage III that children are able to describe routes more coherently, though not always completely accurately, with references to landmarks and directions.

Given Piaget's discussion of the development of route knowledge it would be expected that children's earliest description of routes will generally be made up of indications of directions. However, Siegel and White (1975) proposed a slightly different sequence for the development of environmental knowledge. They based their description on Piaget's, but argued that the first stage of environmental awareness was a knowledge of landmarks, and that only after landmarks were known could they be linked to form a route. Therefore, Siegel and White's theory (in contrast to Piaget's) implies that children's first route descriptions will emphasize individual features or landmarks along the route (for a more detailed and critique of both theories, see Blades, 1991).

Children's earliest route descriptions have received little empirical attention though a study by Spencer and Darvizeh (1983) asked preschool children in Iran and in Britain to describe familiar routes. Spencer and Darvizeh found that some children gave descriptions based on movement and directions (corresponding to Piaget's results) other children provided a list of landmarks (as would be expected from Siegel and White's theory). In other words, the evidence from

Spencer and Darvizeh's study does not distinguish between the relevant theories, but it does demonstrate that young children have some idea of what it means to give directions, even though their descriptions were far from complete and were often ambiguous.

Other studies have focused mainly on older children, but the results from different experiments do not give a completely coherent description of the development of children's ability. For example, Flavell *et al.* (1985) found that seven year olds were unable to detect ambiguous directions when using those directions to follow a route on a map, and this might imply that children of this age do not always have a complete understanding of what constitutes a satisfactory route description. Similarly, in a series of experiments designed to study developmental differences in the ability to communicate information Lloyd (1990, 1991, in press) found that seven year olds were often unable to describe a route from a map. In contrast, Waller (1985, 1986a) and Waller and Harris (1986) showed that slightly older children were competent at giving directions along a short route. Waller asked children to give directions between places in their school or playground and found that eight year olds were not only able to describe the routes, but also altered and simplified their description appropriately when they were talking to younger children.

Studies which have also included older children have shown that after about ten years of age children are able to give adequate directions, but they may not do so in exactly the same way as adults. For example, Lloyd (1991) analysed the strategies which children and adults used when describing a route from a simplified map— adults used a 'directional strategy' (i.e. included references to left, right, and straight on at junctions), but ten year olds relied on describing the landmarks at junctions (and given the simplified map used in Lloyd's study, this was sufficient to identify the correct route). In other words, directions given by older children may differ in content from those given by adults, and to some extent such differences may be due to the children's lack of competence with the component aspects of direction giving—for example, children may not be very accurate or confident at using relational terms like left and right (Boone & Prestcott, 1968; Waller, 1986b).

However, most studies with children have only included two different age groups which limits what can be said about any developmental progression in the ability to describe routes and therefore, to find out more about developmental differences we carried out a small scale study with groups of ten six, seven and eight-year-old children and ten adults (Brewster & Blades, 1989). The subjects were shown an invented map which included typical map features and clearly labelled landmarks, as well as a scale and the cardinal directions. A route which included eight choice points was drawn on the map and subjects were asked to give a verbal description of the route. All the adults did this successfully, but none of the children gave accurate directions, though four of the eight year olds, and two of the seven year olds correctly described at least half of the turns along the route (the six year olds hardly ever described a turn correctly).

The children's accuracy was affected by the nature of the choice point—turns associated with a landmark were easier to describe than turns without a landmark, and turns approached moving 'up' the map were better described than ones where the direction of travel was 'across' or 'down' the map. There was an age related increase in the amount of information included in the directions—the adults provided almost twice as much information as the oldest group of children, and this in itself contributed to the adults' accuracy, because any minor errors or slight ambiguities in their descriptions would easily have been understood by a listener in the context of all the information which was provided. There was also a striking difference in the type of route information given by the adults and the children. Nearly all the adults included descriptions of the road junctions on the map (as crossroads, T-junctions, etc.) but the children never referred to the type of junction. By describing the type of junction the adults were able to use these as landmarks along the route. More importantly, all the adults frequently described the sequence of roads ('first right', 'second right', etc.) which was an effective way of specifying a subsequent turn on the route, and overcame the difficulty of identifying a turn which was not directly associated with an obvious landmark. None of the children ever referred to the road sequence.

Such a difference in the style and content of the children's and adults' directions was unexpected, and it was difficult to explain because in the Brewster and Blades (1989) experiment only a few young children were being compared to mature adults. The difference in style might reflect an age related progression to more sophisticated direction giving, or it might simply have been the case that the older subjects had adopted a different way to describe routes as adults (without any developmental progression being indicated). To investigate this the present experiment was designed to extend the findings from Brewster and Blades, by including a larger number of subjects across a wider age range.

In designing the experiment three other points were also considered. Firstly, it was predicted that children would give directions at turns associated with landmarks more accurately than directions at turns without landmarks (from Siegel & White's 1975, theory). Secondly, we expected that the direction of travel would influence accuracy, and choice points approached travelling northwards on the map would be easier to describe accurately than choice points approached from other directions. This is a common finding from previous studies (e.g. Brewster & Blades, 1989) and in formal spatial tests (e.g. Money, 1965). Thirdly, gender differences in spatial ability have often been reported (for a review see Linn & Petersen, 1985)—for example, Ward et al. (1986) found that male students gave more accurate directions than female students, and other studies with adults have also indicated that males are better map readers (Chang & Antes, 1987). Therefore, the results from the present experiment were also examined for gender differences in ability.

Subjects

Ninety subjects took part in the experiment. There were 18 adults with the mean age of 22 years (range 21–30 years) and 72 children were randomly selected from two schools. The children were divided into four age groups of 18 children, with mean ages of six years one month (range 5–9 to 6–9); eight years two months (range 7–6 to 8–9); ten years (range 9–4 to 10–10) and 12 years (range 11–6 to 12–11). There were approximately equal numbers of male and female subjects in each age group. None of the subjects had received any training about giving directions in school.

Materials

Two maps were especially drawn for the experiment. A diagram of Map A is shown in Figure 1. The map was drawn in colour on a piece of white card (measuring 53 × 81 cm). It was intended to be similar to a conventional road map, though the landmarks were made especially salient and street names were not used because this would have disadvantaged the youngest children whose reading skills were limited. The map included a scale and a compass rose with the cardinal directions marked. A red line, drawn along the roads on the map indicated the route to be described.

Sixteen different landmarks were included on the map (these are represented symbolically in Figure 1, but on the map itself each landmark was drawn as a small picture and labelled). The landmarks were selected so that they would easily be recognized by all the subjects (e.g. a school, a garage, a church and a playground). Six of the landmarks were placed at turns along the route. The map also included traffic lights, pedestrian crossings and round-abouts, but none of these features were directly associated with the route.

The route included 12 turns—these are numbered for reference in Figure 1, but no numbers appeared on the actual map. The route was designed so that half of the turns were at a landmark (turns 1,3,5,7,9,11) and the other half did not have a

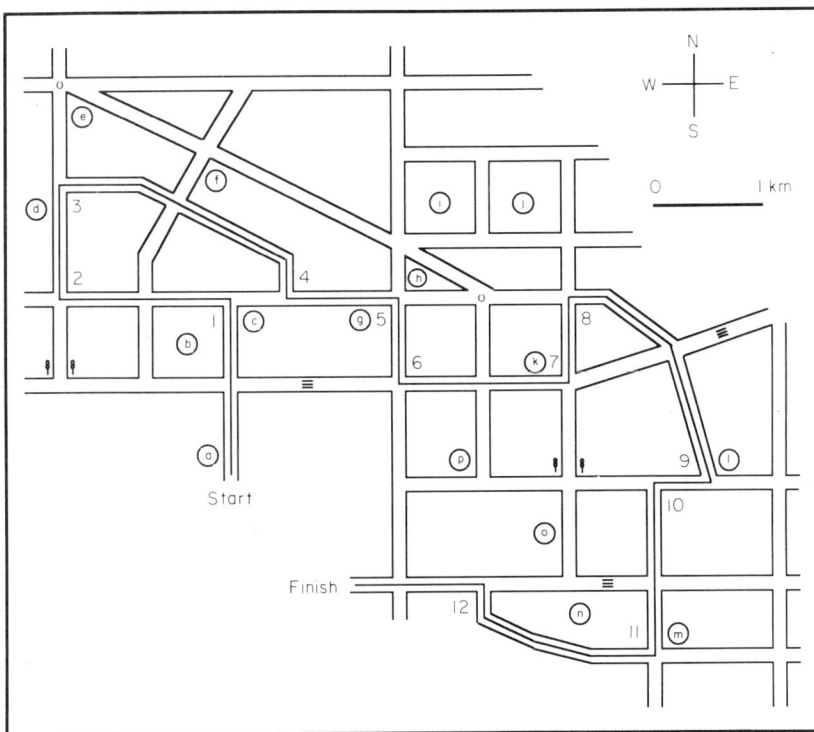

FIGURE 1. Diagram of Map A showing route which was described. Turns on the route are numbered for reference (but the numbers did not appear on the original map). Landmarks, which were included as labelled drawings on the map, were: a, garage; b, house; c, school; d, sweet shop; e, bank; f, telephone box; g, pond; h, library; i, shop; j, post office; k, swimming baths; l, church; m, big house; n, playground; o, cafe; p, doctor's house.

related landmark (2,4,6,8,10,12). Half the turns were at the first junction after a previous turn (3,5,6,8,10,12) and half were at the second junction after a previous turn (1,2,4,7,9,11).

The route was also designed to take into account the direction of travel, so that four turns were approached while travelling north (1,3,8,12); four were approached travelling south (4,6,9,11); and four were approached from the east or west—which will be referred to as 'across' the map (2,5,7,10). In addition, half the turns were right turns and half were left turns; and half the junctions were at crossroads and half were at T-junctions.

As far as possible all these factors were balanced. It was expected that the direction of travel would influence the accuracy of describing route turns, but one the main hypotheses was concerned with the presence (or absence) of landmarks at turns, and therefore to avoid confounding travel direction and landmarks these two factors were strictly counterbalanced. In other words, of the six turns with a landmark two were approached travelling north, two were approached travelling south and two were approached 'across' the map; and the same was the case for the six turns without a landmark.

The second map, Map B, was a mirror image of Map A. This map retained all the features of Map A with the exception that right turns became left turns and vice versa. All other factors were the same as Map A and this provided the opportunity to test the subjects twice with the same route, but on different maps.

Procedure

All subjects were tested individually with both maps. Half the males and females in each age group saw Map A first and the other half saw Map B first. The map was placed on a table in front of the subject (who was not allowed to rotate it).

The roads, compass rose and scale were pointed out and the landmarks were named (and all the subjects were able to recognize all the landmarks). Then it was explained that the red line on the map represented a specific route along the roads. Subjects were asked to pretend that they were describing the route over a telephone to someone who needed to walk the route but did not have a copy of the map. It was emphasized that the route directions should be clear so that anyone using them would not get lost. A toy telephone was provided for the younger children to help them pretend that they were talking to someone at a distance.

After the subjects had completed one map they were then given the alternative map and the procedure was repeated. No help was given, except (for the younger children) occasional prompts to encourage them to complete the task. All responses were recorded on tape and then transcribed.

Scoring

The subjects' route descriptions were scored for accuracy. There are several ways in which a correct turn can be described without ambiguity, for example:

(a) By a combination of landmark and direction. For example 'turn right at the church' or 'turn north after the swimming baths'. Such combinations were the most frequently used turn descriptions and usually the direction was specified in terms of left or right, although two adults, six 12 year olds; three ten year olds; and one eight year old made at least one reference to a cardinal direction in combination with a landmark. Adults also used descriptions of road junctions as effective landmarks in combination with a direction, for example: 'turn right at the T-junction'.

(b) By a combination of landmark and distance. For example 'after the garage you go 1 kilometre and then turn left'. This combination was only used infrequently— two adults; three 12 year olds; and four ten year olds each made at least one use of this combination in their descriptions.

(c) By referring to the sequence of roads. For example, 'the first right, next left and then second left'. All the adults and six of the 12 year olds made at least one reference to the sequence of roads, but it was rare for younger children to do so (see below).

Given the design of the map in this experiment, the above combinations, if applied correctly, are the minimum necessary to specify a turn unambiguously. The route descriptions were scored according to two criteria of accuracy. The first 'strict' criterion assessed whether the description given by the subject would have been sufficient for a person following the directions to make a correct turn at each of the choice points along the route. In other words, a perfectly correct route description on either map would have scored a total of 12.

But it was considered that the strict criteria might underestimate the younger subjects' ability to give directions, because some of the children may have appreciated the need to identify a turn by (for example) combining a landmark and a direction, but made a mistake by giving an inappropriate landmark or describing the direction incorrectly. The latter sometimes happened when children transposed left and right when the direction of travel was down the map. To compensate for errors which might have been due to a confusion of terms rather than an inability to give directions per se, a second 'lax' criterion was also used to assess the descriptions: if subjects (for example) referred to a landmark and a direction to indicate a turn, but gave either a wrong direction or an inappropriate landmark they were scored as correct. If however, both direction and landmark were given inappropriately the subject received no credit. The mean scores for each age group by each criterion are given in Table 1.

In fact, both scoring criteria gave similar results in all the statistical analyses, and therefore only the results from the strict criteria will be reported here. Nonetheless, the similarity between the strict and the lax analyses indicates that the differences between age groups and conditions given below do reflect developmental differences in the ability to understand what is required to give a satisfactory route description. The reported differences are not simply the consequence of younger subjects being less consistent in the use of terms such as left and right, or less precise in selecting landmarks.

Other measures of performance (e.g. the amount of information given in the subjects' descriptions and the number of landmarks or road junctions mentioned) are described in the following section. As explained above there are different ways to describe a route and two individuals might be equally accurate at giving directions, but do so in different ways and therefore the results section also includes a discussion of the different descriptive styles used by the subjects.

TABLE 1
Mean scores for each age group for strict and lax criteria (for definition of criteria, see text). Maximum possible score was 24

	Age group (years)				
	6	8	10	12	Adult
Strict criterion	0·00	0·89	7·00	17·35	23·17
Lax criterion	0·28	3·38	7·92	19·68	23·52

Results

Preliminary analyses showed that there was no difference in performance on Map A and Map B; no effect for direction of turn (left or right) and no effect for the type of junction (crossroads or T-junction). Nor was there an effect for the order of presentation—in other words, there was no practice effect and all the age groups performed equally well with the first map and the second map which they were given. Therefore, the subjects' scores for both Maps A and B were combined (i.e. if subjects described every turn on both maps correctly they would have scored a maximum of 24).

Two analyses were carried out. The first focused on the relationship between the presence of a landmark at a turn and the direction of travel in a 5 (age) × 2 (male/female) × 2 (landmark/no landmark) × 3 (travel direction: north, south, across) analysis of variance.

There was an effect for age, $F = 99·50$; df 4,80; $p < 0·001$. A Tukey HSD post hoc comparison ($p < 0·01$) showed that the adults had a better mean score than the 12 year olds who were better than the ten year olds who were better than the eight year olds and six year olds. There was no difference between the performance of the latter two age groups (for the mean scores of each age group see 'strict' data in Table 1). Therefore, with the exception of the two youngest groups there was a progressive age-related improvement in the ability to give effective directions from the maps.

There was an effect for landmark, $F = 7·09$; df 1,80; $p < 0·01$). The subjects were more accurate when there was a landmark associated with a turn (mean score 5·02) than when there was no landmark (4·66). There was also an effect for direction, $F = 22·24$; df 2,160; $p < 0·001$. Tukey comparisons ($p < 0·01$) showed that when travel direction was northwards turns were better described (mean score 3·70) than when the travel direction was across the map (3·18) or when travel was southwards (2·80). Also, accuracy was better ($p < 0·05$) when direction of travel was across the map than when it was southwards. In other words, subjects found it easier to describe turns correctly when moving 'up' the map than when moving across it, and they were poorest when describing turns as they moved 'down' the map. There was no effect for gender, $F = 0·07$; df 1,80; $p > 0·05$.

A significant interaction was found for age 3 direction, $F = 4·76$; df 8,120; $p < 0·001$. This interaction was because the effect of direction of travel ('up' better than 'across' better than 'down') was generated by the performance of the three intermediate age groups who each showed this trend. But neither the adults (with almost perfect scores for all three travel directions) nor the six year olds (with

TABLE 2

Mean number of items of information and mean number of landmarks included in the route descriptions by each age group. Table also shows the number of landmarks as a proportion of the number of items of information

	Age group (years)				
	6	8	10	12	Adult
Items of information	6·9	18·9	27·6	51·5	69·3
Landmarks	3·9	5·4	6·4	11·8	14·3
Landmarks as proportion of information	0·57	0·29	0·23	0·22	0·20

uniformly poor scores for all three directions) reflected any effect for the direction of travel. A significant interaction was also found for landmark 3 direction, $F = 6·55$, df 2,160; $p < 0·01$. This was the result of performance being more accurate when a landmark was associated with a turn and travel direction was north or across the map. But when travel direction was southwards the presence or absence of a landmark made little difference to the accuracy of the turn description.

The second analysis considered the effect of the sequence of roads. Each subject's mean score for first turn with a landmark; first turn without a landmark; second turn with a landmark; and second turn without a landmark were entered in a 5 (age) × 2 (male/female) × 2 (landmark/no landmark) × 2 (first turn/second turn) analysis of variance.

The significant main effects for age and landmark were described above. The other main finding was an effect for turn, $F = 12·39$; df 1,80; $p < 0·001$. The subjects were better at describing a turn when it was the first turn after a previous turn (mean score 5·07) than when it was the second turn after a previous turn (4·68). There was also a significant interaction for age 3 turn, $F = 12·39$; df 4,80; $p < 0·001$. This came about because all the age groups had similar scores for both types of turn, except for the ten and 12 year olds who were more successful at describing a first turn than a second turn.

As a simple measure of the amount of information provided by the route descriptions a total was made of all the items of information mentioned by each subject (for both maps). An item of information was taken as any reference to a landmark, a feature on the map, a distance, or an indication of direction. These were totalled (without taking into account how accurate or appropriate the information might be) and the means for each age group were calculated. The means indicated a clear age related increase in the total amount of information provided by the subjects—see Table 2.

More specifically, the total number of landmarks referred to by each subject was also calculated and the means for each age group are included in Table 2. This table also gives the number of landmarks as a proportion of all the items of information and shows that the proportion declined with age. For example, names of landmarks represented 57% and 29% of the information provided by the six year olds and eight year olds respectively, but only 20% of the adults' descriptions were references to landmarks.

An assessment was made of the total number of references to landmarks (for both maps and all age groups combined). The six landmarks at turns were referred to most frequently (with a mean of 43·5 references; range 28–54) and the only other landmarks to be mentioned as frequently were a garage (at the beginning of the route) which received 41 references, and a playground (at the end of the route) with 48 references. Four landmarks which were on the route, but not at turns, were sometimes mentioned (with a mean of 15 references; range 4–28). Landmarks which were not associated with the route were almost never included in the subjects' descriptions; one received four references; one received three; one received two; and two received one reference each.

All 18 adults included descriptions of road junctions, usually by stating whether a choice point was a crossroad or a T-junction and they made a mean of 9·7 references (range 2–23) to the type of junction. Eight of the 12 year olds also described the road junctions with a mean of 10·4 (range 3–17) references. But only two of the ten year olds described the junctions (one made seven references and one made three references), and none of the younger children attempted to give information about the type of junction.

An important difference between the adults and most of the children was the adults' use of the sequence of roads (i.e. describing a turn as 'first left'; 'second right', etc.). Without exception all the adults included references to the road sequence—the 18 adults made a mean number of 7·1 (range 2–14) such references. Six of the 12 year olds also mentioned the road sequence (mean number of references, 5·1 and range 2–8). But only one ten year old included the road sequence in his description and referred to it at eight points along the routes; none of the younger children mentioned the road sequence.

The six year olds had difficulty in carrying out the task and a few children were able to do no more than point out one or two unconnected features on the map, and none of the six year olds were able to give a completely accurate description of any of the turns along the routes. Two six year olds mentioned cardinal directions, and one child used the directional term 'right' on a single occasion, but in general, the children relied on vague indicators such as 'over there', 'forward', 'that way' which reflected movement on the map, rather than the directions for actual travel, for example:

You go up and across; up; across; down; across; up; round; across and down; up and round

Six of the children gave descriptions like the above which were just a list of directions and made no reference to the features on the map. Alternatively four of the six year olds simply provided a list of landmarks with little or no indication of direction at all.

Go past house; past sweet shop; past pond; past swimming baths; past church...past playground and you're there

The eight year olds were only a little more accurate than the six year olds, though some of the eight year olds gave accurate directions at individual turns. Eight of the eight year olds mentioned cardinal directions, usually correctly (though of course the cardinal directions were marked on the maps). Nine children included 'left' and 'right', but on only 39% of the occasions when these terms were used were they given correctly. Like the younger age group, one of the eight year olds just listed landmarks without giving any directions, and ten of the eight year olds gave imprecise indications of direction without including any landmarks:

Right; you go up; then turn left; then you go back up; then turn right; then go right again; then go down; then right....

Though a few children did include both directions and landmarks in more detailed descriptions, and with some recognition that it was appropriate to link landmarks and directions together:

First you start at a garage and you go past a house and a school and then turn east and turn a corner and you go past a sweet shop; then you turn a corner again and you go past a bank; then go past a telephone box; then you turn a corner again....

The ten year olds, who were significantly better than both the younger age groups, showed an improvement in giving directions. But only five of the children described more than half of the 24 turns correctly and therefore even the more successful children in this age group only provided partial route descriptions. Five children mentioned cardinal directions (usually correctly) and ten used left and right (but with only 46% accuracy):

You start at the garage; you go past the house; turn left; then you go north; then you turn right...and get to a pond; then you turn south; then you turn left and you come to the swimming baths; you turn again left; then you turn right; go down a hill to the church; then turn left; then south; you come to a playground and then you've finished.

Unlike the younger age groups, four of the ten year olds introduced distance estimates into their descriptions:

You start at a garage and then go up about 1 kilometre; then turn a corner....

The 12 year olds were generally good at giving directions. Sixteen of the children gave accurate directions at more than half the 24 turns, and eight of these children were correct at 20 or more turns. In other words, many of the 12 year olds were able to give nearly complete descriptions of the route. Both cardinal directions and left/right indicators were used with few errors and three children mentioned distances correctly:

You start by the garage; then you go about one and a half kilometres north and you will get to the school; then go left...for another kilometre until you get to a crossroads; then go right and you should pass a sweet shop and then you carry on about a kilometre and you turn right...

As discussed above, six children in this age group also included the road sequence:

...walk along, don't turn until you get to the church; then turn right; then turn first left; walk along and turn second right and walk past the playground....

The adults nearly always gave completely accurate directions, and the few errors were due to confusions of left and right (usually when travelling down the map). Despite their much longer and more detailed route descriptions, only four adults included distance estimates (one subject gave three distances, one gave two and two gave one each) and two adults gave cardinal directions (one gave six references to north or south, and one made a single reference to north). Adults were more likely than children to describe the type of junction at the choice points and quite often used the junction description as a landmark by specifying the type of junction and the turn required without giving any additional information:

...go past the church to a T-junction; turn right at this junction; turn left at the next T-junction; go over another set of crossroads and then turn right at the next crossroads; follow the road round; turn left at the T-junction; then come to the finish.

The adults provided very detailed directions and therefore many of their turn descriptions included several items of information, some of which were redundant because the route turn could have been unambiguously specified with much less information.

Discussion

The results from the experiment demonstrated a clear age related improvement in the accuracy of direction giving. The results showed that the younger children, (the six and eight year olds) had difficulty giving directions from the map. None of the six year olds were able to describe any of the turns accurately, and although some of the eight year old children were able to describe individual turns, they were unable to provide anything approaching a coherent route description.

The finding that six year olds performed poorly corresponds with the results from Brewster and Blades (1989) and confirms that children of this age have difficulty giving directions. In terms of the developmental theories put forward by Piaget et al. (1960) or Siegel and White (1975), children at the age of six years may not have reached the stage when they have a properly formed the concept of a 'route', and if young children's environmental knowledge is limited to a memory for sensorimotor movements through the environment (Piaget) or unconnected landmarks (Siegel and White) then it is not likely that they will be able to verbalise a coherent route. On the other hand there is now much evidence that the descriptions of young children's environmental cognition put forward by Piaget or by Siegel and White have underestimated children's ability (see Blades, 1991), and studies (e.g. Spencer & Darvizeh, 1983) have demonstrated that young children can give at least partial descriptions of a familiar route from memory. The eight year olds also performed poorly in the present experiment, and this contrasts with Waller's (1985) finding that children of a similar age could give directions along an actual route. There are at least two possible reasons for this discrepancy—firstly, the actual routes used by Waller were less complex and easier to describe (in the sense that the choice points on the route were indicated by obvious landmarks). Or secondly, young children may be better at describing actual routes than routes shown on a map. The latter reason, could be part of the explanation for the generally poor performance of the six and eight year olds in the present experiment, and further research could consider differences in young children's descriptions when describing an actual route and a map of the same route.

The two older age groups of children demonstrated a marked improvement in direction giving, though it was only the 12 year olds who were able to give correct directions for all or most of the route. The 12 year olds improvement in direction giving was due to their greater accuracy when using relational terms, and an increased use of road descriptions and road sequences. By describing intersections and naming the road sequence, some of the 12 year old children approximated to the style of the adult subjects. However, it is not clear how children come to adopt this style—it can be assumed that children do not receive specific teaching about how to give directions, and therefore the adoption of an 'adult style' of description may be dependent on imitation or, with greater experience of the environment, a recognition that the combination of certain types of information can provide more economical and less ambiguous route directions. Whether younger children could be taught to give directions in a similar manner remains an open question, but it is worth further investigation because it has important practical implications for teaching children how to give effective route descriptions.

A further factor in the accuracy of the 12 year olds' and adults' was the amount of information which they included in their directions. On the one hand the information which they gave included much redundant detail which was additional to what was strictly necessary to identify a turn, but on the other hand it had the important effect of overcoming any minor errors or ambiguities in the directions. As Flavell *et al.* (1985) have demonstrated, 12 year olds and adults are sensitive to ambiguous route information, but younger children may not notice a potential ambiguity.

Although the cardinal directions and distance information were included on the map, these were comparatively rarely used by any of the age groups, and this was especially true for the adults who, despite the length and detail of their descriptions, hardly ever mentioned distance or cardinal directions. This suggests that route giving proficiency does not depend on the use of these potential aids, and that if anything, the more 'expert' subjects preferred not to use cardinal directions and distance at all.

As expected, the presence of a landmark at a turn usually resulted in a more accurate description of the turn being given, and the landmarks at turns were referred to much more frequently than any of the other landmarks on the map, even those which were placed along the route. Also as expected, the direction of travel was a major factor in accuracy, and when moving across or 'down' the map the need to work out the correct spatial terms proved more difficult than when travelling 'up' the map. This is a common finding for both children (e.g. Money, 1965; Harris, 1972) and for adults (e.g. Farrell, 1979; Shepard & Hurwitz, 1984) and was an inevitable effect of presenting the route in the form of a map.

There were no gender differences in performance and no evidence that the male subjects performed better than the female subjects. Therefore, the (usually weak) effects favouring males which have been found in previous studies with maps or direction giving (e.g. Ward *et al.*, 1986; Chang & Antes, 1987) were not supported.

In summary, the experiment showed that there are important developmental differences in the way that children and adults give directions. When describing routes, young children provided little information and were generally unable to give a coherent set of directions. They focused on landmarks and vague indications of directions (which often related more to movement on the map than the movement in the environment), but the landmarks and directions were rarely combined to give a clear indication of the route to be taken at a choice point. However, by the age of about 12 years many of the children provided more accurate and detailed information and were able to adopt a style of direction giving similar to the style favoured by adults. As well as information about landmarks and the use of left and right directions, all the adults made references to the type of junction being approached and often indicated the sequence of roads in relation to a previous turn, so that the recipient of the directions was told whether the next turn was the first or second road ahead. Details about road junctions and sequences were never included by the children before the age of 12 years, and this was the most marked difference in style between the age groups.

The present experiment has provided an outline of how the ability to give directions develops across the age range, and it has also highlighted the marked contrast in the way that subjects before the age of 12 and subjects after the age of 12 years give directions. Not only were the older subjects more accurate, but there were significant differences in the content and the style of their directions compared to the younger subjects. Further research could consider what factors contribute to and influence the ability to give directions and how children about the age of 12 years come to adopt the style of direction giving most frequently used by adults.

Acknowledgements

We are very pleased to acknowledge the help and support of all the staff at the schools where this research was carried out—in particular, the Headteacher, Mr C. Smith, and Mrs G. Medlicott at Tankersley Church of England School, Barnsley, and the Headteacher, Mr G. Hall, at Lydgate Middle School, Sheffield. We are also grateful to Christopher Spencer and the three reviewers who read and commented on an earlier draft of this paper.

References

Blades, M. (1989). Children's ability to learn about the environment from direct experience and from spatial representations. *Children's Environments Quarterly*, **6**, 4–14.

Blades, M. (1991). Wayfinding theory and research: the need for a new approach. In D. M. Mark & A. V. Frank, Eds., *Cognitive and Linguistic Aspects of Geographic Space*. Dordrecht: Kluwer Academic Press.

Boone, D. R. & Prescott, T. E. (1968). Development of left–right discrimination in normal children. *Perceptual and Motor Skills*, **26**, 267–74.

Brewster, S. & Blades, M. (1989). Which way to go? Children's ability to give directions in the environment and from maps. *Journal of Environmental Education and Information*, **8**, 141–156.

Chang, K. & Antes, J. R. (1987). Sex and cultural differences in map reading. *American Cartographer*, **14**, 29–42.

Farrell, W. S. (1979). Coding left and right. *Journal of Experimental Psychology: Human Perception and Performance*, **5**, 42–51.

Flavell, J. H., Green, F. L. & Flavell, E. R. (1985). The road not taken: understanding the implications of initial uncertainty in evaluating spatial directions. *Developmental Psychology*, **21**, 207–216.

Golledge, R. G. (1987). Environmental cognition. In D. Stokols & I. Altman, Eds., *Handbook of Environmental Psychology*. New York: Wiley, vol. 1.

Gordon, D. A. & Wood, H. C. (1970). How drivers locate unfamiliar addresses—an experiment in route finding. *Public Roads*, **36**, 44–47.

Harris, L. J. (1972). Discrimination of left and right, and development of the logic of relations. *Merrill-Palmer Quarterly*, **18**, 307–320.

Heft, H. & Wohlwill, J. F. (1987). Environmental cognition in children. In Stokols & I. Altman, Eds., *Handbook of Environmental Psychology*. New York: Wiley, vol. 1.

Hill, M. R. (1987). 'Asking directions' and pedestrian wayfinding. *Man-Environment Systems*, **17**, 113–120.

Kirasic, K. C. & Mathes, E. A. (1990). Effects of different means for conveying environmental information on elderly adults' spatial cognition and behavior. *Environment and Behavior*, **22**, 591–607.

Klein, W. (1982). Local deixis in route directions. In R. J. Jarvella & W. Klein, Eds., *Speech, Place and Action*. Chichester: Wiley.

Kovach, R. C., Surrette, M. A. & Aamodt, M. G. (1988). Following informal street maps. Effects of map design. *Environment and Behavior*, **20**, 683–699.

Linn, M. C. & Petersen, A. C. (1985). Emergence and characterization of sex differences in spatial ability: a meta-analysis. *Child Development*, **56**, 1479–1498.

Lloyd, P. (1990). Children's communication. In R. Grieve & M. Hughes, Eds., *Understanding Children*. Oxford; Blackwell.

Lloyd, P. (1991). Strategies used to communicate route directions by telephone: a comparison of the performance of 7 year olds, 10 year olds and adults. *Journal of Child Language*, **18**, 171–189.

Lloyd, P. (in press). The role of clarification requests in children's communication of route directions by telephone. *Discourse Processes*.

Lunn, S. E. (1978). Route choice by drivers. *Transport and Road Research Laboratory Supplementary Report 374.*

Mark, D. W. & McGranaghan, M. (1988). Map use and map alternatives: an experiment in intraurban navigation. *Canadian Geographer*, **32**, 69–75.

Money, J. (1965). *Manual for the Standardized Road-Map Test of Direction Sense*. Baltimore, MD: Johns Hopkins.

Petchenik, B. B. (1985). Facts or values: basic methodological issues in research for educational mapping. *Cartographica*, **22**, 20–42.

Piaget, J., Inhelder, B. & Szeminska, A. (1960). *The Child's Conception of Geometry*. London: Routledge and Kegan Paul.

Psathas, G. (1987). Finding a place by following directions: a phenomenology of pedestrian and driver wayfinding. *Man–Environment Systems*, **17**, 99–103.

Psathas, G. & Henslin, J. M. (1967). Dispatched orders and the cab driver: a study of locating activities. *Social Problems*, **14**, 424–443.

Psathas, G. & Kozloff, M. (1976). The structure of directions. *Semiotica*, **17**, 11–130.

Shepard, R. N. & Hurwitz, S. (1984). Upward direction, mental rotation and discrimination of left and right turns in maps. *Cognition*, **18**, 161–193.

Siegel, A. W. & White, S. H. (1975). The development of spatial representations of large-scale environments. In H. W. Reese, Ed., *Advances in Child Development and Behaviour*. New York: Academic Press, vol. 10.

Spencer, C., Blades, M. & Morsley, K. (1989). *The Child in the Physical Environment: the Development of Spatial Knowledge and Cognition*. Chichester: Wiley.

Spencer, C. & Darvizeh, Z. (1983). Young children's place descriptions, maps and route-finding: a comparison of nursery children in Iran and Britain. *International Journal of Early Childhood*, **15**, 26–31.

Streeter, L. A., Vitello, D. & Wonsiewicz, S. A. (1985). How to tell people where to go: comparing navigational aids. *International Journal of Man-machine Studies*, **22**, 549–562.

Vanetti, E. J. & Allen, G. L. (1988). Communicating environmental knowledge. The impact of verbal and spatial abilities on the production and comprehension of route directions. *Environment and Behavior*, **20**, 667–682.

Waller, G. (1985). Linear organization of spatial instructions: development of comprehension and production. *First Language*, **6**, 137–143.

Waller, G. (1986a). The development of route knowledge: multiple dimensions? *Journal of Environmental Psychology*, **6**, 109–119.

Waller, G. (1986b). The use of 'left' and 'right' in speech: the development of listner-specific skills. *Journal of Child Language*, **13**, 573–582.

Waller, G. & Harris, P. L. (1988). Who's going where?: children's route descriptions for peers and younger children. *British Journal of Developmental Psychology*, **6**, 137–143.

Ward, S. L., Newcombe, N. & Overton, W. F. (1986). Turn left at the church, or three miles north. A study of direction giving and sex differences. *Environment and Behavior*, **18**, 192–213.

Wunderlich, D. & Reinelt, R. (1982). How to get there from here. In R. J. Jarvella & W. Klein, Eds., *Speech, Place and Action*. Chichester: Wiley.

JANINE EBER MAPS LONDON: INDIVIDUAL DIMENSIONS OF COGNITIVE IMAGERY

DENIS WOOD and ROBERT BECK

School of Design, North Carolina State University, Raleigh, NC 27695-7701, U.S.A.

Abstract

In 1971 we accompanied a group of American teenagers on their first trip to Europe, collecting from them more than 300 maps of London, Paris and Rome sketched according to the protocols established in *Environmental A*, an experimental mapping language. Heretofore, with the exception of an exploratory comparison of individual and aggregate imagery (Beck and Wood, 1976b), we have described the results of *group* analyses of these sketch maps (Wood, 1973; Beck and Wood, 1976a; Wood and Beck, 1976). These, however, obscure not only the real behaviors of actual individuals (as called for by, *inter alia*, Canter, 1977, p. 67; Gale, 1983, p. 71 and Walmsley, 1988, p. 50), but also the full power of *Environmental A* to illuminate the development of individual map *surfaces*, with the attendant improvement this engenders in the understanding of cognitive development in the individual. Here, then, we take an essentially empirical look at a single teenager's maps of a single novel environment, as sketched over the period of a week, augmented by both predicted and remembered images. At the same time we take this opportunity to probe the relationship of the *formal* properties of the mapped image to its evaluative, affective, ultimately *meaningful* dimensions, aspects of the image *Environmental A* has proven peculiarly useful in elucidating (Spencer and Dixon, 1983, and Walmsley, 1988, p. 51; but also see Buttenfield, 1986 and Canter and Donald, 1987, p. 1286).

Janine Eber Visits London

In the summer of 1971 Janine Eber, a 17-year-old from a small town in southern New England, in company with thirty-odd other American teenagers on tour, visited London and other attractive parts of Western Europe for the first time. In most respects it was an ordinary tour: ordinary American teenagers saw ordinary tourist Europe in the ordinary way.

Of course, this ordinariness was limited. Most American teenagers don't spend their summers travelling around Europe; but among those that do, these were different in no systematic way (the group was naturally occurring). The situation with respect to Janine was similar. While a unique person with any number of attractive qualities, her history had distinguished her in no startling way. She took accordion in elementary school, wore braces in junior high, remained active in Scouts through high school. When we worked with her she was a college-bound rising senior thinking very vaguely about maybe pursuing a career as some sort of health professional (physical therapy? nursing? medical technology?). During her senior year in high school she studied, among other things, French and Physics; and did sufficiently well to earn a scholarship to Wagner, a small Lutheran-affiliated college on Staten Island in New York. On the other hand, there were some unusual features: nearly every day Janine and the others were asked to fill out questionnaires, draw maps, make checks beside lists of adjectives and do other things to record aspects of the developing images of the places visited. Janine complied with each of our requests for assistance; and if she was in this not alone, she uniquely volunteered an introspective description of her mapping process. It is this in particular that has singled her out for attention here. Among other things that Janine did was to fill out—halfway through her seven-day stay in London—a landscape adjective checklist (Gough and Heilbrun, 1965), that had been created by Ken Craik and his associates (Craik, 1968; 1970). Janine was supposed to check off the adjectives on this list she felt described London, but she did more that this. Janine placed *different numbers* of checks next to different adjectives indicating a hierarchic ordering. One adjective was checked five times, a couple four times, a few two and three times and the rest once. She checked 42% of the 200 adjectives on the list. Table 1 presents the adjectives she checked and the number of times she checked them.

These adjectives allow us to approach (on tip-toe to be sure) Janine's image of London, or one of those many images of London she might summon forth in response to a question like, 'Well, what was London like?' Judging from the list her answer would be something like, 'Well, it was cool, filled with flowers and huge.' These three adjectives, checked four and five times, provided the keys to the rest of the list, for each of them is echoed in the remaining adjectives checked. *Cool*, for example, is echoed in: *invigorating, refreshing, relaxing, tranquil, calm, clean, clear, comfortable, content, crisp, placid, pure, quiet, restful, serene, sunny, warm* and *windy*. Each of these adjectives in isolation tells its own story (like *sunny*), but taken together their evocation is decisive: London is a delightful place, not unlike a seaside resort, or perhaps a small town in New England.

There is a second component, which, while it does not contradict the first, does bring the first more obviously into the context of London. Resonating with *vast*—which Janine checked four times—are: *spacious, big, broad, expansive, large, massive,* and *wide*. While the first component made it clear that London was pleasant, this one makes it clear that London is also large.

The third component we identified ran changes on *flowery*—which she also checked four times: *fresh, green, meadowy, pastoral, wooded, shady, forested, grassy, leafy, lush, natural* and *tree-studded*. Somehow these are all associated in one way or another with chlorophyl. London is delightful, London is large, London is filled with greenery. Note that there are some adjectives that could easily move from group to group without modifying the prevalent feeling. It you object to *sunny* tagging after *cool*, let it tag after *flowery*; and if you would prefer that *shady* and *fresh* run after *cool* instead of *flowery*, we would be glad to go along with you. The attempt is not to segment rigorously Janine's list of adjectives, but to distill from it increasingly easy to deal with—and increasingly powerful—image components.

There is a bunch of adjectives that, while they don't pay any particular allegiance to *flowery, vast* and *cool*, nonetheless support the developing image. They are all very positive: *friendly, happy, lovely, peaceful, active, alive, awesome, beautiful, bright, colorful, free, gentle, good, inspiring, inviting, majestic, nice, old, rich, secure, slow, soft,* and *unusual*. Some of these help to modify the powerful 'outdoorsy' quality of the first three components. Thus, *friendly* and *happy, gentle* and *good, inviting* and *nice, old* and *rich* could apply to Londoners as well as to London's parks, or if not to Londoners themselves, then to the environment built and supported by them. London is vast, but also nice, secure, gentle, and inviting; it is flowery and pastoral, but also old, rich and unusual.

TABLE 1

Adjectives Checked by Janine Eber for London (Numbers at column heads refer to the number of check marks beside each adjective)

5 Checks	4 Checks	3 Checks	2 Checks	1 Check	
Cool	Vast	Fresh	Meandering	Active	Mountainous
	Flowery	Friendly	Open	Alive	Natural
		Green	Pastoral	Arid	Nice
		Happy	Rolling	Autumnal	Old
		Invigorating	Shady	Awesome	Placid
		Lovely	Wooded	Beautiful	Pleasant
		Peaceful		Big	Pure
		Refreshing		Blue	Quiet
		Relaxing		Bright	Remote
		Spacious		Broad	Restful
		Tranquil		Calm	Rich
				Clean	Rough
				Clear	Rugged
				Colorful	Secluded
				Comfortable	Secure
				Content	Serene
				Crisp	Sloping
				Dark	Slow
				Deep	Soft
				Expansive	Springlike
				Forrested	Summery
				Flowing	Sunny
				Free	Tree-studded
				Gentle	Warm
				Good	Wide
				Grassy	Wild
				Hazy	Windy
				Hilly	Wintry
				Inspiring	
				Inviting	
				Isolated	
				Large	
				Leafy	
				Lush	
				Majestic	
				Massive	

The relationship of the rest of the adjectives Janine checked to our three principal components is difficult to resolve and there may be none. There is a handful of adjectives describing rolling terrain (*rolling, hilly, mountainous, rough, rugged, sloping,* and *wild*), another dealing with isolation (*isolated, remote, secluded*), a seasonal quartet (*spring-like, summery, autumnal* and *wintry*) and four others (*blue, dark, deep,* and *hazy*). These may refer to particular places within the city (as may many of the others) or to aspects of the city as a whole that escape us. *Blue, deep* and *dark* could, for instance, refer to the Thames; *isolated, remote* and *secluded* (as well as others in the principal components like *tranquil, calm, placid* and *quiet*) could refer to spots in Hyde or Regent's Park or other rural pockets in the urban maelstrom. Hills surround London and Janine did visit rolling vicinities. Or, this interpretation could be off-base, and the fact that these groups of adjectives don't fit in smoothly could be evidence of that. But assuming that it is not, what sort of image do we have? Janine's London is vast, pleasant and green, a rolling reposeful place that is friendly, happy, gentle and quiet.

Despite a superficial formality, Janine's image is emphatically meaningful. From it we learn not only about the city of London but the personality of Janine Eber, not only what there is in London, but what Janine values, prefers, likes. To sharpen this point, compare the adjectives Janine checked with those checked by Marina Giaconda, another girl on the trip from another small New England town. While Janine checked 94 of the 200 adjectives, Marina checked only 46, of which 28 were *not* in common. Among these the most significant to Marina was *dirty*, the only one checked more than once; but the rest of the adjectives Marina checked that were not also checked by Janine equally conspire to an impression of a big bustling dirty city: *black, cloudy, dangerous, fast, forceful, hard, harsh, imposing, noisy, powerful, rapid, sharp, brisk, bushy, challenging, contrasting, exciting, extensive, flat, living, moving, reaching, reflecting, running, straight,* and *towering*. Is this an image of the same city visited—at the same time on the same tour— by Janine? Marina's city is one of energy, action and power. Her adjectives reek of motion: of running and moving and reaching. Her London is hard, harsh and dirty. It is a city of adventure, danger and challenge. It is an-*ing* city, tower*ing*, impos*ing*, reflect*ing*, inspir*ing*, invigorat*ing*. Almost a third of Marina's adjectives take the gerundive -*ing* form; less than a tenth of Janine's do. Where Janine found repose, Marina found action; where Janine found cleanliness, Marina found dirt; where Janine found hills, Marina found flats; where Janine found security, Marina found danger; what Janine saw as gentle and soft Marina saw as hard and harsh. Against this divergent background, even the adjectives they checked in common are raucous with differences: *awesome, big, cool, deep, large, massive, expansive, wild, old*. In Janine's gentle green London *old* reads *antique*, but in Marina's urban hell, *old* reads *wornout*. Once again, the descriptors used are essentially formal, but once again, the image that results is emphatically meaningful. Janine liked Hyde Park better than anything else in London and hated Madame Tussaud's most of all. Marina liked Madame Tussaud's above all else and saved her opprobrium for the British Museum. When Marina thinks of London the first thing that comes to mind are 'cold people.' When Janine thinks of Londoners she thinks of them as 'friendly.' What are we hearing in these words? Janine and Marina? Or London? The answere is neither alone, for the image invariably reflects both.

Janine draws a map of London

If the choice of adjectives says more about Janine than it does about London, the maps of London that she drew say, on balance, a little more about London than they do about Janine. This is not to suggest that maps are not evaluative. Unquestionably they are: they are highly selective representations, and as such manifest choices—and thus values, goals and intentions—no less than check marks beside lists of adjectives (see Wood and Fels, 1986 for a semiological analysis of the map). But there is this difference (and it is all the difference in the world), that in *mapping* the choices are made among things apprehended in the environment, while in *checking adjectives* the choices are among those found on the list.

Immediately after completing the adjective checklist Janine drew the following map of London (Fig. 1). Its line-and-dot appearance is a consequence of *Environmental A*

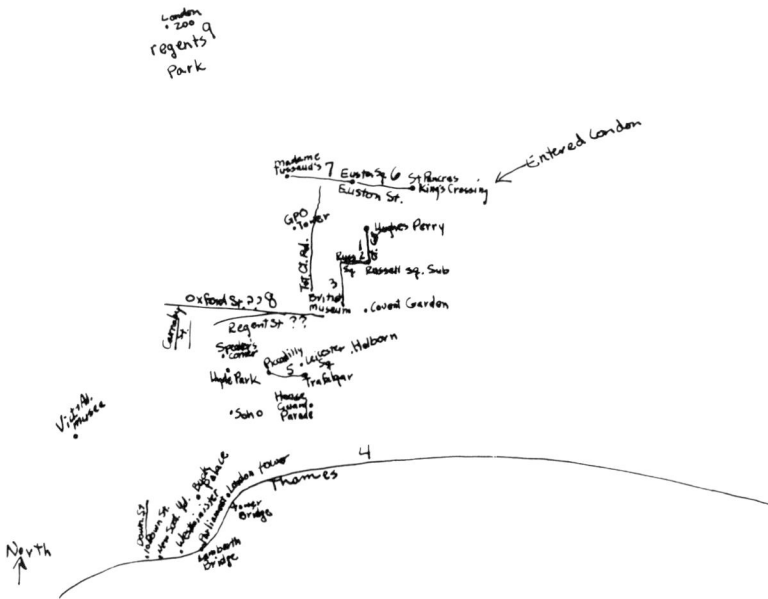

FIGURE 1. Janine Eber's Point-Line Skeleton Map of London, Midway Through Her First Week's Visit.

and all the maps drawn by Janine and her friends had more or less this character. These maps were usually drawn in stages. First a network or *skeleton* of points and lines— ordinarily representing landmarks and streets—was laid down. The order in which this skeleton was created was indicated by arabic numerals drawn beside each line. Next the mapper put a piece of tracing paper over this skeleton and made a series of map overlays (the entire protocol is reproduced in Wood, 1973). In this particular instance, Janine made one overlay showing *areas* in London (Figure 5) and another showing her *feelings* about the places she mapped (Figure 6). In this she was perfectly ordinary: most of her peers produced precisely these two overlays.

The difference between the nature of the information about London found on this map and that found in the adjective checklist is immediately apparent. Janine's point-dot map begins to approach close to the heart of the definition of form, that is, the distribution of the material of a thing in space. While, to be sure, her choice of these precise pieces of London is relevatory of Janine's systems of meaningfulness, there is an important difference between showing the orientation and relative location of Oxford Street and saying Oxford Street is beautiful. The information about the whereness of Oxford Street acquires meaning in the context of the remark that Oxford Street is beautiful, just as the remark that Oxford Street is beautiful acquires formal existence when located in the rest of London. Without the locative information the evaluative remark is unattached to anything, and simply floats; but without the evaluative remark the locative information is similarly barren. And to the extent that this is true for Oxford Street, so it is true for the whole of London. It is nice to know that London is friendly, green and big, but it becomes useful when we know *how* London is. And similarly, knowing how London is, is enhanced by knowing that it is friendly, green and big. An image that is not whole provides a limited basis for intelligent human action.

That this is true does not mean that the image cannot be taken apart. It can be taken apart, and Janine's point-dot skeleton is just such a part. But it can be taken even

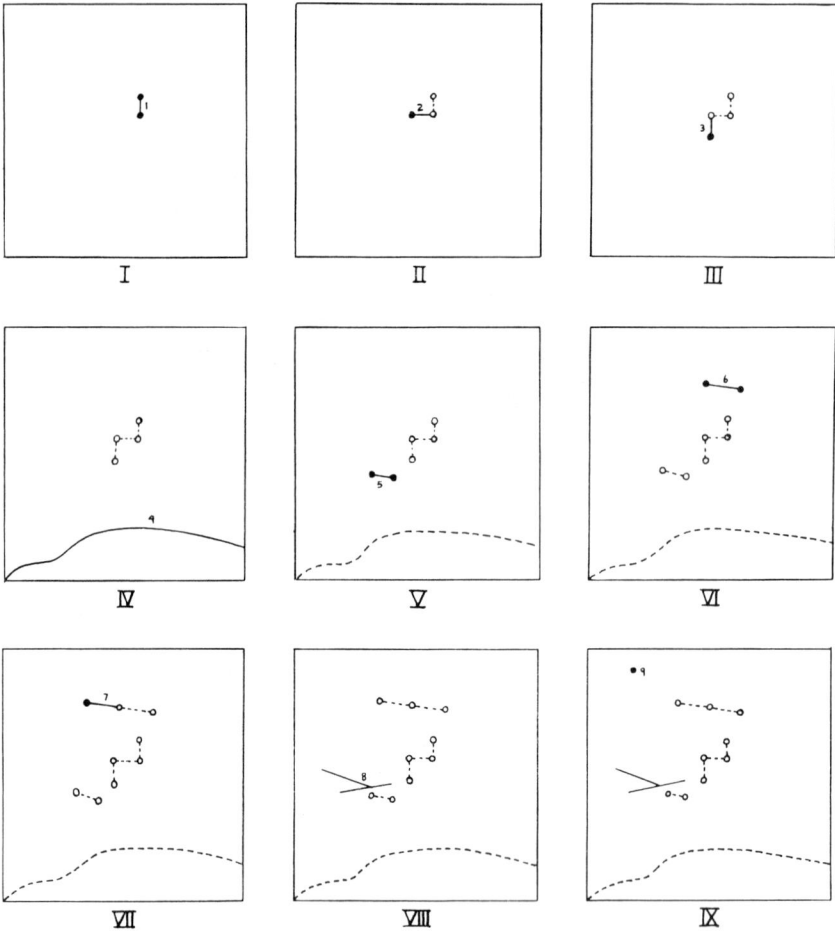

FIGURE 2. Sequence of Construction of Figure 1.

farther apart, a dissection that will be more than helpful in understanding how the image is put together in the first place. Note that each line on Janine's map is numbered. These refer to the sequence in which they were drawn. Figure 2 shows how Janine drew her map, how she composed her image. Janine commenced drawing with the dormitory in which she was staying. That is, she started mapping what she knew best. This initial gesture—as an initial gesture—is in and of itself meaningful rather than formal, hinting at the deep inter-relationship between these two aspects of the image. Her first three lines keep her in the vicinity of the dormitory, in, that is to say, familiar territory (as expected: see Herman, 1980; Wapner, Kaplan and Ciottone, 1981; Stegel, 1982). But then, having apparently exhausted her large-scale knowledge, she leaps to the south and draws the Thames. The number of places along the Thames suggests (and is confirmed by the actual itinerary of the guided tour taken by the group on its first full day in London) that this region, while distant, is nonetheless familiar. The fact that these two regions—the dormitory and the Thames—were placed on the map without being connected indicates that Janine did not yet know how to connect them. Her next step confirms this: she locates her fifth line *somewhere* between the Thames and dormitory regions. This line is a fictitious street but serves to indicate the connection between Piccadilly Circus and Trafalgar Square. What is important is that while she knew something about the local connections in this region, she was unable to connect it to her two anchoring spaces. In fact, Janine's mapping was an act of *creation*, in that as she mapped she was learning about the limits of her knowledge. Thus her map was not simply a graphic representation of something she knew, but a process of discovery in its own right. In her next foray she heads north, to sketch in something of another locally construed region whose connections to other regions are not mapped. Finally she zooms far to the northwest to add Regent's Park.

What we have in these incremental drawing gestures are a number of individual schema, indicative of internalized—but relatively fragmentary—experience. As Janine visited Madame Tussaud's or the Thames she locally modeled her routes, internalizing memorable connections. These *schema* would be valuable in allowing her to find her way back to a starting point, for example, or to strike out on new paths within local regions. The schema—these models of her movements—materialize on her map as relatively well-defined local areas capable of being sketched by virtue of the existence of the schema, this time directed, not toward retracing her steps, but representing them on a piece of paper. The connections among these schema, however, were never themselves schematized, either because they were made in the underground (subway), where— because of (1) lack of landmarks other than stations; (2) the subtle turns that do not reflect street patterns and which are nearly impossible to apprehend, and (3) the dark— schemas never developed; or because they were made on a tour bus, directed by a guide. In either case (subway or bus) there was little question of active navigation or of becoming lost and hence little motivation for learning what would amount to nearly useless or redundant information. (Once the mapping became habitual the students took close note of where they were going, even on the guided tours, so that they would be able to reproduce even these connections on the next set of maps. This was especially noticeable during the first day's tour in Rome eleven days later.)

Janine's map is then, an aggregation on a simultaneous (*all at once*) surface of a number of events that took place sequentially in time. Sequential events have been synchronized (Beck and Wood, 1976a, pp. 213–215). These local events, internalized as operable schema, become manifested on the map (symbolized) as local regions of relatively dense detail. Their relations in space (as opposed to their historical ordering) are merely stabbed at, but this does not lessen the actuality that the map is an attempt at a purely spatial ordering of events that transpired essentially in a temporal order. This two-dimensionalizing of four-dimensional experience is fraught with peril whenever the four-dimensional data are incomplete, as here, due to the underground trips and the trips on the tour bus. Incomplete data result in the inability of the mapper to project accurately the relationships among the data in hand. Figure 3 is a transformation of Janine's map into a grid which removes all the data, preserving only their spatial relationships. Had Janine had a complete and veridical set of data (such as is in theory available to a cartographer) she could have produced a map for which the transformation would have shown an equally spaced orthogonal grid. When the grid lines on the transformation of her image fail to cross at right angles, or fail to be equally spaced, it is because she has incorrectly located two or more items with respect to each other. Thus if she places two items too close to each other with respect to the other items displayed, the grid lines are drawn closer together; contrariwise, locating them too far will cause the lines to spread apart. Other errors will have similar if more

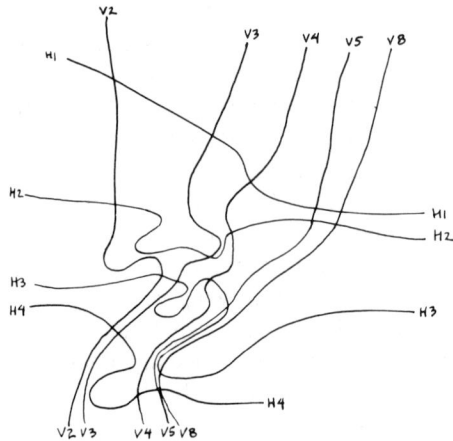

FIGURE 3. Grid Transformation of Figure 1.

noticeable consequences (Tobler, 1966; Wood, 1973, pp. 532–588; Golledge, 1987, p. 161).

Notice that Janine is most accurate in the north and northwest of London where she includes the least information, and least accurate in the centre of the map where she includes a lot of detail she was unsure of. In the extreme south, along the Thames, she actually made only two, essentially trivial, errors, but errors whose consequences are enormous. In the first place she located the Tower of London a couple of miles too far west. This has caused the entire southern half of her map to be squashed together in a mental funnel, dragging much of central London with it. Secondly, Janine has overestimated the distance between the dormitory and the Thames, particularly in the vicinity of Parliament. Both are problems of distance estimation. In the centre her error was to locate Trafalgar Square west of Piccadilly Circus rather than south. This is an error of orientation. A similar problem in orienting Regent Street has been ignored in the grid analysis because Janine indicated her lack of assurance about this location with the use of question marks. Janine has a generally useful picture of London. Distortions are in distance estimation or otherwise local, not wholesale confusion about what's where. A sense of how her map compares with others can be acquired by contrasting her grid transformation with those of Wood, Gray and Lincoln in Figure 4. The type of error that shows up in both Janine's and Gray's map was fairly common and can be attributed to the organization of the first day's guided tour (see further Wood, 1973, 565; Beck and Wood, 1976a, pp. 213–214).

Janine's areal overlay for this map (Fig. 5) includes designations for twelve areas, only eight of which are bounded. The eight small bounded areas are those portions of her skeleton rich in detail. Again, these reflect the elaborated local schemata worked out in Janine's actual experience in the field. The lack of connections among the schemata —that we saw on the skeleton—show up here as area names that are unbounded: West End, East End, South Bank, North Bank. These are names of regions that Janine realized are part of London, perhaps parts of London which she has visited, but they are only vaguely understood. In them, as known islands in an unknown ocean, float the areas of the city about which she has developed knowledge in which she feels confidence. Given Janine's image of London as green and tree-studded, it is not surprising that Regent's Park accounts for 34% of the surface area demarcated or that

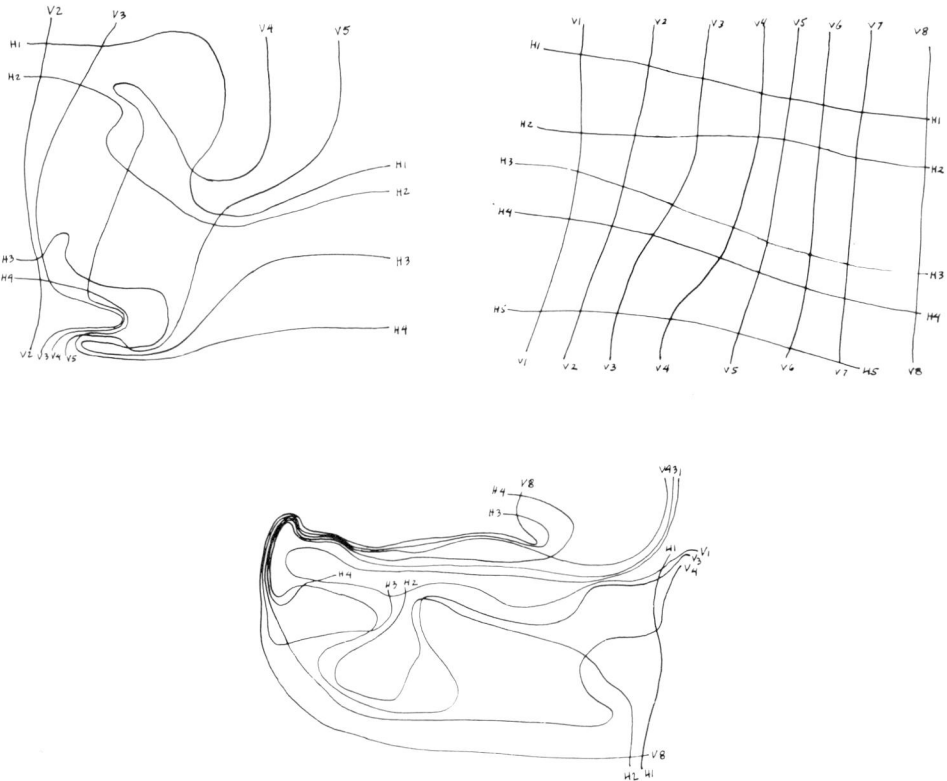

FIGURE 4. Grid Transformations of the Maps of Others Produced on the Same Occasion as Figure 1.

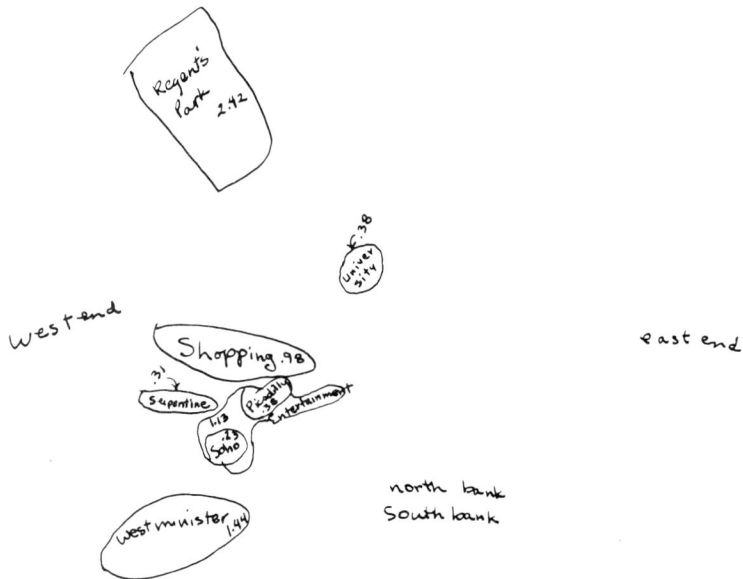

FIGURE 5. Areal Overlay for Figure 1.

FIGURE 6. Attributive Overlay for Figure 1.

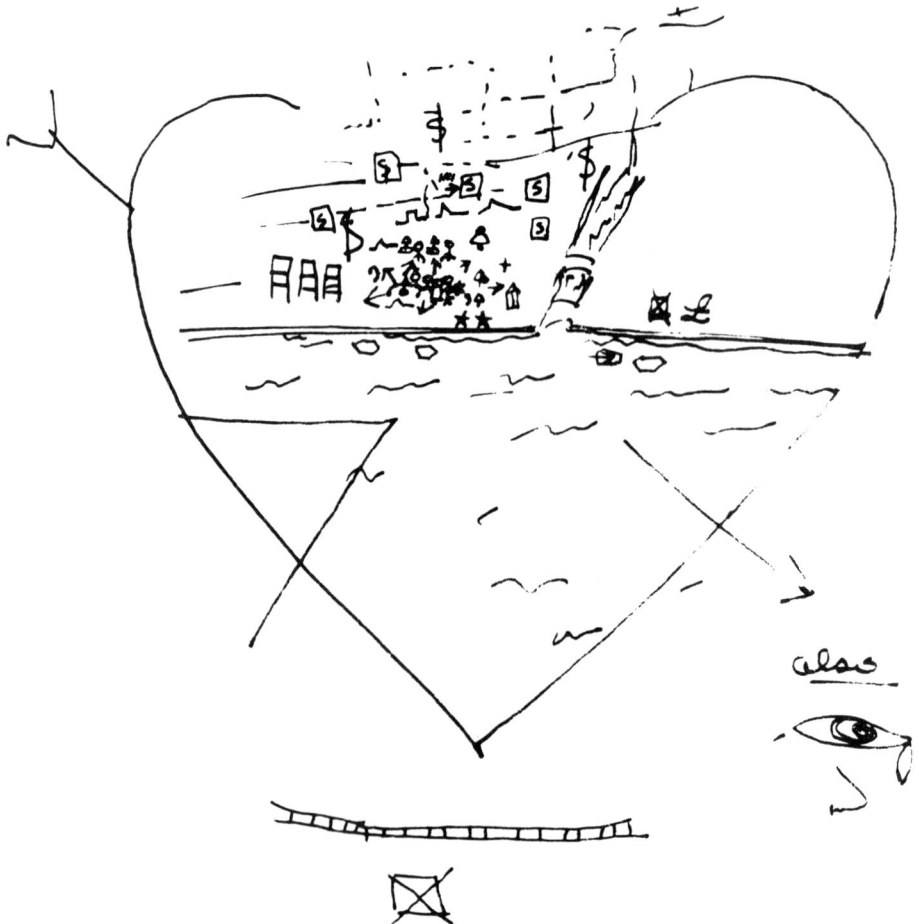

FIGURE 7. Janine's Overlay of Her Second Remembered Map of Venice.

the Serpentine (the large lake in Hyde Park) accounts for another 3%. That 37% of her knowledge of London areas should be comprised of parks is still another indication of the indissolvability of the union of meaning and form, schema and figure.

In Figure 6 we can look at Janine's other overlay. These overlays—termed interpretive or attributive—were composed by the mappers with recourse to a small vocabulary of graphic signs contained in a small dictionary available to them at all times (Wood, 1973, pp. 589–603; Murray and Spencer, 1979; Spencer and Dixon, 1983; Walmsley, 1988, p. 51). Most of the symbols that Janine used on this overlay are purely interpretive; that is, they interpret the dots and lines of the skeleton indicating which dots are underground stations, which lines are streets, and so forth. Others, however, are attributive; that is, the mapper used them to attribute feelings to specified places or things. Thus the *crescent moon* sign indicates that it was cool, the *parentheses*, that an intensely personal experience occurred, and the *circle of arrows pointing inward* a feeling of being crushed or tightly enclosed. This overlay reinforces all we have come to understand of Janine's image of London and of her value system. Contrast the three crescent moons (completely defined in the dictionary as 'cool, like shady parks, shady avenues, breezy'), the comma (defined as the 'the pause that refreshes, quiet in a crazy city, the personal touch in a mechanical world'), the parentheses (and its positive personal experience) and the bandstand—all drawn over the Serpentine Lake in Hyde Park—with the skull-and-cross bones ('sense of hostility'), the minus sign (to indicate a strongly negative impression), the multiplication signs ('up-to-the-minute, swinging, with-it'), the inward pointing arrows (for a sense of constriction) along with the symbols for bars and discotheques—all drawn over the Piccadilly Circus area. This is Janine Eber all right—these are her values—but here they have been given urban *form*. We not only know how she *feels* about things, but how these feelings are *distributed over the space* of the entity called London. Here we read not simply 'I felt good!' but 'I felt good HERE!' Or, to put it the other way, not simply that 'the Serpentine is here,' but that 'the Serpentine is here and I love it!' Janine's ultimate expression of this sentiment, applied to an entire city, is her overlay of Venice, where a smitten heart embraces the whole city (Fig. 7).

The dynamism of cognitive images

Obviously we cannot discuss the manner in which Janine's ability to structure cognitively a city matured over seventeen years (see the introduction to Canter's discussion of his mapping subject, Nick, in Canter, 1977, p. 67), but we can catch a glimpse of this process as it might have been recapitulated during the short period of her experience with the tour, for the map we have just examined was the third of five maps. Janine drew her 'first' map of London several weeks before she ever saw the city. The idea behind this 'predictive map' was to establish a sort of pre-image—an anticipatory image—that would establish a base-line against which to make comparisons. Janine's pre-map and its attributive overlay—that's all there was to this map—can be seen in Figure 8 (Canter's Nick also drew a pre-map of London. Comparison of his—Canter, 1977, p. 68—with Janine's demonstrates the extent of her *idealization*.) Unlike the London Janine was going to visit—with the Thames running along the south of the tourist city—her pre-London was equally disposed on either side of the Thames, along which at the very heart of the city, lie the things the tourists come to see: the Tower of London, Buckingham Palace, Parliament, Trafalgar Square, Hyde

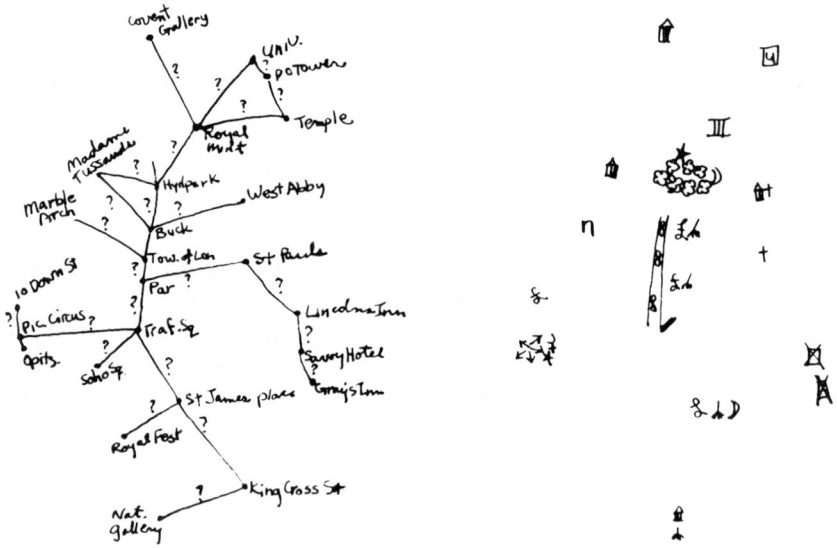

FIGURE 8. Janine's Map of What London Might Be Like. Point-Line Skeleton at Left; Overlay at Right.

Park... Look at the attributive overlay. Over Hyde Park lies a forest of trees, and already—weeks before she goes there—there already lies a crescent moon. Brooms, the symbol of cleanliness ('so clean it sparkled' reads the definition), are everywhere. There are plenty of museums and a fair share of monuments. At the right and left extremes of her map are commercial spots, a couple of hotels to the right (one of these, Gray's Inn, Janine will discover is not an inn after all), some activity to the left over Piccadilly Circus and the Opera House. (In this cluster of symbols there appears a nose—defined as an overwhelming smell either good or bad—which either indicates that that she was

FIGURE 9. The Point-Line Skeleton of Janine's First Map of London, Drawn After Two Days' Experience.

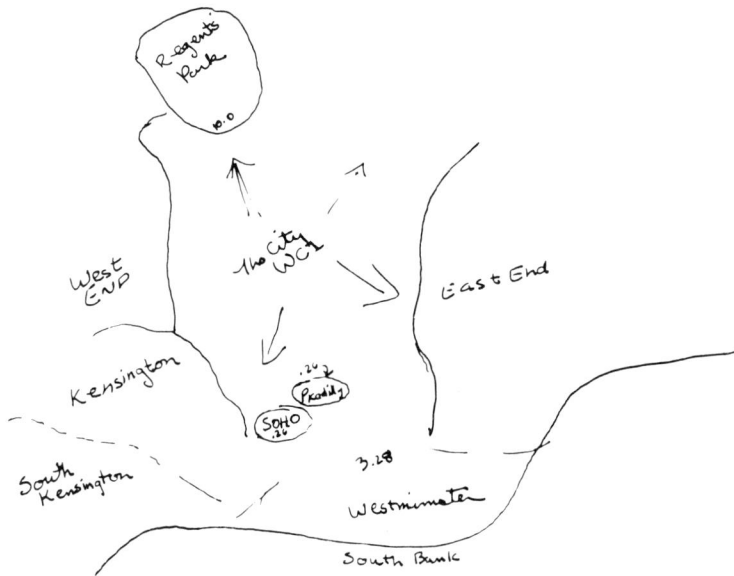

FIGURE 10. Areal Overlay for Janine's First Map of London (Fig. 9).

prepared not to appreciate the discotheques and bars—as actually turned out to be the case–or that she thought that Piccadilly Circus was actually going to be a circus, an unlikely possibility, but worth mentioning, especially since Canter's Nick—Canter, personal communication—similarly expected Oxford Circus to be a circus.)

This isn't *exactly* the London that Janine found, but it is not exactly not the London that Janine found. It is the city she hoped to find and it consequently exerted a powerful force over everything she experienced, and everything she mapped.

A couple of days after arriving in London, Janine produced her *first* map of the city. It consisted of the two portions—a skeleton and an areal overlay—exhibited as Figures 9 and 10. A grid transformation of her skeleton can be seen in Figure 11. This

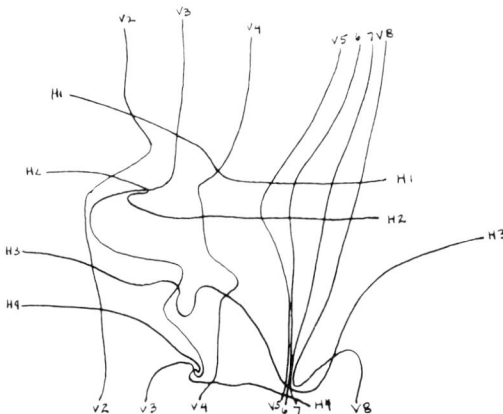

FIGURE 11. Grid Transformation of Janine's First Map of London (Fig. 9).

first map is surprisingly well organized, certainly more integrated and better connected than her second map (again, compare the first map of Canter's Nick–Canter, 1977, p. 68). It was hard labour that won this prize. Although they do not reproduce, signs of erasure cover her map. Practically everything shown had been previously located elsewhere on the map. The whole of the Thames has been moved to the south, and the London Zoo (located within Regent's Park) has been moved to the north. An erasure under the north arrow indicates that this was the first thing she put on the map. If we follow the Arabic numerals (on Figure 12) we can again reconstruct a portion of the sequence in which the map surface was created. She commenced with the dormitory, moved south to Russell Square and then again south to the British Museum. This is precisely the same sequence she followed in beginning her *second* map. At this point, however Janine did not leap to the Thames, but rather tried to work out the central portion of the map, beginning with Piccadilly Circus and working out from there to Leicester Square, Hyde Park Corner (with a stab to Speaker's Corner) and Trafalgar Square. Soho Square floats in limbo. Her comprehension of these elements is much better on the first map than it is on her second. Things are quite well-oriented both with

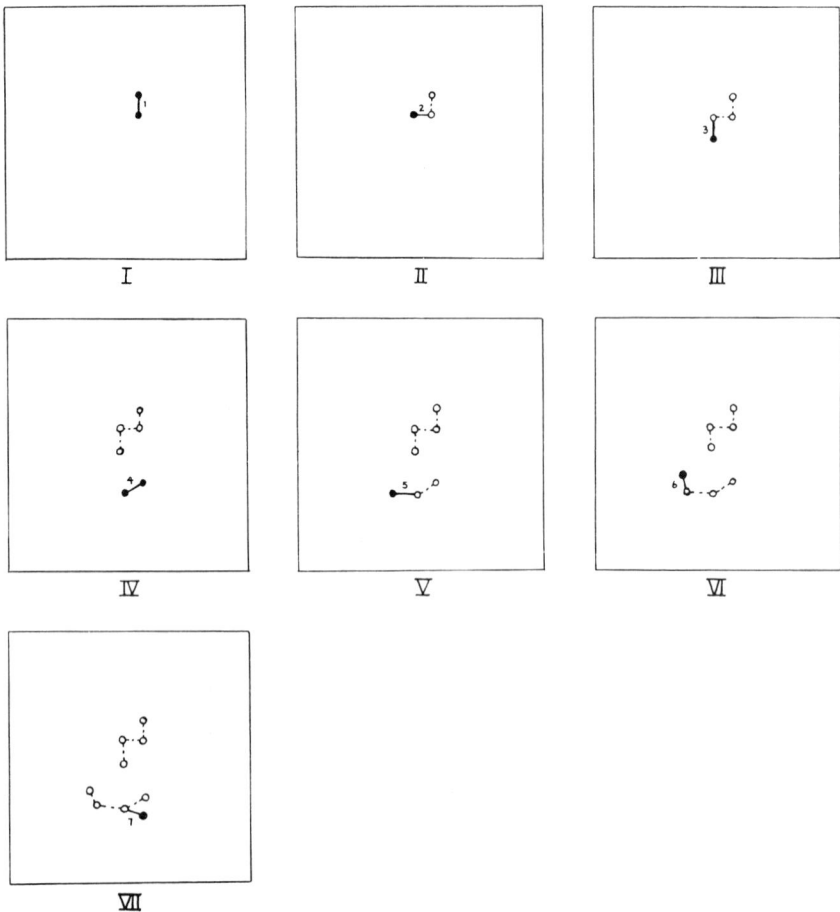

FIGURE 12. Sequence of Construction of Janine's First Map of London (Fig. 9).

respect to each other and to the map as a whole, although her distance estimation leaves something to be desired. The lack of any more numbered sequences does not allow us to continue the reconstruction past this point, but if the second map is any guide at all, Janine then did the Thames, the Euston axis to the north of her dormitory (HP Hall on this map), London Zoo, and her attempts at Oxford Street-Regent Street. In essence, the order in which the second map was completed likely reproduced the order in which this first map was produced. This suggests that her earliest learning experiences resulted in hard and fast schemata that provided a real foundation for the reconstruction of her subsequent experiences.

Not only is the map well-connected, but as the grid transformation shows, it is quite veridical. Because Janine spread her Thames-related landmarks along the river, her map preserves the east-west dimension of the city. The most obvious disasters are the location of the bend in the Thames to the east, rather than south of her dormitory, and the close proximity of Parliament and the Tower. These two errors produced a southern drag in the map as a whole and a constriction or funnel between the Tower and Parliament. Her other errors are local: a misapprehension of the location of Soho and a horrifying misunderstanding of Regent Street.

Nonetheless, a comparison of this first map (Figure 9) with her second (Fig. 1) reveals apparent deterioration in Janine's ability to organize the space of London. Increasing confidence in moving around the city caused her to give up her map study despite her awareness that we would be asking her to draw another map, and the insecurely developed images acquired by map reading faded rapidly. This left only the strongly and experientially developed schema as map-making material (the cluster or 'mini-spatial-representations' of Schadler and Siegel: see Siegel and White, 1975, p. 41). Thus, much of the deterioration is only apparent. What is actually revealed on the second map is less glibness, greater honesty and increased knowledge (something apparent in the differences between Nick's first and second maps—Canter, 1977, pp. 68–69—as well). On her second map the lack of connections is not evidence of an *unlearning* of connections once known, but rather an *admission* that these were never as well established as she suggested they might have been on her first map. This results in the fragmentation of what on map one was a single connected sequence into a collection of eight discrete aggregations and free-floating points. But in the end this results in greater accuracy. Thus, for example, her first map shows Tottenham Court Road running into Russell Square, an error rectified by a breaking of that connection on the second map. Similar remarks could be made with respect to the rest of the disconnections. When we come to the Thames we see that she has moved Westminster onto the bend in the river (veridically) but, unable to break the Parliament-Tower connection, has consequently dragged the entire map into an east-tending funnel. This alone accounts for the greater insanity of the grid transformation of the second map.

An important breakthrough on map two shows up on the areal overlay. Here the surface is much more differentiated: the large blob called 'The City' that occupies the centre of the first overlay is replaced by shopping and entertainment areas on the second. In fact, all of the areas with the exception of the four unbounded ones, are better defined on map two than on map one.

In general, the image at large seems to be developing—over a period of days—in a fashion similar to Janine's construction of a single map. If we find erasures galore on map one, we find mental erasures in comparing maps one and two. Those things omitted on the second map may be viewed as admissions of earlier error. The entire production of the map is still highly experimental, and the formal externalization of the image is still very much a learning experience, though this is more true for the city as a whole than for its parts. Thus, the small nuggets of detail that appeared on map one (dormitory to Russell Square underground station to Russell Square proper) are retained on map two. These schematized relations seem to be enduring; it is *their* relations that are undergoing substantial revision. One significant development should be noted: Janine included an interpretive-feelings overlay for the first time with her second map.

Janine's third map of London reveals another aspect of the process. A glance at Figure 13 shows that much *new* information has been added to the map, that certain original connections have been remade, and that new connections have developed. Adjustments have also been made in the scale of presentation of certain features. Note, for example that the Serpentine, the large lake in Hyde Park, is now represented at a seventh its former size. On the other hand, the Horse Guards Parade has increased in size from a mere point on the second map to an incredibly large area on the third. Many things are going on (equally apparent in Nick's third and final map—Canter, 1977, p. 69—which drawn after three *weeks'* experience is less comparable to Janine's third map than were his earlier ones).

The sequence in which this map was created is set out in Figure 14. As is amply demonstrated, the old schemata are still holding sway. The initial three lines are the same as on maps one and two. This time, however, presumably as a result of surer information, Janine works steadily out from her origin rather than leaping about the map. Finally she makes the connection to the Euston axis to the north of her dormitory, with assurance, but—indicating the strength of the initiating schema—what results as a straight line connection between Euston and the Russell Square

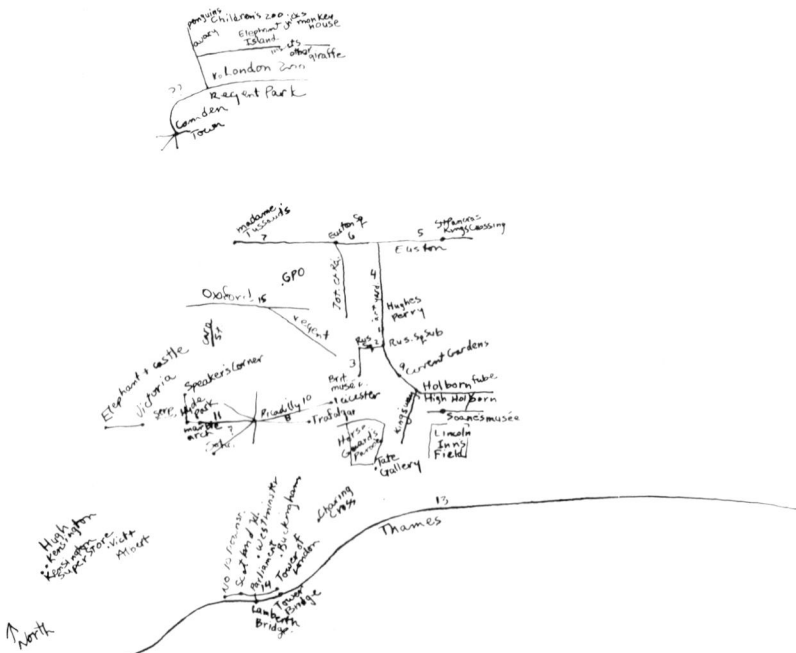

FIGURE 13. The Point-Line Skeleton of Janine's Third Map of London, Drawn After a Week's Experience.

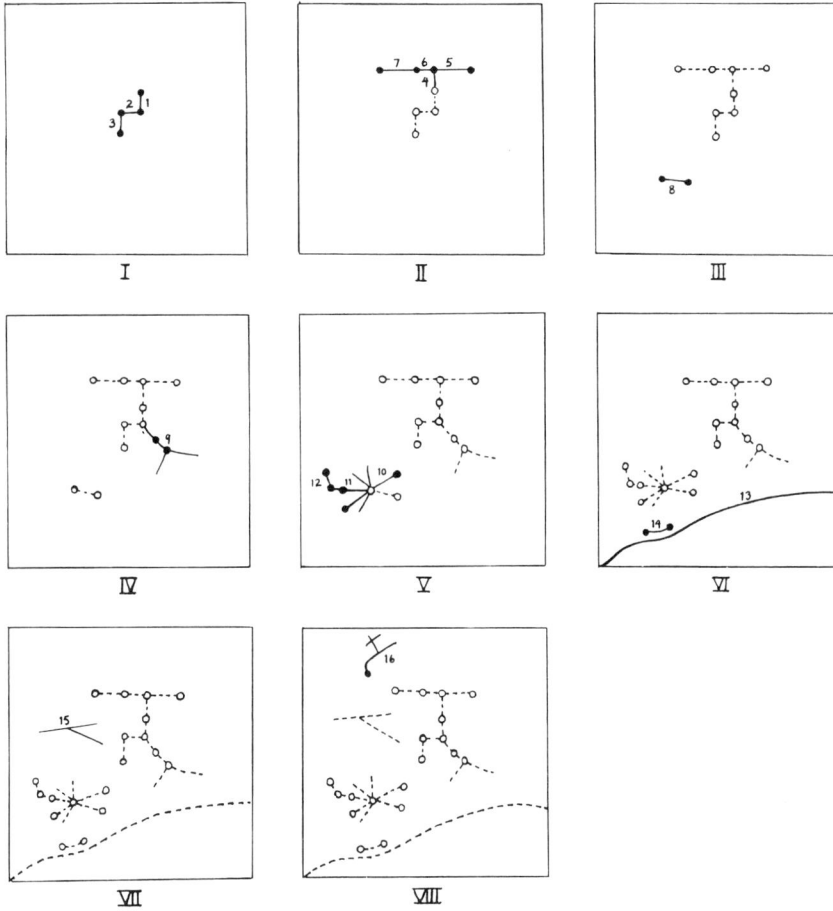

FIGURE 14. Sequence of Construction of Janine's Third Map of London (Fig. 13).

underground station is actually drawn in two segments, south from the dormitory to the underground and north from the dormitory to Euston. From this we would argue that the image is still fragmentary, but that now the fragments link up at critical junctions. Thus we feel that were Janine to visualize the route from Euston to the Russell Square underground, she would do so in two 'scenes', one encapsulating the dorm-to-Euston portion, another the dorm-to-station portion. These are important things to understand about environmental images: they exist in suspension, as it were, precipitated at need.

From her home territory Janine moves out into central (not southern) London. Her representation of the Piccadilly–Trafalgar Square connection is a simple repetition of her performance on the second map; that is, another instance of a powerful schema holding sway. The ninth connection (shown in the fourth diagram in Figure 14) is a representation of the one that was inadequately and inaccurately portrayed on the first map and erased on the second. Janine is now sure of herself and now knows that this particular route takes her toward the Inns of Court and the Strand, not toward Piccadilly Circus. But it took her one relocation—on the first map—and one erasure—on the second map—before she was able to do what she has done here. Returning to

Piccadilly Circus she now makes the connection to Hyde Park via Piccadilly. This also had been attempted on the first map and discarded on the second. The third map is an apparent return to the level of quality of the first map, but once again this is mostly apparent, for what it really represents is the schematization of what had on the first map been exclusively figural attempts at the organization of London. She still doesn't understand how to get from central London to the Thames, and she still squashes everything along the Thames together, and she no more understands the location of Regent Street than she ever did, but increasingly this is a map of London that might have been drawn by a Londoner living in Bloomsbury. In fact, in his discussion of our work David Ley refers to what has happened here as *place learning* (Ley, 1983).

This increase of detail and connectivity is not reflected in the grid transformation (Fig. 15). What this means is that among this new material there are no significant errors. The errors that destroy the grid are the primeval errors that have distorted Janine's grids from the beginning. The family resemblances among her grids are an important indication of the strength and permanence of the underlying schema that organize her image of London's form. Uncorrected errors made early in her visit manage to survive increasing experience in the environment to say nothing of repeated attempts at mapping the environment; so do early *correct* understanding. Early initial *Urforms* are the bedrock on which Janine's London is built, and these *Urforms* consist of nothing but a hierarchic organization of better or poorer articulations of motile schemata. During her stay in London these representations of schemata were locally elaborated and worked over, and the higher order schemata relating the local schema underwent perpetual if relatively subtle adjustment. This accounts for the bulk of change seen on the three London maps. The balance is due to the representation of new local schemata (like Camden Town or the stuff southeast of Covent Garden) and the attempt to incorporate these within the higher level organizations of London, as a whole. This process shows up on the maps as:

1. A steady increase in the number of things and places represented (similarly observed by Canter—1977, p. 70—across Nick's maps of London);
2. A continuous process of incorporation and deletion of given elements as their reality is questioned or confirmed by experience;
3. Increased differentiation of London into regions;
4. Increased integration of detail into regions;
5. An increase in interconnections from the first to the third map (also observed by Canter—1977, p. 70—across Nick's maps);
6. A gradual reduction of the veridicality of the representation of the relations among the parts of the city as revealed in the grid transformations.

We maintain that this last manifestation would be reversed in the long haul; and that it would have already reversed itself were London smaller or less complex. The tourist is not ordinarily trying to memorize the street pattern of a city, but see as much of it as possible in what is usually little time. Under the continuous impact of new sensations, and in the absence of the practice effects that would result from permanent residence, there is little opportunity to acquire the experience necessary to rectify early errors (but a great opportunity to examine the 'Main Sequence' of Siegel and White—1975—, and to probe development 'micro-genetically'). To the extent that these errors reduce the utility of the image for moving through the environment on a long-term basis they will

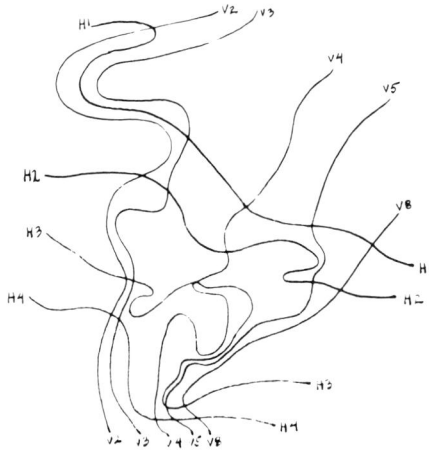

FIGURE 15. Grid Transformation of Janine's Third Map of London (Fig. 13).

be more or less corrected. Similarly, over the long haul, the other manifestations of increasing intregration and differentiation, increasing detail and interconnectivity, will slow down, and finally plateau. (Canter—1977, p. 67—observed a similar plateau in precisely these dimensions of Nick's maps of London; but see also the group analysis and Janine's place in it in Wood, 1973, especially pp. 447–495, for the number of elements represented and their connectivity.)

But this is not all that's been happening. On the interpretive level (see Fig. 16), Janine is beginning to generalize about London. The mess of shops and activity with which she first described Oxford and Regent Streets has resolved itself into a commercial district dominated by middle-class department stores. Danger has disappeared from the

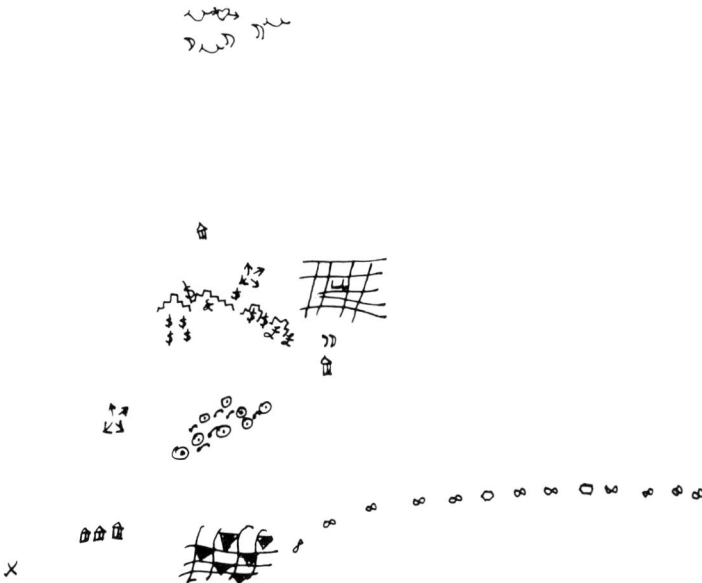

FIGURE 16. Attributive Overlay for Janine's Third Map of London (Fig. 13).

entertainment district. The University region is shown to have a right-angle street
pattern, while Westminster has been generalized into large buildings on irregularly
patterned streets. Hyde Park appears as nothing more than open space (although
Janine's latest love—the London Zoo—is covered with smitten hearts, smiles and
crescent moons!). Gone is the finicky detail—individual subway entrances, individual
shops—of initial experience. In its place is an embryo of a general comprehension of
the city's economic and social organization. This is very much in keeping with the more
mature skeleton Janine produced on this trial: she is very much getting a handle on the
physical, social, cultural and economic morphology of London as an integrated whole.
The day after drawing this map Janine Eber left London.

Remembering London

Several months after Janine returned from Europe she produced another map of
London (Fig. 17) in response to our request. It is completely numbered so that it is
again possible to reconstruct the sequence in which it was created, but it is
accompanied by notes that permit us yet *further* insight into the process of its
construction. In this, as noted previously, Janine was unique.

Whereas each of her in-London maps began with the dormitory and the almost
reflexive route down toward the British Museum, her remembered map begins with the
Thames (Fig. 18). This is a totally novel approach. Whether it has resulted from a re-
evaluation of the significance of the Thames in the form of London, from a desire to
anchor the map from the start with its most obvious feature, or from a radically reduced
importance for the dormitory now that she is no longer living there, we do not know.
Suffice it to say that after sketching in the Thames, Janine immediately reverts to form
and returns to the dormitory for her second mark on the map. From the dorm she

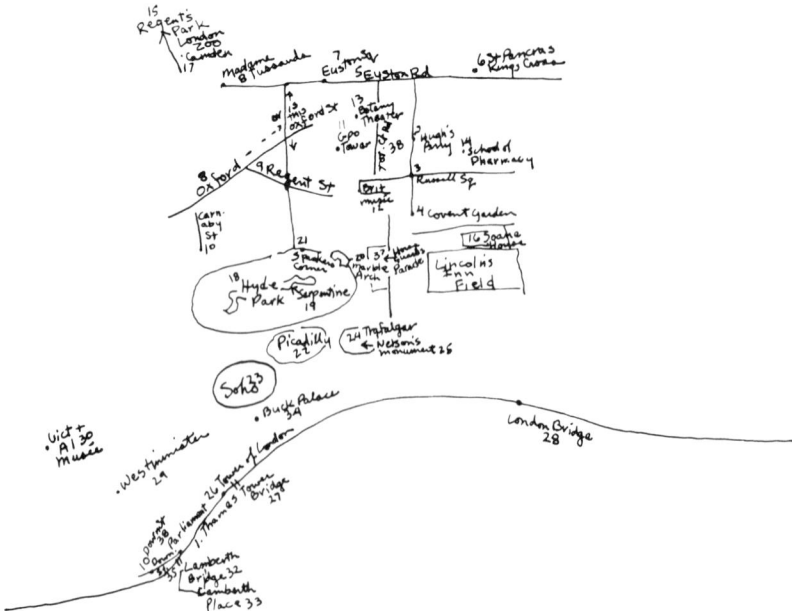

FIGURE 17. Janine's Remembered Map of London.

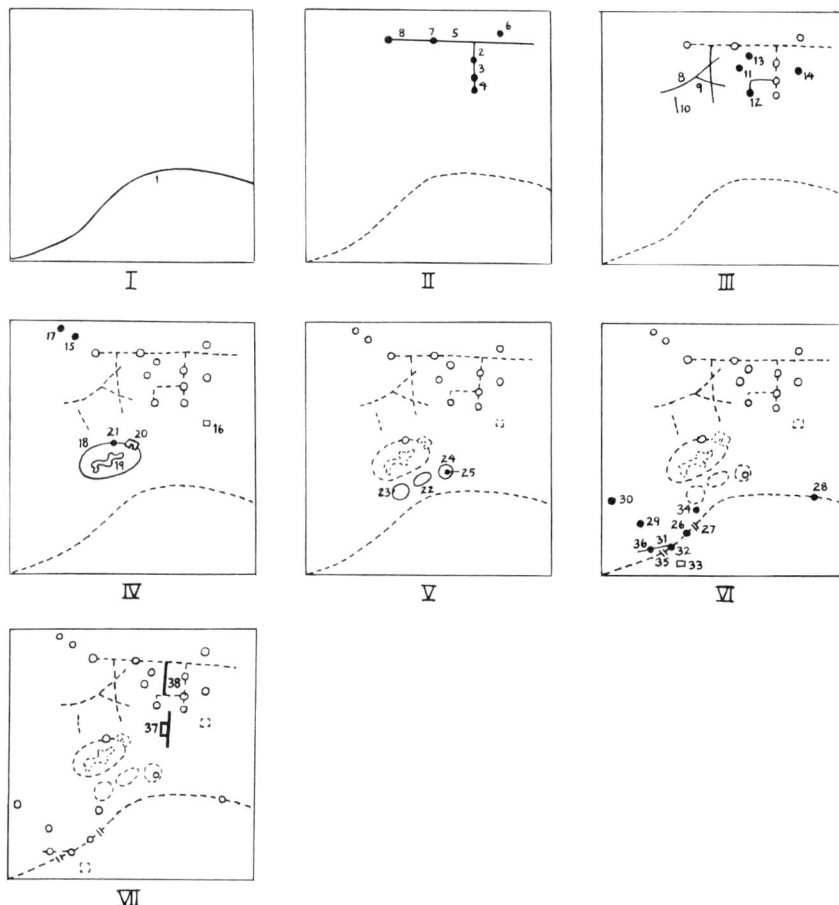

FIGURE 18. Sequence of Construction of Janine's Remembered Map of London (Fig. 17).

heads south *all the way to Covent Garden* with two strokes of the pencil, returning to add the route to the British Museum. This initiatory schema, never mind its being the second thing drawn, retains its role as the core of the formal image. It has, however, been modified, by taking in more southerly material, and by failing to include the British Museum (which will appear later as the twelfth mark). Sticking to the approach pioneered on her third map, she now tackled the Euston Road axis. Thus, by the time she has made her eighth mark she has completed the frame for most of London in the north and south. She now proceeds to adjacent and little understood material. Two Oxford Streets appear with the question: 'Or is this Oxford Street?' attached to what should be Regent Street. She has never known how these streets worked and she knows it:

> First I went through the list of check off points I knew the location of. Now I draw the map. It occurred to me that Oxford Street ran into Euston, right? Or it comes close, so I don't know how to draw it. I never could get the stuff by the river right! This is a terrible map—I feel like I'm just putting down places. I'm not sure of most of them. I just looked at a map of London to see how I did and yuck! It's horrible. Of course, I never did know where the Tower of London, Westminster, and Oxford and Regent Streets were...!

Whatever it was in her very early experience that confused these streets for her has been a powerful influence on her cognition of London's form. This is not the first time Janine has looked at a map of London, but all the looking she has done has had a minimal impact compared to her veritable and confusing experiences in the streets. (Nor was Janine alone here. Though many of her peers finally did get it figured out, confusion between Oxford Street and Regent Street, and/or among Piccadilly Circus, Trafalgar Square and Leicester Square, was all but universal on first and/or second maps; and many of our mappers never got it straight. See the detailed discussion in Wood, 1973, pp. 505–511 for Oxford Street; and pp. 556–557 for Oxford-Regent and Piccadilly-Trafalgar-Leicester.)

From this endeavor Janine continues her orderly progression south to add Hyde Park to the map. Hyde Park has never had so full a treatment: the park is bounded, the Serpentine is included and Speakers' Corner and the Marble Arch appear for the first time as integral to the park. Hyde Park reminds her of Regent's Park and she zooms north for this. To the east she adds Soane's House Museum. Still to the south she adds Soho, Piccadilly and Trafalgar. She has never been very accurate about the location of these either, but she seems not to be aware of this. Here they are presented as regions. Finally, along the bottom of the map she adds detail along the Thames. The production of this map has been essentially orderly. With the exception of the skip to Regent's Park and the ultimate backfilling of the Horse Guards Parade and Tottenham Court Road (# 37 and # 38), there has been no leaping hither and yon. The map was framed north and south and then filled in working south in a highly systematic fashion. It is as if her formal image of London has been stewing and had reached the point from which Janine could finally visualize the *map* she was about to create. The order in which she produced the map contradicts her suggestion of 'just putting down places.' The way she has put them down locally is the way she has always put them down: the proximity of the Tower and Parliament, Soho southwest of Piccadilly, her Regent-Oxford confusion her initiatory schema, the tripartite representation of Euston: each of these has been presented the same way from the first. Maybe 'just putting down places' refers to the automatism involved. But these very powerful local schema have finally been subordinated to an overriding conception of London's form that allows her to produce them systematically, north to south, as opposed to the more actively creative maps she produced *in* the city, jumping here and there as need and knowledge prompted.

In a sense this is Janine's best map of the city. There is much more detail, and while she has not attempted literal connections, she has located places so closely as to suggest connections infallibly. The final separation of the Tower Bridge from Parliament suggests that although the river is still unclear in her mind, it is not as unclear as it was. Other students also found this remembered map easier to draw and better than any of their others. David Abrams, who also produced his best map of the city on this occasion, reported that he did so in one-fourth the time it had taken him *in situ* and that he found the operation much less painful. (But David was our best mapper—on measures of richness, integration and veridicality Janine fell somewhere in the middle—on almost all dimensions. Compare his remembered map of London in Wood, 1973, p. 637.) From the evidence gathered of these remembered maps it seems that time was the vital digestive that aided the integration of the environment into reasonable wholes. It might be just this time to digest that is the operant difference between maps produced by residents and maps produced by tourists.

The effect of memory on the image was not to dim, but to add up experiences into wholes. This does not preclude flashes, visions or fleeting memories. Janine experienced these. Six months after returning she wrote:

> It's strange—I can be typing, or reading, or cleaning, and all of a sudden I recall a hotel, or a place, or a restaurant. Sometimes I don't know where it is and I have to pause and think; and finally I remember. WOW! It's really weird!

Around the same time, she described feeling 'funny' whenever she saw some place she'd seen on television or in books or magazines. The discrete hunks of experience that constituted the European tour are reaching higher and higher levels of organization where form and meaning are gradually being remapped onto each other. Refer back to Janine's overlay of her remembered map of Venice (Fig. 7), for example, or consider the following monologue she wrote to accompany her remembered map of Innsbruck. As she attempts to structure the form of the city, she finds herself narratively reliving her experiences, which *mutatis mutandis* uncover the form of the city. As she maps she remembers, and as she remembers she has more to map. The process she describes is more than reminiscent of the navigational procedures employed by many primitive peoples whose migratory routes are intimately linked with the mythic tales embodying them (for example, Blakemore, 1981).

> Here's my running commentary on the checklists, maps and bus charts so you'll know what I'm going through (!): First of all I sat and thought about Innsbruck. We never did maps on the city—don't remember doing a checklist either. We came in from the north–down that big mountain—saw several danger signs, a car avec a trailer which didn't make it, then to the hotel—built in 1452 or 1453 or some year like that. It just occurred to me that because I didn't go all the way up the mountain I didn't remember Porter's escapade, but only *heard* about it. But I do remember the hassle about the drugs (Scene 1 of 'Was It Fate' or 'The History of Group L'—a play in five acts…)
>
> Anyway back to the city, I remember the park where my group had a picnic and a water-fight (I was not drunk!!!): going up the mountain by cable car and train; the Inn River; the walk back from the mountain trip avec Sven, Betty, Claire, Susan, Nybia (??—no I guess not), Vanessa—maybe Erica and Karl. I'm not sure. We stopped at a covered bridge (remember that?)—Hey, that's the bridge that Cliff, Vanessa and I had to cross to be with the rest of "our group" and go to another park…

Janine goes on from here for several pages to detail minutely her memories of Innsbruck. The catches in her narrative ('Hey, that's the bridge…') when she fits two memories together are beautifully captured and manifest the integrative process that has been going on. In memory she is recreating the entire experience, probably putting it back together for the first time. What we are seeing here in action is the process of synchronization (Beck and Wood, 1973a, pp. 214–215): the map Janine is making is the narrative she is telling *drained of the temporal dimension* characterizing its narrative form, but in such a way that the narrative may be plausibly reconstructed *from the map*. It is a two-way street: map-space is not merely Meredith's (1970) 'synoptic anytime', but that 'time' void of distinctions among past, present and future in which uniquely may past narratives be reconstituted and future events lawfully unfold. Narratology has much to offer here (for example, Prince, 1982 and Bal, 1985), for each of these connections is another clue as well, though when she says 'Hey, that's the bridge…' it is not merely another step in the narrative, but the recovery of another *formal* fact that will further the construction of her map.

> It's strange—I remember only being able to see the whole valley from the north side (when we went up by cable car) and only the village from the ski jump. Therefore, I had to draw the River first, holding the paper south to north (upside down) and then turn it right side up to do the rest of it.

She is clearly operating out of visualizations here, one of the village from the ski jump and another of the whole valley from the north side. But the form she will produce on the map will reflect neither visualization, but an integration of the two of them, *mechanically accomplished by the rotation of the paper*. The form of the city is being born as a whole unit for the very first time before our eyes.

> Just decided to do most of the map upside down, 'cos we always went south (across the river) to shop, sightsee, etc. Oh no! I have to start all over again. I don't have room east of what I put on the map already for more stuff. Besides, turning it rightside up and positioning myself and the map... Ugh! Hope it's okay if I just add the rest on another sheet instead of drawing the whole thing smaller. Well, I guess I'd better make the whole thing smaller and cancel the added sheet idea.
>
> Oh yes, the school where we have those horrible lectures–except for the one on music—Hey! Those were the ones where Odin feel asleep. Now, boy, I wish you were here to answer my questions. Do you want me to turn all the names around to make them right side up??? This has already taken me 45 mins and I haven't started the overlays yet...

The map of Innsbruck Janine produced (Fig. 19) isn't really important. It looks in fact very much like the sort of map you have come to expect from Miss Eber. What is important is the description of the process she had to go through to draw the map and what this reveals about the way places and things are remembered. It didn't have to be Innsbruck. Janine could have been recalling the car she rode in on her first date or her first bicycle. The process of recall—and the process of creation—would have been similar, inevitably dealing in both form and meaning, place dredging up story, story place, the feeling of being in love illuminated suddenly by the vision of the place of being in love, a vision of that place suddenly enriched by the feeling of being in love, and on and on, usually less dramatically, sometimes more so, but always this dual involvement of signified with signifier, signifier with signified. There is no form without

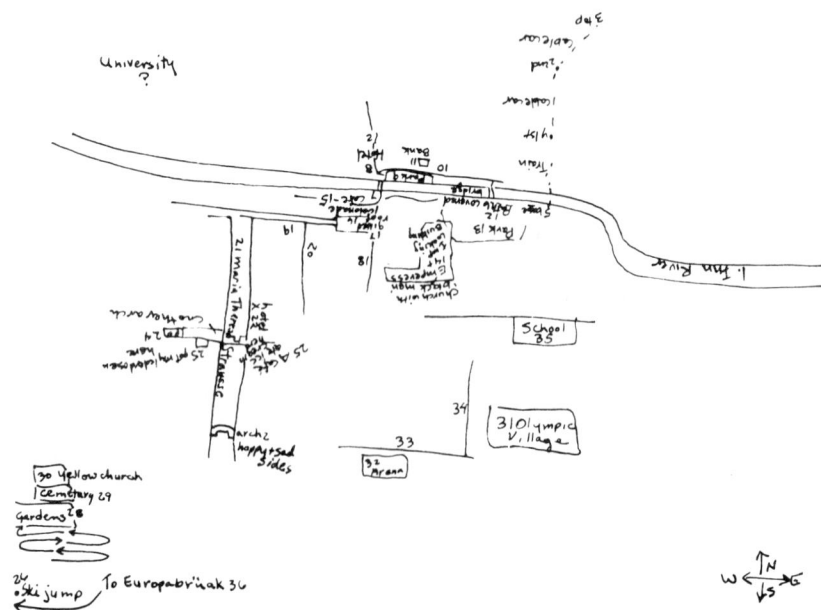

FIGURE 19. Janine's Remembered Map of Innsbruck.

meaning, no meaning without form, no image without both. Forms of things are not existential isolates—free of meaning—and may not be manipulated at will. As carriers of meaning, forms underwrite, make palpable, shape, and give shape, to that meaning. Changes in form result in changes in meaning. Similarly, changes in meaning result in changes of form—cognized form, of course, but existential form as well—for as meaning is not independent of form, so form is not independent of meaning, of role, of function, of need, of end and of purpose. Meanings justify form, make sense of form, create places for forms in our lives. London fulfilled Janine's need for parks and Janine in turn created an image of London filled with, dominated by parks. Neither the dominance of her maps by parks nor her love of London can be understood in isolation, for it is her image of London as park-filled that helps her love it, and her love of London that helps her fill it with parks. And these, working together, take her to the parks—St. James Park, Hyde Park, Regent's Park, Green Park—over and over again and this experience, translated from movement and being into schema, fills her with the park-related detail with which she fills her map, and this familiarity takes her back to the parks.

Over and over again. So the image mediates between the world as apprehended and the world as acted upon.

References

Bal, M. (1985). *Narratology: Introduction to the Theory of Narrative*. Toronto: University of Toronto Press.

Beck, R. and Wood, D. (1976a). Cognitive transformation of information from urban geographic fields to mental maps. *Environment and Behavior*, **8**, 199–238.

Beck, R. and Wood, D. (1976b). Comparative developmental analysis of individual and aggregate cognitive maps of London. In G. T. Moore and R. G. Golledge (eds), *Environmental Knowing: Theories, Research and Methods*. Stroudsburg, Pennsylvania: Dowden, Hutchinson and Ross, 173–184.

Blakemore, M. (1981). From way-finding to map-making: the spatial information fields of aboriginal peoples. *Progress in Human Geography*, **5**, 1–24.

Buttenfield, B. P. (1986). Comparing distortion on sketch maps and MDS configurations. *Professional Geographer*, **38**(3), 238–246.

Canter, D. (1977). *The Psychology of Place*. London: Architectural Press.

Canter, D. and Donald, Ian. (1987). Environmental psychology in the United Kingdom. In D. Stokols and L. Altman (eds), *Handbook of Environmental Psychology*. New York: John Wiley, 1281–1310.

Craik, K. (1968). The comprehension of the everyday physical environment. *Journal of the American Institute of Planners*, **34**, 29–37.

Craik, K. (1970). Environmental psychology. In *New Directions in Psychology* **4**. New York: Holt, Rinehart and Winston, 3–121.

Gale, N. (1983). Measuring cognitive maps: methodological considerations from a cartographic perspective. In D. Amadeo, J. B. Griffin and J. J. Potter (eds), *EDRA 1983: Proceedings of the Fourteenth International Conference of the Environmental Design Research Association*. Lincoln, Nebraska: University of Nebraska-Lincoln, 65–72.

Golledge, R. G. (1987). Environmental cognition. In D. Stokols and I. Altman (eds), *Handbook of Environmental Psychology*. New York: John Wiley, 131–174.

Gough, H. G. and Heilbrun, A. B. (1965). *The Adjective Check List Manual*. Palo Alto: Consulting Psychologists Press.

Herman, J. F. (1980). Children's cognitive maps of larger-scale spaces: effects of exploration, direction, and repeated experience. *Journal of Experimental Child Psychology*, **29**, 126–143.

Ley, D. (1983). *A Social Geography of the City*. New York: Harper and Row.

Meredith, P. (1970). Developmental models of cognition. In P. Garvin (ed.), *Cognition: A Multiple View*. New York: Spartan Books.

Murray, D. and Spencer, C. P. (1979). Individual differences in the drawing of cognitive maps. *Transactions of the Institute of British Geographers*, (New Series), **4**, 385–91.

Prince, G. (1982). *Narratology: The Form and Functioning of Narrative*. Berlin: Mouton.

Siegel, A. W. (1982). Toward a social ecology of cognitive mapping. In R. Cohen (ed), *New Directions for Child Development: Children's Conceptions of Spatial Relationships, no. 15*. San Francisco: Jossey-Bass.

Siegel, A. W. and White, S. H. (1975). The development of spatial representations of large-scale environments. In H. W. Reese (ed), *Advances in Child Development and Behavior, Vol. 10*. New York: Academic Press.

Spencer, C. and Dixon, J. (1983). Mapping the development of feelings about the city: a longitudinal study of new residents' affective maps, *Transactions of the Institute of British Geographers*, (New Series), **8**, 373–383.

Tobler, W. R. (1966). Medieval distortions: the projections of ancient maps. *Annals of the Association of American Geographers*, **56**, 351–360.

Walmsley, D. J. (1988). *Urban Living: The Individual in the City*. New York: John Wiley.

Wapner, S., Kaplan, B. and Ciottone, R. (1981). Self-world relationships in critical environmental transitions: childhood and beyond. In L. S. Liben, A. H. Patterson and N. Newcombe (eds), *Spatial Representation and Behavior Across the Life Span*. New York: Academic Press, 251–282.

Wood, D. (1973). *The Genesis of Geographic Knowledge: A Real-Time Developmental Study of Adolescent Images of Novel Environments (London, Rome, Paris)*. Worcester: Clark University Cartographic Laboratory. (mimeo.) (Available from University Microfilms).

Wood, D. and Beck, R. (1976). Talking with *Environmental A*, an experimental mapping language. In G. T. Moore and R. G. Golledge (eds), *Environmental Knowing: Theories, Research and Methods*. Stroudsburg, Pennsylvania: Dowden, Hutchinson and Ross, 351–362.

Wood, D. and Fels, J. (1986). Designs on signs: myth and meaning in maps. *Cartographica*, **23**, 54–103.

SEX DIFFERENCES IN HOME RANGE AND COGNITIVE MAPS IN EIGHT-YEAR OLD CHILDREN

PAUL WEBLEY

University of Exeter, England

Abstract

This study investigated the hypothesis that there is a sex difference in the extent of children's home ranges and therefore a corresponding difference in their home area cognitive maps. Cognitive maps were elicited using a road construction kit and the resulting maps scored for embedding, extent and detail. Home ranges were investigated using a photographic recognition test and by children indicating home range extent on aerial photographs. The results supported the hypothesis of a sex difference. However, these sex-related mapping differences disappeared when children were required to create maps of areas to which both sexes had limited and controlled exposure. Thus, the original sex-related differences appear to be associated with differences in the size of familiar territory rather than with any male superiority in spatial cognition or map-building ability. The latter was shown to be related to general intelligence as reflected by scores on Raven's Progressive Matrices.

Introduction

Most research into children's cognitive maps has been concerned with the developmental aspects of the 'navigational geometry' that children use, and has been firmly grounded in Piaget's theoretical description of fundamental spatial cognition and the child's use of reference systems (see, for example, the reviews by Hart and Moore (1973), Siegel and White (1975) and Siegel *et al.* (1978)). Briefly, Piaget's scheme is as follows; pre-operational children use an action-centred egocentric system, which means that routes and the position of landmarks are thought of in terms of a child's actions. A concretely operational child uses a fixed reference system. This means that his cognitive map is partly co-ordinated by the use of landmarks. These can be related to each other in discrete local areas but the totality of relations between them is not appreciated. The cognitive map of a pre-operational or concretely operational child will tend to be a route map (a mental tracing of a route through an area). Finally, the formally operational child uses a co-ordinated system of reference, each single part of his cognitive map being related to all other parts. These maps will tend to be survey maps (mental configurations of the relative location of places).

This concentration on structural development in cognitive maps has meant that although many researchers have commented on the need to relate environmental behaviour and environmental cognition, this issue has rarely been investigated in children. One notable exception is Hart's (1979) intensive study of children living in a New England town. He found a positive relationship between the accuracy and extent of children's sketch maps and the extent of their home range. Girl's ranges were more constrained and this was reflected in their maps being smaller and less accurate than those of boys.

Hart, however, used sketch maps as an index of environmental cognition, and despite their extensive use with children (e.g. Ladd, 1970; Maurer and Baxter, 1972; Bishop and Foulsham, 1973; Waldvogel, 1974; Partridge, 1978; Catling, 1978; Hart, 1979) their use is almost certainly inappropriate. From their study of children's drawings of simple objects, Kosslyn et al. (1977) concluded that children's cognitive representations cannot realistically be inferred from their drawings. Rothwell (1976) found a significant correlation between accuracy of children's floor plan sketches and their scores on the Draw-a-Man Test. These two studies led Evans (1980) to conclude that 'children's drawings are not a suitable technique for probing their cognitive representations of large-scale environments'. One aim of this study has thus been to examine the sex-differences in large scale spatial representations that Hart (1979) reports, using a more satisfactory method (a constructionist technique rather than sketch maps).

Hart did, however, study children's knowledge of a real large-scale environment and not a simulated environment (the choice of most researchers). Siegel et al. (1978) have pointed out that one reason why the quality of children's cognitive maps may have been underestimated is that 'children's knowledge of large environments has only rarely been tested in a layout of the same scale'. A second aim of the study was, therefore, to examine the quality of children's cognitive maps of familiar and novel large-scale environments.

Sex differences in spatial cognition have been frequently reported; males are superior to females on most, although not all, spatial tasks (Maccoby and Jacklin, 1975). Some have suggested that this difference is genetically determined; others that there is a hormonal involvement in spatial skills (Waber, 1977) and yet still others that the spatial superiority of males derives from differential experiences. Munroe and Munroe (1971), for example, tested the proposition that extensive movement through the environment might lead to enhanced spatial ability and found that children observed further away from home were more skillful at a spatial task. This result was replicated by Nerlove et al. (1971) using three spatial tasks. Munroe and Munroe suggest that the sex-difference in environmental experience results from an innate tendency for males to investigate more and also from child training methods which encourage this tendency. Given the age range that Munroe and Munroe studied (three to seven) this would suggest that by the age of eight, differences in certain aspects of spatial ability (e.g. 'the capacity to perform a set of behaviours ordered sequentially in space') would be firmly established.

As well as sex, there is another variable that seems likely to affect the size of a child's home range and thus the extent and accuracy of his or her cognitive map: the distance from school at which a child lives.

This study, therefore, had the twin objectives of (i) examining the hypothesis that there are differences in the extent of children's home ranges which are related to sex and the distance a child travels to school, and corresponding differences in home area cognitive maps; (ii) exploring the origins of the sex-difference in cognitive maps by giving children a controlled exposure to a novel large-scale environment.

Study 1

Method

Sample

The sample consisted of 16 males and 16 females with a mean age of eight years seven months, ranging from eight years three months to nine years one month. The children were of average intelligence and were residents of Hayes, Middlesex, England. Half of the children lived in the area immediately surrounding their school (North Hayes) and half lived in a separate estate half a mile away (Hayes End).

Materials

(1) *Road Construction Kit*: this consisted of lengths of thick grey card, 1·2 cm wide with a white dotted line down the middle. These were of various lengths from 1·2 to 15·3 cm and in addition to the straight pieces there were curves of diverse arc and length, cross-roads, off-set crossroads and T-junctions. Red and green plastic houses were included. It was possible to construct a perfect map of the area with this kit.

(2) *Photographic Recognition Test*: a complete list of landmarks in the area was compiled by the author and two local inhabitants. The area around the school was then divided into 75 approximately equal segments and a landmark chosen and photographed from 35 of these (see Figure 1).

(3) *Home Range Map*: a composite vertical aerial photograph of the area around the school was formed from nine 19 × 19 cm Aerofilm photographs (scale, 1:4224). This covered a rectangle of approximately 3·3 by 1·5 km, mainly to the north of the school.

(4) *Raven's Standard Progressive Matrices* (Raven, 1960): this is a measure of general intellectual capacity which requires the comparison and manipulation of geometrical patterns.

Procedure

(1) *Road Construction Kit*: all the elements of the kit were put on a large table and an example of each type shown to the child. The experimenter then said 'Now, I would like you to make me a map of round here starting from the school (a red plastic house was placed on the table). If this is the road outside (a 5 cm road strip was placed next to the 'school') could you carry on from there?' These two elements of the map were initially *orientated correctly* by the experimenter. When the child began to falter, the experimenter would probe further. He would point to an open end on the child's map and ask 'What if you go up there?' This was repeated continually until all of the open ends had been extended to the point where the child professed ignorance. He would then be asked to name all the roads, indicating what towns would be reached by travelling along the main exit roads and if anything was missing or inaccurate.

(2) *Photographic Recognition Test*: the slides were projected in a random sequence. The child was instructed to say 'yes' when he recognized a slide. These responses

were probed to establish that they were genuine, by asking questions to establish whether or not the place had been seen, for example 'What is opposite it then?' In cases of doubt the photograph was marked as not recognized.

(3) *Home Range Map*: the child was shown the composite aerial photograph, and the school, houses and roads pointed out. He then identified other places and each was labelled. After the composite photograph had been studied for five minutes it was explained that the experimenter wanted to know where the child had and had not been. The child indicated this on the photograph and this information was recorded. When the spontaneous statements were finished, the experimenter went through the composite photograph section by section to check the extent of home range reported with the child.

(4) *Raven's Progressive Matrices*: these were administered on an individual basis following the instructions given by Raven (1960).

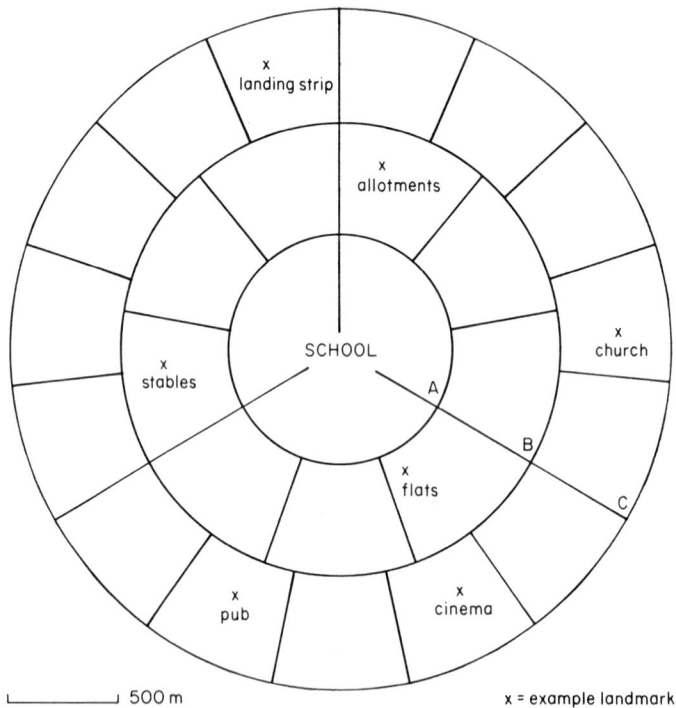

FIGURE 1. Area around school showing segments and example landmarks.

Results

Scoring
(1) *Road Construction Kit*: the constructed maps were scored for embedding, extent and detail. The embedding score was a measure of the extent to which the child's cognitive map of his home area was anchored in a wider network. The criterion used

was whether a child knew what towns were reached by travelling along the three main exit roads. The extent score was a measure of how large an area a child's map represented. Each constructed map was transformed into a true scale map which was then overlaid with a grid. The number of squares which the transformed map covered was the extent score (1 square equalled approximately 10 hectares). Detail scores included road names known, the number of buildings used and the number of roads included in a designated core area around the school.

(2) *Photographic Recognition Test*: the photographs simply scored as recognized or not.

(3) *Home Range Map*: an extent score was derived for the home range map using a similar overlaid grid to that used for the road construction kit.

(4) *Raven's Progressive Matrices*: (total number of problems solved correctly).

Analyses

The scores derived from the constructed maps were analysed using a mixed model analysis of variance (sex, fixed factor; place of residence, random factor). There were no significant effects of home area or sex on embedding or detail, but a highly significant effect of sex on the extent of constructed maps; boys' maps covered, on average, an area 40% larger than the girls.

TABLE 1
Sex differences in the extent of constructed maps

Place of residence	Boys ($n = 16$)	Girls ($n = 16$)
North Hayes	22·0	14·5
Hayes End	27·0	19·5
Both*	24·5	17·0

* $F = 14297$, $d.f. = 1,1$, $P < 0·01$, Mixed Model Analysis-of-Variance.
Figures are mean extent scores, 1 unit equals 10 hectares.

Both the Photographic Recognition Test (P.R.T.) and the Home Range Map extent scores revealed significant sex differences. For each separate ring around the school and for all of them together boys claimed to recognize about 40% more photographs than girls (see Table 2).

That this difference reflects a genuine difference in home range was supported by the home range maps derived from children's recollected behaviour as indicated on the aerial photograph. These showed that boys reported having visited about 50% more of the locality than girls (see Table 3).

These two indirect measures of home range, which depend upon the accuracy and honesty of the respondents, can be cross-validated. From the home range map it is possible to predict whether a child should recognize a photograph from the P.R.T. For example, if according to a child's home range map he or she had never been past Charville School one would predict that they would not recognize the photograph of Charville School in the P.R.T. In the area covered by the home range map were ten of the photographs included in the P.R.T. These were used to make predictions of recognition. Overall 82·5% of the predictions and actual recognitions

TABLE 2
Total number of claimed recognitions of landmark photographs

	Maximum possible	Boys ($n = 16$)	Girls ($n = 16$)	p(t-test)
Ring B	9×16	75	50	< 0.005
Ring C	15×16	63	40	< 0.01
Rings D and E	11×16	32	26	< 0.1
All Rings		170	116	< 0.001

TABLE 3
Sex differences in the extent of reported home range

	Boys ($n = 16$)	Girls ($n = 16$)	
North Hayes	22·93	12·81	
Hayes End	19·06	16·07	
Both*	20·99	14·44	$* P < 0.01$, t-test

Figures are mean extent scores, 1 unit equals 10 hectares.

corresponded. Only two photographs had matching scores lower than 75%, and one of these illustrates an inherent problem in the construction of the P.R.T. According to their home range maps, none of the children had set foot in Heinz's grounds, and so predicted recognitions were nil. But 11 of the 32 children reported recognizing the building, as it can be seen from two nearby roads. Clearly it is not necessary to visit a place to know what it looks like. Nonetheless the fact that predicted and actual recognitions generally correspond is encouraging and that similar conclusions can be drawn from disparate methods increases one's confidence in the results.

The data, therefore, suggest that eight-year old boys have more extensive cognitive maps and more extensive ranges than girls. The difference in maps appears to be the result of the difference in range; scores on the P.R.T. and the extent scores of the constructed maps correlated significantly, rho = 0·62 ($P < 0.02$).

The type of map that is elicited by the construction technique is illustrated in Figure 2. These maps are topologically accurate although some roads are missing. They are strikingly superior to sketch maps produced by children of the same age in previous studies. Bishop and Foulsham (1973) for example describe their nine-year old respondents maps as the 'archetype children's map' with school and home at each end of a route map with a combination of plan and elevation.

Discussion

These results provide support for the hypothesis that boys have a more extensive home range than girls and a corresponding difference in their cognitive maps. Other indices derived from the constructed maps show no significant differences between the sexes. Children who live further away from the school also have more extensive home ranges and cognitive maps.

However, although these sex differences in home range and cognitive maps are likely to be causally related, the data presented above are only correlational. Study 2 was carried out in order to examine directly the effect of home range on cognitive maps.

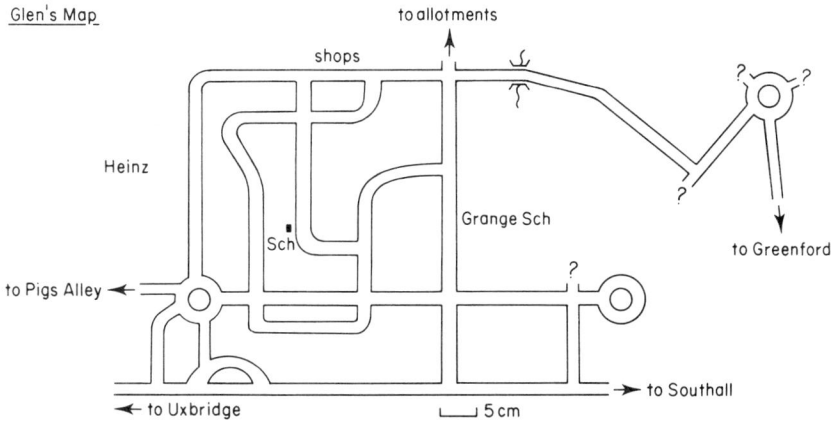

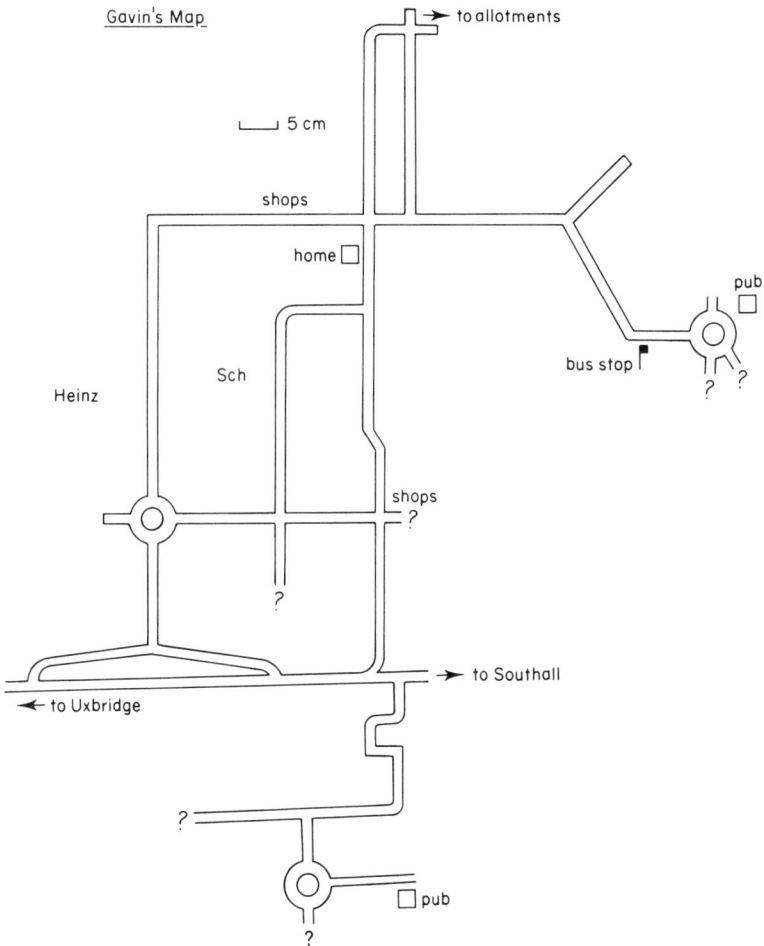

FIGURE 2. Typical examples of constructed maps.

Study 2

Method

Sample
The 16 children from North Hayes.

New area
An area unknown to the children that was as similar as possible to their home area was chosen. This was Northolt Village, Middlesex.

Procedure
Children were taken by car to Northolt (on four different days) in four groups of four, two boys and two girls in each, selected to cut across friendship ties. Before leaving the school they were told 'We are going to an area which you do not know. What I want you to do is imagine that you have moved house and want to find out about your new home area quickly. We will drive around the area first and then we will park in front of your new house and walk around a bit. When we return I will ask you to make up a map'.

Northolt Village would then be visited, following a *standard* route. The car was driven at 30 k.p.h. and the route was walked at 4 k.p.h. The whole trip took approximately one hour ten minutes. On their return the children were given a quiz which asked them to list the names of roads and shops they had seen and what buildings and other interesting things they had seen. This was intended to help those children who could not make maps immediately by giving them access to previously recalled information later.

Then, one at a time, the children constructed maps, elicited as in Study 1. The order used on the four days was counterbalanced for sex.

Results

Analysis of the scores derived from the Northolt constructed maps reveals only one sex difference, and that is that girls show a better recollection of road names than boys (see Table 4). The girls' scores on the other three measures are slightly, but not significantly higher.

The important result is that when exposure to the environment is controlled, the sex differences found in home-area constructed maps disappears. What is more, all the large variation in the children's maps of Northolt cannot be the result of different environmental experience. The correlations between Raven's Progressive Matrices scores and the various map measures suggest that intelligence may be an important factor.

TABLE 4
Constructed maps of Northolt: mean scores on all measures

	Boys (*n* = 8)	Girls (*n* = 8)	
Number of buildings included	3·25	3·50	N.S.
Number of road names	0·875	1·875	$P < 0.05$, *t*-test
Road detail	19·43	20·62	N.S.
Extent (in hectares)	103·7	117·5	N.S.

It is instructive to compare the correlations between Raven's Progressive Matrices and the various map measures found in the two studies (see Table 5). The correlations are all lower in the children's home area, although only two are considerably so. This suggests that intelligence is an important determinant of constructed map scores and becomes more important when range is controlled.

Discussion

This investigation indicated that sex differences in home area cognitive maps are probably the result of differences in the extent of home range. When the latter is controlled a similar sex difference is no longer found.

The findings are more illuminating, however, if looked at in a complementary way. All the large variation in the children's maps of Northolt cannot be the result of differences in exposure to the environment and must be due to other factors. The most obvious candidate is intelligence. Raven's Progressive Matrices scores correlate highly with detail scores in both home area and new area maps. It appears, therefore, that range is more likely to affect the extent of a cognitive map than its detail. Another factor affecting the maps of Northolt may be habitual style of interaction with the environment. Some children's maps reflected only the route they had walked and therefore covered a small area. Others reflected mainly the information derived from the car route whilst the more detailed, accurate and extensive reflected information from both sources.

TABLE 5

Correlations between Progressive Matrices scores and map indices

	Maps of Northolt (n = 16)		Maps of Hayes (n = 32)	
	r	P	r	P
Number of road names	0·64	< 0·01	0·57	< 0·01
Number of buildings	0·57	< 0·05	0·08	N.S.
Extent	0·78	< 0·01	0·24	N.S.
Detail	0·64	< 0·01	0·44	< 0·01

The present results, as far as they go, are at variance with the conclusions of the Munroes (Munroe and Munroe, 1971). They suggested that the more extensive movement through the environment of boys leads to superior spatial ability. However, in this study differences in the extent of range did not produce differences in spatial ability, at least as reflected in constructed maps of a new area. This does not imply that the processes identified by the Munroes are not operating; it is more likely that in western society the effects of experience of the large-scale environment on spatial ability are swamped by the effects of other relevant experience. Especially important in this context may be scale toys and manipulative toys.

The technique used in this study to elicit spatial representations has given rise to maps that show previous research to have underestimated the quality of children's representations. The attainment of survey maps has usually been considered an achievement of early adolescence; however, all but one of the eight-year old children produced comprehensive and detailed survey maps which were topologically accurate. This suggests that Evans (1980) was right to warn against using sketch maps to study children's environmental cognition. If an 'ability amplification' technique such as the road construction kit is used, a far better index of children's cognitive maps can be obtained.

Acknowledgements

The author wishes to thank Julie Dawick and Dr Beryl Geber for many helpful suggestions. The constructive comments of three anonymous referees were of great assistance in improving the clarity and accuracy of this paper.

References

Bishop, J. and Foulsham, J. (1973). Children's images of Harwich. *Architectural Psychology Research Unit Working Paper No. 3*, Kingston Polytechnic.
Catling, S. (1978). Cognitive mapping and children. *Bulletin of Environmental Education*, **91**, 18–22.
Evans, G. W. (1981). Environmental cognition. *Psychological Bulletin*, **88**, 259–87.
Hart, R. A. (1979). *Children's Experience of Place*. New York: Irvington.
Hart, R. A. and Moore, G. T. (1973). The development of spatial cognition: a review. In R. Downs and D. Stea (eds), *Image and Environment*. Chicago: Aldine.
Kosslyn, S. M., Heldmeyer, K. H. and Locklear, E. P. (1977). Children's drawings as data about internal representations. *Journal of Experimental Child Psychology*, **23**, 191–211.
Ladd, F. D. (1970). Black youths view their environment: neighbourhood maps. *Environment and Behaviour*, **2**, 74–99.
Maccoby, E. E. and Jacklin, C. N. (1975). *The Psychology of Sex Differences*. London: Oxford University Press.
Maurer, R. and Baxter, J. C. (1972). Images of the neighbourhood among children. *Environment and Behaviour*, **4**, 351–88.
Munroe, R. and Munroe, R. (1971). The effect of environmental experience on spatial ability. *Journal of Social Psychology*, **83**, 15–22.
Nerlove, S., Munroe, R. and Munroe, R. (1971). The effect of environmental experience on spatial ability: a replication. *Journal of Social Psychology*, **84**, 3–10.
Partridge, G. M. (1978). The development of children's ability to represent space. MSc dissertation, London University.
Raven, J. C. (1960). *Guide to the Standard Progressive Matrices*. London: H. K. Lewis.
Rothwell, O. (1976). Cognitive mapping of the home environment. *Dissertation Abstracts International*, **36**, 4758A.
Siegel, A. W., Kirasic, K. C. and Kail, R. V. (1978). Stalking the elusive cognitive map. In I. Altman and J. F. Wohlwill (eds), *Children and the Environment*. New York: Plenum.
Siegel, A. W. and White, S. H. (1975). The development of spatial representations of large-scale environments. *Advances in Child Development and Behaviour*, **10**, 9–55.
Waber, D. P. (1977). Sex differences in mental abilities, hemispheric lateralization, and rate of physical growth at adolescence. *Developmental Psychology*, **13**, 29–38.
Waldvogel, C. (1974). Children's knowledge of the city. In R. White, (ed.), *Geographical Studies of Environmental Perception: Research Report 61*, Evanston; Illinois Northwestern University.

CHILDREN'S SPATIAL REPRESENTATION OF THEIR NEIGHBOURHOOD: A STEP TOWARDS A GENERAL SPATIAL COMPETENCE

ANDERS BIEL

Department of Psychology, University of Göteborg, Sweden

Abstract

The ability of six-year-old children to combine perspectives of spatial layouts in their own neighbourhood was assessed. The children were asked to imagine themselves as being at a particular reference site and to decide which of two landmarks was closer to the site. From a theoretical point of view, one should not expect six-year-olds to have acquired projective and Euclidian concepts. Nevertheless, the judgements were internally consistent and tended to be veridical. This sensitiveness to relative and absolute accuracy was interpreted as a step towards a general cognitive capacity, governing the representation of spatial relations. Distances to the home were underestimated relative distances to other landmarks. Also estimations from the home tended to be more accurate than estimations from other reference sites. These results suggest that the home acts as a central reference point in children's mental representation of the environment.

During the last few years, there has been an increasing interest in the investigation of space by young children (e.g. Hart and Moore, 1973; Siegel and White, 1975; Hardwick *et al.*, 1976; Hazen *et al.*, 1978; Anooshian and Young, 1981). Almost all researchers refer to works by Piaget and his coworkers (Piaget *et al.*, 1960; Piaget and Inhelder, 1967). The developmental trends to be found are either interpreted in direct relation to the three developmental stages proposed by Piaget; topological, projective and Euclidean (Hart & Moore, 1973; Acredolo *et al.*, 1975), or in terms related to Piaget's model (Siegel and White, 1975; Hardwick *et al.*, 1976; Anooshian and Young, 1981). Although Piaget's theoretical model has served as a guide for later studies, there seems to be a consensus among many of these researchers that the use of small scale models of space is a limitation with Piaget's own empirical work (e.g. Acredolo, 1977).

Part of this critique of Piaget could probably be attributed to a different focus of interest. Piaget and his associates were mainly concerned with a general cognitive ability to structure space, a capacity to be aware of the abstract concepts, such as distance and projective relations governing all spatial representation. Whether these were applied to tabletop models or large-scale space was not their main concern. Other authors see the study of children's knowledge of the large-scale environment as their essential field of study. The reason for studying large-scale environment as a separate topic is motivated by the fact that it cannot be perceived from one location but requires observations over a period of time. Furthermore, this fact is taken as a basis for postulating that the representation of environment as compared to small-scale space may follow different developmental sequences (see, for example, Hazen *et al.*, 1976).

The present study focuses on the large-scale environment. The choice of scale is made in order to study children's knowledge of the everyday environment. However, there is no attempt to introduce a state of opposition between the study of large-scale space and children's knowledge of the everyday environment, on the one hand, and the use of small-scale models as a tool to understand children's representation of space, on the other hand. Children's knowledge of the everyday environment is regarded as a necessary precondition for a general spatial competence, as proposed by Piaget. The problem with table-top models is rather that they implicitly presume that children should use geometrical concepts. As Piaget and Inhelder (1967) and Laurendeau and Pinard (1970) have shown, children of six years of age do not apply operations associated with projective space. Neither have they incorporated Euclidean concepts. However, it does not follow that children's spatial representations lack internal consistency and are non-veridical. With a task better adopted to measure children's practically acquired knowledge, our understanding of their spatial abilities will hopefully increase.

The aim of the present study was to examine if young children's representation of the location of landmarks in their everyday large-scale environment are internally consistent and whether they coincide well with their actual spatial locations. The general procedure followed was a pairwise comparison of the distance from two landmarks to a reference site. The reference site could either be the home or a landmark previously mentioned by the child. The children were asked to imagine themselves as being at a particular place in their neighbourhood and to decide which of two landmarks was closest to the 'anchor-point' (see Figure 1). This procedure is similar to one of the conditions in a previous study by Hardwick et al. (1976). These authors asked subjects to align a tube towards three target objects in a room from a particular 'anchor-point'. Four target objects were chosen to define the spatial layout of the room, one of these being the 'anchor-point'. When this was completed, subjects were told to imagine that they had moved to another 'anchor-point'. They were then asked to point the tube towards three target objects as if actually standing at the imagined 'anchor-point'. The task was performed under two conditions. One group performed the task while in direct visual contact with the spatial layout of the room (unobstructed), while the other group had their view obstructed by a screen. Thus, the obstructed condition corresponds to the procedure used in the present study. The unobstructed condition, on the other hand, is similar to the co-ordination of perspective taking task used by Laurendeau and Pinard (1970). An unobstructed task requires the child to free itself from its own view and coordinate it with other perspectives. However, when the view is obstructed, interference between one's own and other perspectives is reduced. Thus, one would expect superior performance in the obstructed condition, at least for younger children. This is also what Hardwick et al. found. Hence, the task used in the present study with a combination of children's every-day environment and no need to co-ordinate perspectives was thought to make it easier for children to demonstrate their practically acquired knowledge.

Some factors were expected to influence the children's answers. In the exploration of the neighbourhood, the home serves as a point of departure. In accordance with Schouela et al. (1980), who, from a number of studies, conclude that some relatively stable reference point is required as a basis for establishing a cognitive organization, the distance comparisons from the home to other landmarks were ex-

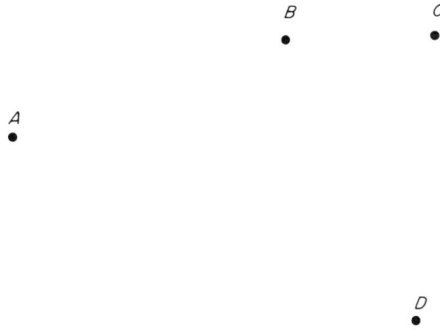

FIGURE 1. A hypothetical spatial layout for four landmarks. Each landmark served as reference site in a randomly chosen order.

pected to be more accurate than estimates from other landmarks. In a study by Evans and Pezdek (1980) where adults were asked to compare distances, the authors suggested that 'each intrapair distance was imaged and then directly compared to one another' (p. 17). In addition, one could expect distances more often traversed to be more easily accessible from long-term memory. As a result, the total time to scan familiar distances would be shorter. This, in turn, can lead to an underestimation of these distances compared to less familiar ones. In order to investigate this line of thought, the present study compared intrapair distance judgements to the home with intrapair judgements to other landmarks. It is assumed that distances to the home will be underestimated relative distances to other landmarks since roads to the home are used more often.

Estimation given by the children were also compared to a cartographic map in order to see if their judgements were sensitive to actual distances in the environment, and not only internally consistent. If the former is true, one would expect judgements diverging from factual distance relations to be few in number. Furthermore, when factual distances to the reference site were nearly equal, it would be more common to reverse the estimations.

Finally, the distance judgements were compared to landmark locations which the children indicated on a piece of paper.

Method

Subjects
A total of 23 six-year-old children from a nursery school and a local day-care centre participated in the study.

Procedure
The task was performed individually in a room at each place. The experimenter had been at each place for two days beforehand in order to get acquainted with the children. All the children agreed to have the session tape recorded.

First, the children were asked where they lived and, if possible, to specify whereabouts in the city their homes were situated.

Second, they were asked to give a short description of the area surrounding their house, what it looked like and what could be found there. From this description, four or five places, including the home, were chosen by the experimenter to be used in the distance estimation task. This procedure was followed to ensure that landmarks were included in the child's cognitive map.

Third, the distance estimation task was performed. The child was asked to imagine him/herself standing at a certain reference site and to decide which one of two landmarks was closest to the site without being given any guidance on how this judgement should be performed. This paired comparison was carried out from each place to all other places. It was ascertained that each child knew what 'closest' meant before the task was performed, and they gave an affirmative answer when asked if they really could imagine themselves standing at the reference site. The instructions did not cause any confusion.

After completing the task, the child was asked to describe how he or she performed the distance estimations.

Finally, the child was given a blank piece of paper and asked to locate the chosen landmarks in accordance with their geographical positions.

Information was collected from 12 children about the actual spatial locations of the landmarks. This was done in company with their mothers after the testing.

Results

Consistency within and between reference sites
With four reference sites, the number of possible comparisons added up to twelve and with five sites there were 30 comparisons for each child. In the case where four places were utilized, the number of comparisons from each place were three with one possible intransitivity, i.e. a total of four potential intransitivities. An example of intransitivity would be that $AD < AC$, $AB < AD$ and $AC < AB$. With five reference sites there were six comparisons from each site with three possible intransitivities, fifteen intransitivities altogether.

It was found that five children made one, and one child two intransitivities. The potential number was 158 but of course it is highly unlikely that all the children would be totally inconsistent.

Even though the children remained consistent in their judgements within each reference site, there was a possibility that they imagined different relations between landmarks when they changed from one site to another. If the distance AB is judged to be shorter than AC from A, and BC to be shorter than BA from B, then CB would be shorter than CA from C in order for the estimation to be consistent, (see Figure 1). To determine whether the relative distances were compatible regardless of where the child imagined him/herself standing, the number of intransitivities between reference sites were counted. Once again, five children with one intransitivity and one child with two, were found. This time the potential number was 128.

The results for consistency clearly indicated that the children maintained the location of landmarks from different places. Yet, one may still wonder if this relative accuracy embraced systematic distortions or corresponded well to the factual locations in the environment. In order to shed some light on this question, estimates from 12 children were compared to landmark locations on a map.

Map correspondence

In all, these 12 children made 162 paired comparisons. Twenty-three out of these, or 14%, diverged from factual relations, i.e. the shorter distance was judged to be longer or vice versa. The low number of divergencies suggests that the judgements tended to be veridical at least on an ordinal level.

Deviations were also expected to be more common the closer the ratio between judged distances got to one. In Figure 2, the judgements are ordered in categories with respect to relative distance relations. Also, the divergent estimates are arranged. Testing for the assumption that errors would be more common the closer the intrapair distance ratio got to one, gives a *P*-value of ·05, from a Kolmogorov-Smirnov one-tailed test. Thus, one can say with a fair degree of confidence that the children were sensitive to actual distances.

Another way to present the overall agreement between estimates and the map was obtained by looking at rank orders for landmarks from each reference site. These ranks were compared to corresponding rank orders from the map. For 31 out of 49 reference sites, there was a perfect agreement with a mean rank correlation of $r_s = 0.79$. If the judgements had been made by chance only as compared to the map, a $r_s = 0$ would be expected. All children performed better than chance, $P = 0.003$, (one-tailed sign test).

Accuracy in estimations from the home were considered to be superior compared to those from other places. Out of 39 judgements from the home, 2 deviated from factual distant ratios. The corresponding figures for estimates from remaining places to landmarks besides the home were 45 judgements and 8 deviations, respectively. Although the difference was in the expected direction, it was not significant, $P = 0.07$ (binomial test).

A check was also carried out to verify whether the children underestimated the distance to the home in relation to the distance to another landmark. In those 28 cases where the home was farther away from the reference site than the compared landmark, for 11 judgements the positions were reversed. On the other hand, when a compared landmark was farther away than home, which was the case for 50 judgements, only two discrepancies were found. This difference was significant, $P = 0.007$, McNemar test for significance of changes. There was no difference between the average distance ratios for the two considerations, $t (76) = 0.27$.

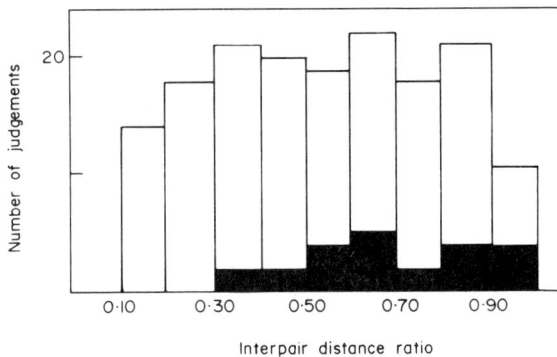

FIGURE 2. Distribution of distance judgements with respect to interpair distance ratio. Shaded areas correspond to non-veridical estimates.

The children were also asked to reproduce the locations of their chosen landmarks on a piece of paper. The same type of rank comparison that was made between estimates and the map revealed that 28 of 49 judgements showed perfect agreement. The mean rank correlations was $r_s = 0.62$, thus slightly lower than for the estimates.

Sometimes rank orders differed between the map and the reproduced locations. When this occurred, it was tested as to whether rank orders of paired comparisons were closer to the map or the reproduced locations. Fifteen out of 22 relevant cases were in favour of the map. This difference was not significant, $P = 0.13$, two-tailed sign-test.

Finally, how did children generate their judgements in order to ascertain which of two landmarks was closest to a reference site? Seventeen children replied that they had walked, or imagined they had walked, the distances and knew that one was shorter/longer. Four children replied that they just knew or could imagine, one said that she had seen the landmarks and, finally, one child said that her mother had told her. The last two children lived downtown and were not allowed out on their own which could explain their answers. By and large, thinking of own movements from the reference site to the landmarks in question seemed to serve as a basis for the judgements.

Discussion

The analysis of the distance estimations to and from the home revealed the unique position of the home in children's mental representation of their environment. With growing cognitive capacity, this exceptional position will probably diminish since more deliberate strategies can be applied. The suggestion that relative under-estimation of the distance to the home depends on the ease by which it is represented could be amplified. The estimates of the other landmark of the pair, besides being judged from the reference site, could also be affected by its location in relation to the home. Even though a certain landmark is not far away from the reference site, being far from the home may influence its spatial representation. Thus the feeling of security that children associate with their home may lead to systematic distortions in their spatial representations. The present data do not allow for a test of this hypothesis but the question of influence played by the home is worth further investigation.

Through interaction with the environment, children acquire knowledge of spatial locations. This knowledge has been described as a number of mental representations from various viewpoints. In the present study, care was taken that landmarks used to define these viewpoints were part of the children's spatial representations. Under these circumstances, it could be shown that six-year-old children were able to combine perspectives and represent projective relations. A similar result was reported by Anooshan and Young (1981) for children round the age of eight, although these authors preferred to call it a general level containing wholistic, co-ordinated representations of the ordinal or relative relationships between objects in space, an explanation adopted after Hardwick et al. (1976).

The children were not only internally consistent. By relating the distance estimates to a map, it was possible to show that these judgements were sensitive to actual distances.

The fact that projective and perhaps Euclidean properties are well reproduced should not be interpreted such that the children had incorporated projective and Euclidean concepts. The tentative answer on how comparisons were generated underline that these children were strongly dependent on their own movements. An Alternative interpretation would be that mental representations studied here are close to projective and Euclidean concepts but that the children did not apply operations associated with projective and Euclidian space.

In future studies it would be interesting to compare strategies used by children and adults when organizing a spatial representation of the same large-scale environment. The reliance on own transactions in the environment is salient also for adults as indicated by earlier studies (Evans *et al.*, 1981; Gärling, 1980). However, being too dependent on own movement patterns makes the system vulnerable. With an extended activity range, more parsimonious strategies are called for. In order to relate different parts and landmarks in the environment to each other, reference systems independent of the individual's own transitions must be learned. Children's spatial competence is limited in this respect. This is evidenced in studies by Piaget and Inhelder (1967), Laurendeau and Pinard (1970), and Hardwick *et al.* (1976) where it was found that young children did not apply operations associated with projective and Euclidian space. Also, children's ability to abstract information from external sources like maps or to choose perceptually salient landmarks (see Allen *et al.*, 1979) are inferior to adults' ability. However, as the present study has indicated, own experience is a strong factor in acquiring a spatial representation of a familiar environment. This sensitivity to relative and absolute accuracy can be regarded as a step towards a more general cognitive capacity governing all spatial representation.

References

Acredolo, L. P. (1977). Developmental changes in the ability to coordinate perspectives of a large-scale space. *Developmental Psychology*, **13**, 1–8.

Acredolo, L. P., Pick, H. L. and Olsen, M. G. (1975). Environmental differentiation and familiarity as determinants of children's memory for spatial location. *Developmental Psychology*, **11**, 495–501.

Allen, G. L., Kirasic, K. C., Siegel, A. W. and Herman, J. F. (1979). Developmental issues in cognitive mapping: The selection and utilization of environmental landmarks. *Child Development*, **50**, 1062–70.

Anooshian, L. J. and Young, D. (1981). Developmental changes in cognitive maps of a familiar neighbourhood. *Child Development*, **52**, 341–8.

Evans, G. W. and Pezdek, K. (1980). World distance and local information. *Journal of Experimental Psychology: Human Learning and Memory*, **6**, 13–24.

Evans, G. W., Marrero, D. G. and Butler, P. A. (1981). Environmental learning and cognitive mapping. *Environment and Behavior*, **13**, 83–104.

Gärling, T. (1980). *Perception av miljöers spatiala organisation under förflyttning.* (In Swedish). Swedish Council for Building Research, R159: Stockholm.

Hardwick, D. A., McInture, C. W. and Pick, H. L. (1976). The content and manipulation of cognitive maps in children and adults. *Monographs of the Society for Research in Child Development*, **41**, 1–55 (3, Serial No. 166).

Hart, R. A. and Moore, G. T. (1973). The development of spatial cognition: a review. In R. M. Downs and D. Stea (eds), *Image and Environment*. Chicago: Aldine.

Hazen, N. L., Lockman, J. J. and Pick, H. L. (1978). The development of children's representations of large-scale environments. *Child Development*, **49**, 623–36.

Ittelson, W. H. (1973). Environmental perception and contemporary perceptual theory. In

W. H. Ittelson (ed.), *Environment and Cognition*. New York: Seminary Press.

Laurendeau, M. and Pinard, A. (1970). *The Development of the Concept of Space in the Child*. New York: International University Press.

Piaget, J., Inhelder, B. and Szeminska, A. (1960). *The Child's Conception of Geometry*. New York: Basic Books.

Piaget, J. and Inhelder, B. (1967). *The Child's Conception of Space*. New York: Norton.

Schouela, D. A., Steinberg, L. M., Leveton, L. B. and Wapner, S. (1980). Development of the cognitive organization of an environment. *Canadian Journal of Behavioural Science*, **12**, 1–18.

Siegel, A. W. and White, S. H. (1975). The development of spatial representations of large-scale environments. In H. W. Reese (ed.), *Advances in Child Development and Behavior*, Vol. 10. New York: Academic Press.

URBAN EARLY ADOLESCENTS, CROWDING AND THE NEIGHBOURHOOD EXPERIENCE: A PRELIMINARY INVESTIGATION

FRED J. VAN STADEN

Department of Psychology, UNISA, Pretoria

Abstract

This interview study focused on urban early adolescents' conscious experiences of high density conditions. Thirty children residing in apartments in two boroughs of New York City (Brooklyn and Manhattan), participated in this study. Control variables included income level, age, family constellation, and length of residence. Sex, apartment density, building size and neighbourhood density served as classification variables. The interview lasted approximately forty minutes and addressed the following issues.

Personal conceptualizations and definitions of crowding.
Evaluation of the children's perceived neighbourhood.
Peer relationships.
Perceived social and physical opportunities and constraints.

Interviews were tape recorded, transcribed and content analyzed.

The children's conceptualizations of crowding emphasized large numbers of people, spatial restriction and an aversive experience of the situation as relevant components of the crowding construct. The restriction of behavioural choice as it relates to goal postponement and the child's efforts to come to terms with larger organizational social structures, underlined the interpretation of crowding as a socio-physical stressor. While information overload was depicted as another relevant characteristic of crowding situations, its aversive experience was mediated by the presence of known others. Even though this diversity on social and physical levels was often experienced as stressful, it was also positively viewed in situations where interpersonal contact or a sense of familiarity were not overridden by other situational demands.

Boys tended to spend more time in their neighbourhood engaged in a broader range of activities than girls. In this sample, boys also took less part in structured group activities and rated their neighbourhoods somewhat more positively than girls. Girls experienced less freedom and more social restrictions in exploring and actively interacting with their extended surroundings than boys.

The issues of movement and situationally imposed behavioural restrictions were central themes in children's conscious experiences of high density situations. Emphasis was placed on physiological arousal linked to the perception of aversive social qualities of high density situations. It appears that an increase in the need to act out is more a function of the child's awareness of having to conform to external norms, or experiencing a loss of perceived social control in situations where vigilance is required, than a clear cognisance of spatial restrictions.

The investigation of the impact of high density conditions on the child's socio-cognitive development and identity formation is viewed as potentially fruitful directions for future research.

Introduction

While persons of all ages function in a variety of high density conditions, only a small fraction of presently available research on high density and stress is concerned with children. Research on children's experiences in high density environments provides a fragmented perspective which focusses primarily upon selected aspects of the residential setting, the day care/school environment and institutional settings (Murray, 1974; Saegert, 1980; Wolfe, 1975). Contexts such as children's experiences of crowding in their home neighbourhood have as yet received no direct attention.

It seems likely that the levels of interpretation and generalizability of results in this field would be enhanced by extending research investigations to include experiences of crowding in secondary environments as well.

Literature review

Spatial and social variables which have been taken into account in density studies of children, involve high rise vs. low rise dwellings and social, room, apartment, building and classroom density, with each factor providing it's own unique constraints and potential contributions to the child's experience of crowding. Recognition of the physical context primarily involved ratios of the number of people per space unit. This density quotient has generally been accepted as a baseline in differentiating between high density and low density conditions and along which individual or group experiences of crowding is measured (Dean *et al.* 1975; Epstein and Baum, 1978).

Kaminoff and Proshansky (1981) review a number of papers indicating that high rise living may inhibit play activities associated with young children's normal developmental growth. Partly because of the constraints posed by high rise living, difficulties may arise in parents' monitoring of their children's outdoor activities which can result in the child's play activities being largely confined to indoor settings and influenced by concerns such as noise and neighbours' complaints. The extent of parental controls may therefore have a significant impact upon children's growth and development in high density settings. While noise appear to be an integrated quality of high density living, questions about children's exposure to community and apartment noise have led to some investigations. Cohen *et al.* (1973) found that children who lived in noisier apartments showed greater impairment of auditory discrimination and reading ability than did those who lived in quieter apartments. Social class variables were ruled out as possible explanations for differences. Cohen and his co-workers (1980) found that children living and attending school under the air corridor of a busy metropolitan airport performed more poorly on puzzle solving tasks than their counterparts living in quieter neighbourhoods. Similarly Heft (1979) reported that children from homes which were described by their parents as either noisy or quiet suggest poorer post-stimulus task performance on the part of children from noisier homes. While none of the above studies incorporated high neighbourhood density as a relevant variable, it seems likely that noise may be one of various forms of children's experience of overload arising from extended high density conditions.

Cognitive orientations to the relationships between perceived complexity and high density in field situations have provided some evidence that children's functioning in physical and social structures which are organized beyond their levels of comprehending, may initiate stress reactions which are related to the experience of cognitive overload. Saegert (1980) reported that children's estimates of the number of apartments in their building were significantly lower than the actual number for large buildings. This finding interacted with a heightened lack of clarity about the tenant population and fragmented interpersonal relations. In comparison with their counterparts living in low rise buildings, children who reside in high rises also perceived other tenants as less friendly and less willing to assist others. The sample consisted of elementary school age children from low income households. Controlling for socio-economic class, high density was also related to poorer performance at school, with girls performing somewhat better than boys. In this context, Booth and Johnson (1975) also found decrements in intellectual development.

In a laboratory study Chandler *et al.* (1978) utilized physiological, behavioural and affective measures and noted an occurrence of heightened sweating and behavioural disruptions whilst children also reported more frequently negative experiences as a function of heightened social and physical complexity.

In terms of affect, high density households seem to be experienced as qualitatively more negative by children who reported more frequent feelings of anger, aggression, anxiousness and nervousness, than those living in low density households (Gasparini. 1973; Loo, 1972; Saegert, 1980). Acting out strategies occurred more frequently, perhaps in part because of lesser options to withdraw, so that a child may feel pressed to express feelings of discomfort more overtly rather than dealing with it in other ways.

While it seems likely that the development of environmentally related dispositions such as field-dependence or locus of control are directly influenced by the 'chronic' experience of high density, little research evidence is available. Saegert (1980) argues that children's development of field-independent qualities may be jeopardized by a lowered incidence of self-determination and frequent exposure to overload conditions. Employing a more limited construct, it seems likely that long term exposure to high density conditions may enhance a child's development of stimulus-screening abilities. While stimulus screening abilities have been reported to be relatively successful strategies in coping with overload situations, they may also have a direct influence upon a child's development of social skills and behaviour. Based on these speculations, it seems that a child's history of, and exposure to, varying density situations may be a potentially significant variable in determining the extent to which density conditions have an impact upon social and cognitive development, and the extent to which it contributes to the experience of stress.

It appears that children living in high density conditions are more likely to portray behavioural problems, e.g. hyper-activity and anti-social behaviours, than their low density counterparts. Booth and Johnsons (1975) and Saegert (1980) reported behaviour ratings of children from high density homes as being less behaviourally adjusted. While Murray (1974) reported that boys from these conditions had more records of criminal offences, Gillis (1974) noted that children living in high rises were more frequently arrested for juvenile delinquency. Also, Saegert (1980) noted that children from high rises described fewer guilt feelings about vandalism and fighting. It seems that these indicators of social strain may in part be enhanced by extended high density conditions.

Because of the interactional nature of daily experiences, it is likely that household problems would extend beyond the home environment to school situations and vice versa. While school environments differ in its nature, and composition (both inter-

and intra-school) as well as in relation to household settings, the relevance of density may in part also be a function of complex interactions between home and school environments which may serve to ameliorate or exacerbate stress experiences. Epstein (1981) notes that while a number of studies have attempted to relate class size with pupil achievement (with differential results) none have investigated the differential processes which may occur in high density vs. low density classrooms. Sex also appear to serve as a relevant differentiator between experiential styles in children. Boys in high density conditions are reported to be more frequently hyperactive, aggressive, angry and anxious, while girls reported more feelings of loneliness and avoidance behaviours compared with children from low density homes (Loo, 1972; Loo and Smetand, 1977; Saegert, 1980). Saegert (1980) also depicted a number of sex by density interactions within families, which lifted out changes in parents' behaviour styles when children are present. Fathers from high density apartments reported spending more time at home when boy children were at home, and doing more things with boys, while mothers shared more time with girls at home. Girls living in high rises spent more time inside, while boys spent more time outside when the apartment was crowded. Boys also perceived that they generated more conflict for others than girls in high density apartments.

It seems likely that these interactions may become more complex when other differentiating factors of family constellation (such as single parent households and number of children) are considered. The recognition of age requires specific attention, especially when children are the subject of study. Chandler *et al.* (1977) note that in density/stress research virtually no direct attempt has been made to relate children's experiences to age within a developmental context. Loo (1977) argues that the failure of personal space to act as a significant variable in preschoolers' density experiences could be indicative of undifferentiated needs as a function of age. A relevant developmental factor indicated by Saegert (1980), involves elementary school children's attribution of negative experiential density variables to interpersonal conflicts rather than to the environmental constraints of a high density living setting, which may also reflect an undifferentiated level of cognitive development with regard to conflict or stress experiences and its environmental concomitants.

The complex nature of high density situations and the few available studies on children's experiences within these situations probably constitute the major reasons why developmental models have not yet emerged within this field. By verifying and broadening the base of research investigations to all age groups, a gradual integration of the literature may give rise to more comprehensive theoretical perspectives.

The Present Investigation

Indications from studies with young children express the relevance of density factors to the child's developing levels of cognitive, social and affective differentiation, changes in identity formation, the experience of stress and the development of coping skills (Chandler *et al.*, 1978; Fagot, 1977; Loo, 1977b; Marshall and Heslin, 1975; McGrew, 1970; Rodin, 1976; Saegert, 1980).

Since the existing literature does not contain any investigation with regard to early adolescence, it seemed useful to explore a few aspects of this age group's conscious experiences of high density situations. A major goal of this pilot study involved the generation of hypotheses which may serve to stimulate further investigations.

Sample Selection

To ensure some variability within the urban context of New York City, participants were recruited from two different areas: Clinton Hill (Brooklyn) and Murray Hill (Manhattan). The initial interviewees provided names and addresses of other children living in their area in a networking process. Parents were approached for consent if their children met the selection criteria. A denial rate of 27·6% was recorded. Reasons why this group of parents refused permission for their children to participate were not ascertained. However, no immediately apparent characteristics seemed to differentiate this group from those parents who gave permission for their children to participate in the study.

A total of 30 children (15 in each area) were selected by controlling the following variables which crystallized from the literature review as being potentially significant factors: income level, age, family constellation, and length of residence. Sex, apartment density, building size and neighbourhood density were introduced as classification variables.

Income level

At least one parent per household was required to have a professional or white-collar occupation, typically an occupation for which at least a high school diploma was needed, and in most cases there was some form of higher educational training as well. This sample then basically consisted of middle-income families. Because the literature on field research reflects an emphasis on children from low-income families (Murray, 1970; Gillis, 1974; Booth and Johnson, 1975; Saegert, 1980), it was reasoned that children from middle-income families might provide a somewhat different set of perspectives and experiential patterns.

Only middle-income families were included in this sample since the introduction of income level as a fourth classification factor would have further complicated the interpretability of the results. However, future investigations in which groups from different income levels are contrasted may be well worthwhile.

Age

Children's ages ranged from 11 to 13 years, with a mean age of 12·35 years. The differentiation of early adolescence from other stages of development is still in an early phase of conceptual definition (Lipsitz, 1980). While puberty and early adolescence in themselves may be regarded as a stressful stage which involves biological changes, intensification of emotional experiences and a heightened differentiation of the self from parental or family bonds, a stronger need for privacy, stating territorial boundaries and control may also occur. Peer group identification also emerges as a central theme (Elkind, 1975; Konopka, 1973; Lipsitz, 1980).

Structure containing the child's household

While the larger body of New York City housing structures involve apartment dwellings, it also seemed likely that marked experiential differences would occur between children living in apartments and those residing in single-family houses, specifically with regard to available space and the child's identification with the house and neighbourhood as a home environment. For this reason, housing structures contain-

TABLE 1
Building structure containing interviewees' apartment

Apartment building	Number of floors	Mean number of apartments per building	Interviewees living in Murray Hill	Interviewees living in Clinton Hill	Total
High rise	14–33	192	10	4	14
Medium rise	8–12	53	5	2	7
Low rise	4–6	4	—	9	9

ing only a single family were excluded from the sample. However, single-family houses (e.g. row houses) which had been converted into multi-family apartments were included in order to provide some variability as far as high rise/low rise apartment living is concerned. Fourteen interviewees lived in high rise buildings, seven resided in medium rises and nine children lived in low rise apartments (see Table 1).

Time living in the apartment
Children's subjective experiences of the socio-physical context in which they live are partly dependent upon the opportunity over time to familiarize themselves and to define relationships with their surroundings. In order to limit potential confounding factors related to recent residence change, interviews were limited to respondents who have been residing in their present dwelling units for a period of at least two years.

Sex and ethnic composition
The sample was racially mixed, with a total of 13 Afro-American respondents, 16 Anglo-Americans, and one Chinese-American.

On the basis of research evidence suggesting sex differences in people's reactions to and experiences of different density situations (Sundstrom, 1978; Altman, 1978), it was decided to include an equal number of boys and girls in the sample, and to introduce sex as a classification factor.

Definitions and Classification Variables

Density is defined on two levels: (1) Socio-physical aspects of the setting where emphasis is placed on objective physical characterizations; and (2) Subjective experiential qualities of the physical situations whereby the individual's perceptions and interpretations are recognized as relevant components of the individual's conscious experience of density factors (Rapoport, 1975; van Staden, 1983).

The objective socio-physical characterization of density was divided into three categories: apartment or household density; building size; and neighbourhood density.

Apartment density
Apartment density was measured by the number of people (living in the apartment) per room. In this analysis, bathrooms were given full weight as rooms.

Apartment densities ranged from 0·44 to 0·86 persons per room, with a mean density level of 0·75. Differences between the Manhattan and Brooklyn sub-groups were negligible, and an overall median split seemed inappropriate in view of the small sample size and the small range of density levels.

In comparison with samples used in other field studies (Booth and Johnson, 1975; Gillis, 1976; Saegert 1980), this measure expressed medium to low levels of household densities for this sample. Being cognisant of this sample characteristic, it seemed useful to consider bedroom density as an additional indicator of the density situation within the apartment. Overall, 13 children shared a bedroom with one other person. In all cases, the other person was a sibling.

Building size

Building size included three levels: high rise (14–33 floors, $N = 14$); medium rise (8–12 floors, $N = 7$); and low rise (4–6 floors, $N = 9$). Table 1 gives a more detailed account.

However, building size not only involves the number of floors, but is also a function of the size and number of apartments, the latter being especially relevant in the case of low-rise apartment buildings. In order to minimize variation in low-rise building structures, the low-rise buildings in this sample consist only of single-family row houses which have been converted into multi-family apartment houses.

The medium- and high-rise buildings, on the other hand, were originally constructed to provide apartment living space for large numbers of families, and therefore include such features as elevators, entrance halls, and basement laundry rooms, features which are not included in the configuration of this sample's low-rise apartment buildings.

Neighbourhood density

A survey of the neighbourhood densities in the two selected areas involved such aspects as residential building structures and number of residences, business activities and services, public services, vehicle traffic and recreational opportunities. While space restrictions do not allow a detailed account of the survey data, a comparison of the two neighbourhood areas yielded the conclusions given below.

Both areas can be described as predominately middle class residential urban neighbourhoods. The Murray Hill area contains somewhat greater diversity with regard to building structures, children's recreational opportunities and businesses than does the Clinton Hill section. The larger number of high- and medium-rise residential buildings, office buildings, street front businesses, traffic lanes and a concentrated public bus service in Murray Hill, would seem to be indicative of generally higher density levels (that is, number of people per space unit), larger fluctuations in these density levels, and the occurrence of a greater variety of public activities than that which occur in the Clinton Hill section.

The Clinton Hill area on the other hand, contains more public parks, schools, churches and a stronger homogeneity in its architectural composition, elements which would seem likely to enhance a sense of community.

In each area, the children lived within a few blocks of each other. The survey included only those areas in which the children reside, and not the neighbourhoods in general. Whereas the Clinton Hill section consisted of 14 blocks, the survey area in Murray Hill comprised of 12 blocks. In terms of geographical surface area the two neighbourhood sections were comparable in size.

While differences between the two neighbourhoods also occur as a function of locational differences between the borough of Manhattan and the borough of Brooklyn, it should be recognized that they nevertheless form part of the same larger city context.

Interviews

Measurements of the participants' conscious experiences of density situations were obtained by means of an interview questionnaire which consisted of two parts: (1) the meaning of crowding; and (2) density issues related to the perceived neighbourhood.

Since crowding is considered to be an extreme experiential experience of density, it was thought to be useful to obtain some data on the meaning which young urbanites attach to the crowding concept. This first part of the interview addressed the following issues.

(1) Personal conceptualizations and definitions of crowding.
(2) Personal experiences of crowding in terms of affective, physiological and behavioural reactions.
(3) Examples of situations which are generally experienced as crowded.
(4) Reports of the social construction of crowded situations encountered.

The next section covered aspects regarding the perceived neighbourhood.

(1) Description and personal conceptualizations of the perceived neighbourhood.
(2) Likes and dislikes regarding the neighbourhood.
(3) Constraints and opportunities in spending free time in the neighbourhood.
(4) Usual activity patterns and peer relationships.
(5) Aspects about the neighbourhood which participants would want to change.
(6) School location, and experiences of going to and from school.

Because of the relatively unexplored nature of this field the application of a less structured method was preferred to more structured approaches. However, in order to keep the interviewees' responses focussed on the objectives of the questionnaire, a number of open ended and fixed-alternative items were included (Robinson, 1981).

The use of a fairly unstructured interview method stands in contrast with the available research studies in this field where emphasis is placed on the analysis of aggregate data or the application of standardized scales (Booth and Johnson, 1975; Gasparini, 1973; Gillis, 1974; Murray, 1974). While these studies involve more rigorous research designs, limits are also placed on the scope of information which is gathered. Consequently it was reasoned that the utilization of a descriptive method would yield a more holistic view of how early adolescents experience life in an urban environment.

Procedure

The order in which the different sections of the questionnaire were covered was randomly interchanged in an attempt to equalize the possible effects of fatigue on the nature of the answers in any one section.

The children were interviewed at home and the interview lasted approximately 40 mins. In most cases, at least one parent was at home and within hearing distance of the interview situation.

In order to establish some rapport between the interviewer and interviewee, approximately ten minutes were spent discussing city life in general, interests and hobbies before focussing on the questionnaire. Subjects appeared to relate and respond to the questions with ease.

The interviews were tape recorded, transcribed and content-analyzed for each question topic, respectively. Categories were developed for each sub-section of the questionnaire by identifying and tabulating the issues mentioned by the interviewees.

Results

The analysis of qualitative data is based upon nominal measurements which is only accessible to descriptive correlational statistics. Results should therefore be regarded as tentative hypotheses which need to be developed and verified using more rigorous research designs. Results of this study will be presented separately for each section of the questionnaire. A synthesis of issues which appear across the various sections is provided in the conclusion.

Interpretation of crowding
Children's personal interpretations of crowding involved five main themes. In general, crowding was depicted as an aversive experience of large numbers of people consisting mainly of unfamiliar others. The restriction of movement and issues relating to information-overload were also emphasized. Table 2 reflects the categories which emerged from the children's answers.

TABLE 2
Interviewees' interpretations of crowding

Themes	Number of children reporting issues with relevance to each theme		
	Boys	Girls	Total
Aversive experiential aspects	13	15	28 (93%)
Large numbers of people	11	14	25 (83%)
Group consisting of unfamiliar others	10	12	22 (73%)
Restriction of movement	10	7	17 (57%)
Information overload	6	9	15 (50%)

Aversive experiential aspects
Virtually all participants viewed crowding as a negative or aversive experience. Reportedly, crowding involved feelings such as being mad, angry, frustrated, scared, uncomfortable or pushed. Four participants (three girls, one boy) also emphasized feelings of "smallness". This rather vague term seems not only to involve a heightened awareness of bodily limitations (e.g. height) but also to express feelings of ineffectiveness or inadequacy during the experience of being crowded.

Physiological changes reported by children involved stress or arousal symptoms such as feeling hot, tight, nervous, restless, cramped and tired. Probes indicated that some children (four boys, three girls) were not aware of any physiological changes which might be linked to their experiences of being crowded. It seems that physiological symptoms are in general more affectively interpreted without clear differentiation, an aspect which may in part be a function of chronological age and level of cognitive development.

Large numbers of people
When asked to define crowding, most children's immediate responses focused on large numbers of people within secondary environments. Examples of places where participants have personally felt crowded, involved transportation settings, recreation-settings, departmental sidewalks, and schools. A summary of their responses is provided in Table 3.

Some sex differences between boys' and girls' responses were noted. Where five boys mentioned crowding when going to or leaving a sports stadium, eight girls emphasized department stores as examples of places where they usually feel crowded.

The issue of waiting at bus stops, in elevators and in queues also merits attention. The emphasis which young people placed on waiting as a relevant part of their experience of crowding reflects the relevance of personal goal blocking or postpone-

TABLE 3
Summary of children's examples of crowded places

	Murray Hill (Manhattan)	Clinton Hill (Brooklyn)	Total
(A) *Secondary environments*			
1 Transportation (city bus, subway, getting on or off a train, and elevators)	15	10	25 (83%)
(2) Sidewalks	10	5	15 (50%)
(3) Recreation (waiting in queues for tickets, leaving theatre or stadium and children's recreation clubs)	7	4	11 (37%)
(4) Rush hour in Manhattan	9	1	10 (33%)
(5) Stores (departmental stores, e.g. Macy's or Bloomingdales)	4	6	10 (33%)
(6) Specific locations in Manhattan (e.g. Times Square, c/o 34th Street and Broadway)	6	1	7 (23%)
(B) *Primary environments*			
(1) School			
Hallways when changing classes	9	12	21 (70%)
Classroom	7	5	12 (40%)
Lunchroom	1	4	5 (17%)
(2) Homes in low income neighbourhoods	2	3	5 (17%)

ment and its interaction with the young person's efforts to come to terms with social-organizational processes.

None of the children's examples were related to their own home environment, an issue which corresponds with the fact that the sample consisted of children from objectively low to medium household densities. On the other hand, it may also be that crowded situations (however frequently or infrequently they may occur) within the home environment are not interpreted as a function of crowding, but instead as a function of interpersonal conflicts of relationships (Saegert, 1980). Some support for this view is provided by children's reports of the social composition of crowding situations. The section on group composition has some relevance to this issue.

Movement

An aspect of crowding which was emphasized by early adolescents from all subgroups involved the issue of movement and the restriction thereof. The expression of the need to act (or to move) formed a large component of answers pertaining to the conceptual meaning of crowding and the affective, physiological or behavioural experience thereof.

For example, crowding was conceptually described as 'no place to walk' or 'people pushing you and you can't move'. Eight boys and three girls integrated this issue of physical constraint in their descriptions.

Affectively, answers such as 'I feel nervous—like I have to hurry up and do something, but that I have to wait at the same time' express aspects of conflict which are interpreted on a behavioural level as the frustration and constraining of movement. Also statements like these appear to indicate that the physiological arousal which takes place is interpreted by some as a heightened need to act behaviourally. However, the simultaneous perception of the loss of behavioural choice completes the basis of the conflict experience. Most children's immediate reactions to crowding are then expressed by a strong need to 'get out of it', in an attempt to limit their exposure to this form of frustration.

Information overload

Participants' experience of information overload is expressed on two levels: (1) their conceptualizations of crowding; and (2) reports of behavioural reactions.

Descriptions of crowding frequently included words like 'a *lot* of traffic, people, grown-ups, stores, lights'. The descriptions emphasized a wide range of interpersonal, societal and technological aspects, while references to the 'natural' world did not occur.

The relevance of information overload perspectives appears to be related to the children's comments about the abundance of human-produced stimuli, which might be perceived as conflicting, overwhelming, senseless, unstructured, or irrelevant. Crowding situations which are perceived in this way may, in turn, provide strong challenges to the child's cognitive and social abilities to interpret and to understand the complexities within these situations (Chandler *et al.*, 1978; Loo, 1977).

Behavioural reactions to crowding included reports (seven boys, six girls) of 'palliative' reactions, whereby the children attempted to minimize external input. As one boy puts it: 'I try to stay with myself, and be with myself'. However, a need for relevant information through which the situation can be more satisfactorily structured also occurs. Answers like 'I wish I was taller' and 'I watch my pockets and the people closest to me' not only express constraints which are tied with the age-relevant factor of height, but also a stronger awareness of potentially adverse social aspects of some crowding situations.

Group composition
Children's reports of being crowded occurred in situations where; (1) they didn't know anyone else, and/or (2) in situations where they did not feel free enough to talk to other people (12 boys, 10 girls).

It appears that being able to talk to somebody can be a significant mediator of crowding experiences for this age group (22 responses). The presence of friends or family members was interpreted as alleviating crowding stress (e.g. 'Sometimes there's somebody I know, whom I can talk to, and then it's O.K.').

On a cultural level, two participants indicated that a heterogeneous ethnic composition might also contribute to their experience of crowding.

Finally, the invasion of privacy as a component of the crowding experience was mentioned by only one child: '... You can also be crowded when you want to be alone and other people bother you'.

The relative absence of direct privacy-related comments from this sample may reflect a developmental factor where social interaction, the development of interpersonal relations, and the acquisition of social skills and knowledge override the need for privacy. However, it should be remembered that children's depiction of places where they usually feel crowded focused primarily upon public situations where privacy may be viewed as secondary to more central themes, e.g. stimulus seeking, behavioural constraints and goal attainment.

These results also appear to correspond with Loo's (1977) findings that personal space did not act as a significant variable in young children's experiences of high density situations.

The perceived neighbourhood
A discussion of participants' responses about their neighbourhood can be organized around the following topics.
  (1) Descriptions of the perceived neighbourhood.
  (2) Constraints and general activity patterns.
  (3) Peer relationships.
  (4) School location and experiences of going to and from school.

Descriptions of the perceived neighbourhood
While the children's personal evaluations of their neighbourhoods ranged from 'pretty bad' to 'one of the best' in both Clinton Hill and Murray Hill, the children from Clinton Hill gave, on the average, more positive and less mixed statements (i.e. answers which contained both positive and negative qualities) than their Murray Hill counterparts (see Table 6). Clinton Hill children also provided shorter and less varied descriptions of the people in their neighbourhood.

The participants' descriptions of what they perceived as the boundaries of their neighbourhood generally focussed on nearby street names, distinguishing features of the area (parks, ethnic composition, buildings), and areas which are frequently

TABLE 4
Participants' perceptions of their neighbourhood: likes and dislikes

Likes	Murray Hill (N = 15)	Clinton Hill (N = 15)	Total (N = 30)	Dislikes	Murray Hill (N = 15)	Clinton Hill (N = 15)	Total (N = 30)
Friends are nearby	6	10	16	Some older children	8	9	17
Parks	7	9	16	Noise	11	4	15
Stores and shops in general	8	2	10	Dirt, litter, garbage	9	5	14
School is nearby	4	5	9	Crime	5	6	11
Restaurants (including 'fast food' places)	7	—	7	Disagreeable people in general	6	3	9
Movie theaters	7	—	7	Graffitti	4	3	7
Trees	2	5	7	Vandalism	2	4	6
Video arcades	5	—	5	Air quality	3	1	4
Library	—	4	4	Bad smells	1	—	1
Block parties	1	3	4				
Grass	—	3	3				
Neighbourhood garden	1	—	1				

used by them. The following statement is a typical example of the answers regarding the boundaries of their neighbourhood:

'There are certain blocks that make up your own neighbourhood. It's probably not the real neighbourhood, but what I would call my neighbourhood is only a couple of blocks. And it has a couple of stores that I know well, and some restaurants. My neighbourhood is about three blocks up, and three blocks down, from here'.

Statements such as the above would seem to indicate that the young urbanites' perceptions of areas and aspects which constitute their neighbourhoods are partly a function of familiarity and personal feelings of security. In nearly all cases (28 participants) the home formed the centre or the primary reference point of their perceived neighbourhood, while areas which were well known, frequently used, or which contain significant personal or symbolic impressions (e.g. the Empire State Building or a specific pizza den) formed reference points in their descriptions. Rather than giving fragmented or scattered impressions, children tended to present their neighbourhood as one integrated area.

Participants' likes and dislikes with regard to their neighbourhood are summarized in Table 4. The two subgroups' responses indicate a somewhat stronger uniformity in their dislikes than their likes.

While the Clinton Hill sample emphasized the availability of friends and open spaces as positive aspects of their neighbourhood, the children from Murray Hill viewed the variety of stores and entertainment centres as positive features of their neighbourhood. It seems possible that the differences and variation in the children's responses are in part reflective of the general socio-physical differences between the two areas.

TABLE 5
Average free time spent within the perceived neighbourhood

		Winter		Summer	
		Usual week day (h)	Usual weekend day (h)	Usual week day (h)	Usual weekend day (h)
Murray Hill	Boys	2–3	4–5	5–6	Over 6
	Girls	1–2	1–2	3–4	3–4
Clinton Hill	Boys	1–2	1–2	3–4	5–6
	Girls	Under 1	1–2	3–4	4–5

Early adolescents' awareness (and dislike) of qualities such as noise, dirt, crime, vandalism and air pollution express some of the neighbourhood frustrations experienced by both subgroups. These aspects have all been linked to the urban situation as partial symptoms of high density living (Cohen *et al.*, 1973; Epstein, 1981; Kaminoff and Proshansky, 1982).

Interpersonally, both boys and girls expressed a strong dislike of 'some older kids' who are perceived as antagonistic and aggressive towards them. This issue appears to be specifically relevant to this age group where the development of a broader social consciousness becomes a central theme. The children also reported dislikes with regard to people in general who are perceived as 'strange, weird or mean'—an aspect which expresses a growing awareness of the diverse nature of society outside the primary home environment.

Constraints and general activity patterns

Table 5 represents a summary of children's estimates of the free time they usually spend outside their home and within their perceived neighbourhood.

Boys tended to spend more free time within their neighbourhoods than girls, while all groups spent more time in their neighbourhoods during the summer and on weekends than during the winter or on weekdays. Murray Hill boys consistently spent somewhat more time in their neighbourhood than did Clinton Hill boys. This pattern was not repeated in the girls' responses.

While the incentives to spend leisure time within the neighbourhood may depend significantly upon the early adolescent's perception of socio-physical *opportunities* and *constraints*, it does not fully contain the process-like basis of experiencing one's extended home environment. An element of the 'neighbourhood experience' and its conceptualization can also be found in experiencing it as a 'transitory' environment, that is, a place through which one moves in order to get to and from other places. Children from all subgroups frequently used the concepts 'walking to ...' or 'going to ...' in describing their leisure time neighbourhood activities. It seems possible that the process of *going to* a friend's house (for example) represents in some respects a more direct experience of the neighbourhood than being at the friend's house.

Places where children usually spent their leisure time included parks (26 responses), stores (20), playgrounds and courts (16), homes of family members or friends (13), video arcades (6), boys' club (5), and library (2). Activities which were not confined to specific places were cycling (9 responses), roller skating (6), 'hanging around' (6). No clear differences emerged from any of the subgroups except for Murray Hill boys expressing a preference for video arcades and the boys' club.

An aspect which all the recreational activities mentioned seem to have in common, is that they generally occur within larger social situations where other people are always present (regardless of the child's choice of a more solitary or socially interactive form of activity). While the significance of others' presence varies in its relevance or importance, the urban early adolescent's neighbourhood activities appear to involve the presence of other people as a constant element.

Related to this issue are three aspects with significant social ties which were perceived by participants as constraints or aspects which limit their access to their neighbourhood: (i) fear of being mugged (26 responses); (ii) crime (12 responses); and (iii) 'the people' (10 responses).

Children's fears of being mugged appear to be partly founded upon personal experience and partly on reports from peers, family or acquaintances. Three children (two in Clinton Hill, one in Murray Hill) reported personal experiences of having been mugged.

While the children's mention of crime as a deterrent to spending their free time in their neighbourhood is related to their fear of being mugged, their reasons for mentionining crime were generally vague and not clearly based on factual knowledge of crime occurrences within the neighbourhood, an aspect which seems to be partly reflective of an awareness about their own vulnerability outside the home environment.

The fear of crime or of being mugged tends to affect children's activities in the following ways.

(i) Going out in the evenings is limited, and generally occurs only when friends and/or family members are present.

(ii) There is an emphasis on organized and group activities (especially in the case of girls).

(iii) A heightened vigilance and monitoring of others occurs, not only as potential assailants but also as potential helpers (e.g. doormen, police, etc.).

(iv) There is an awareness and avoidance of places which are labelled as unsafe.

Children's views of the constraints experienced in their encounters with other people in their neighbourhood involve two components which tend to be quantitative and qualitative in nature. Quantitatively, four children from Murray Hill viewed the number of people on the sidewalks as interfering with recreational goals ('You can't ride your bicycle', for example). On a more qualitative level, seven children (five girls, two boys) felt intimidated by 'people who stare at you' or 'some creepy people' or 'nasty people'.

Other reported constraints less explicitly related to high-density urban phenomena were as follows:

(i) bad weather (15 responses),

(ii) homework (14 responses),

(iii) organized activities such as ballet, Girl/Boy Scouts, sports activities at school, etc. (9 responses) and

(iv) watching televison (7 responses).

Parents were generally not perceived as placing constraints upon children's choices to spend time in their neighbourhood during the daytime.

Finally, children were asked to rate (on a 10-point Likert type scale) their neighbourhood and the block containing their home, in terms of worst-best neighbour-

TABLE 6
Children's evaluations and ratings of their neighbourhoods and their blocks

		Evaluative statements of neighbourhoods			Mean score of ratings on a 10-point Likert type scale	
		Positive responses only	Mixed responses	Negative responses only	Neighbourhood	Street block containing child's home
	Boys	6	1	0	6·125	7·5
Clinton Hill (Brooklyn)	Girls	4	3	1	5·8	7·3
	Total	10	4	1	5·963	7·4
	Boys	4	4	0	6·6	7·0
Murray Hill (Manhattan)	Girls	2	3	2	5·5	6·25
	Total	6	7	2	6·05	6·625

hood/block for them to live in. Table 6 provides a summary of the average scores on this rating scale.

Children consistently gave their blocks higher ratings than their neighbourhoods, and generally viewed their neighbourhoods as somewhat better than average. Boys were slightly more positive (on the average) in their evaluations of their neighbourhoods and blocks than girls. While this difference does not appear to be strongly significant, it may be reflective of emerging sex differences where boys experience more freedom in their explorations and transactions with their surroundings. (Boys tended to spend more of their free time in their neighbourhood than girls, and were also engaged in a somewhat larger range of activities in their neighbourhood than girls).

Peer relationships
Children's contact with close friends ranged from 'once every two weeks' to 'every day'. Meetings with friends appear to be mostly spontaneous and generally take place at school (26 responses), at church or Sunday school (12 responses), at other organized activities such as ballet, Boy Scouts, drama, etc. (7 responses), or in passing by (17 responses).

Four Murray Hill boys reported meeting their friends in the Murray Hill boys club, while five boys and two girls from both areas sometimes meet their friends in nearby parks.

Regular visits from friends at participants' homes (or vice versa) were relatively low (eight responses).

The formation and maintenance of friendships seem to be significantly based upon mutual activities and interests which frequently take place within socially structured situations. This was slightly more the case with girls than with boys. While a tentative hypothesis regarding emerging sex differences may be found in boys' tendency to function somewhat more autonomously in their neighbourhood than girls do, the differences between the responses of these two subgroups were weak.

Another aspect affecting adolescents' peer relationships may involve the proximity of friends' homes. Although the school environment generally provided the most direct opportunities for making friends, it is likely that the extent of friendship experiences is also affected by the availability of peers residing within the same apart-

TABLE 7

Participants' usual mode of transportation to and from school, compared with perceived school location

| Transportation and from school | School location | | Total |
	Inside perceived neighbourhood	Outside perceived neighbourhood	
Walk	10	—	10
Public bus	3	13	16
Subway	—	4	4
Total	13	17	30

ment building or in the nearby vicinity. However, this factor is also coloured by the extent to which activities are shared, such as going to the same school, taking the same routes and taking part in the same recreation activities.

More generally, it seems that this age group's peer relationships are focused on mutual interests rather than lasting interpersonal relationships.

School location and experiences of going to and from school

All participants who usually walk to school viewed their school as part of the perceived neighbourhood. On the other hand, taking a city bus or a subway train generally led to the exclusion of the school environment from the perceived neighbourhood environment.

Table 7 reflects participants' usual mode of transportation to and from school, as compared with their classification of the school as located within or outside the perceived neighbourhood.

Differences between the subgroups (male–female, Clinton Hill–Murray Hill) were too small to indicate any clear trends, except for Clinton Hill participants who did not at all make use of the subway system as a mode of transportation to school. However, it should be remembered that the small sample size on which this preliminary data is based may not adequately reflect significant existing variations such as cycling or school bussing.

When comparing children's comments about their usual mode of transportation, each seems to contain different experimental facets which may not only affect the extent of the child's familiarity with (and conceptualization of) his/her neighbourhood, but also have an impact upon the nature of his/her interaction with others.

While children from all groups felt bothered by noise (17 responses), dirt (14), and other people (13), each group also expressed dislikes which were directly tied to their way of travelling.

Children who usually walk to school complained about litter—especially dog litter (4)—and having no one to talk to (2). Comparatively, however, they expressed fewer complaints about the experience of going to school than did bus or subway riders.

City bus riders were most frequently bothered by issues related to crowding, which involved aspects such as being pushed or squashed (6), 'no place to sit' (4), too many passengers (4), or people crowding parts of the aisles and not moving to the back of the bus (1). Other dislikes involved being bothered by other kids (3), waiting for the bus (4), and the bus passing one's stop without picking up passengers (1).

Like bus riders, children who take the subway trains to school also complained about being crowded (4). Additional dislikes involved foul air or bad smells (2) and vandalism (1).

For all groups (walkers, bus riders, subway riders) the single most positive aspect about going to school and returning home involved meeting and talking to friends (11 girls, nine boys). On a more passive level, participants also enjoyed watching or observing other people (two bus riders, four walkers). While participants' meetings with friends occurred mostly spontaneously, their emphasis on the positive nature of these experiences may also express a developmental element in the child's socialization process.

As expected, bus and subway riders liked getting a seat (6), and not having to wait for the bus or a train (4). One subway rider also viewed subway riding as a fast means of getting to school.

Children walking to and from school enjoyed 'getting out' (2), and 'hanging around' on their way home (3). Two participants also liked buying and having breakfast on their way to school. Walking seemed to provide participants with a less externally controlled basis on which to form transactions with their surroundings than did taking a subway car or a city bus. Thus, walking appeared to provide a less socially controlled experience of the neighbourhood in which opportunities to explore and familiarize themselves with aspects of their socio-physical surroundings are more readily realized.

Children's experiences when going to school and returning home in the afternoons provide a regular and semi-structured basis from which they perceive, experience and explore different sociophysical systems (transportation systems, residential and work environments, etc.). The nature of these systems involve various degrees of complexity and interaction, which may be organized beyond an abstracted or clear comprehension by the early adolescent (Chandler et al., 1979). However, this age group appears to perceive and act within these situations in increasingly autonomous ways.

High-density urban environments inevitably involve complex fusions of socio-physical structures in which the urban early adolescent must function, and is likely to develop a differentiated awareness of larger socio-physical contexts and their changing and dynamic nature. Where this awareness may be directly or indirectly filtered by culture, race, sex, previous experiences, familiarity, socio-economic status, residential location, etc. it nevertheless remains an increasingly relevant and dynamic aspect of the young urbanite's psychological development.

Conclusion

Some issues which emerged from this interview study as potentially significant aspects of the early adolescent's experience of urban density conditions can be summarized under the following headings.

Impact of heterogeneity.

Sex differences.

Movement and behavioural constraint.

Cognitive style.

Heterogeneity

The variance in children's responses appears likely to be positively correlated with the extent of perceived heterogeneity within their socio-physical environment Children living in an urban neighbourhood with diverse housing situations, consumer services and businesses, also appear to develop an awareness of the social diversity which occurs within such a situation. Even though this diversity on social and physical levels is often experienced as aversive or stressful, it is also positively viewed in situations where interpersonal contact or a sense of familiarity are not overridden by other situational demands.

Another positive quality within the more heterogeneous context was found in the existence of more diverse recreational opportunities. Heterogeneity is expressed by means of multiple (and often competitive) stimulus outputs, which place demands upon the child's ability to select and interpret relevant stimuli. While children develop various strategies to deal with overload conditions, they appear to make more use of palliative methods (e.g. stimulus screening, avoidance tactics, superficial friendship commitments) than active or instrumental methods. The development of these strategies may be supported in part by complex socio-physical structures within the child's life world which enhance the experience of personal goal deferment, an awareness of ambiguity and stimulus overload. This notion also appears to be supported by the findings of Chandler *et al.* (1978) that children are likely to experience social and physically complex situations as stressful.

With regard to environmental stimulation, children emphasized two adversely experienced aspects, intensity and lack of clarity. Where the latter may be manifested in feelings of insecurity and an awareness of possible threat or harm, the former is partly expressed by strong dislikes of city noise, litter and dirt. Inasmuch as these dislikes are expressed as attitudes, it is likely that they are also influenced by the socialization process at home and at school. Nevertheless, the work of Cohen *et al.* (1973, 1980) noting impairment of auditory discrimination and decrements in task performances of children living in noisier neighbourhoods, serves as a reminder of the potentially negative effects of overloaded situations.

Two fundamental hypotheses can be formulated.

The occurrence of socio-physical diversity in the early adolescent's everyday life world is positively correlated with an awareness of variation and ambiguity within his/her environment.

The more diversified available options are in satisfying the early adolescent's needs, the less emphasis will on the average be placed upon any single source of satisfaction.

Sex differences

Sex differences in children's experiences can be summarized as follows.

Boys tended to spend more time in their neighbourhood being engaged in a larger range of activities than girls. In this sample, boys also took less part in structured group activities and rated their neighbourhoods and apartment buildings somewhat more positively than girls.

In general, it seems that girls experience less freedom and more social restrictions in exploring and actively interacting with their extended surroundings. While this state of affairs can in part be socio-culturally defined, its expression also appears to be socio-physically supported by high-density urban situations, which give rise to perceptions of ambiguity, challenge and threat, and feelings of insecurity. These findings appear to be supported by a number of studies (Booth and Johnson, 1975; Loo, 1972; Loo and Smetand, 1977; Murray, 1974) relating more actively-oriented styles of interaction to boys and more passive styles to girls.

While it can be posited that boys are inclined to meet the demands of high-density urban conditions more actively (in an overt behavioural sense) than girls, indications are that girls tend to develop a more differentiated perception of the constraints and opportunities embedded within their socio-physical life world.

The development of sexually differentiated orientations in children's interactions with their socio-physical surroundings is also supported by Saegert's (1980) research with younger children. Though neither orientation is clearly successful in ameliorating stress effects or urban living, it seems that children's experience of urban stress is a significant developmental force through which different strategies are employed in exploring their potential or capacity to comprehend and act within these situations.

However, where the results of this study illuminated the possibility of some sex differences, it is also likely that a clear labeling of male and female experiences and strategies would be reductionistic in its disregard of other interactional properties.

Movement and behavioural constraints
In discussing their experiences of crowding, children emphasized the issue of movement and situationally imposed behavioural restrictions. The loss of behavioural choice and the postponement of personal goal realization appear to be central themes in children's conscious experiences of everyday crowding situations.

While fluctuations occur in the interaction between physical and social characteristics of urban situations, it appears that children's emphasis on movement and the need to act out generally express a state of physiological arousal which is often linked to the personal awareness of negative social aspects in the situation. It seems that an increase in the need to act out is more a function of the child's awareness of having to conform to (informal/formal) external norms, or experiencing a loss of social control in situations where more vigilance is required, than it is of him/her being clearly cognisant of spatial restrictions. Children's discussions of places they like and dislike tend to support this comment. For example a number of the places or situations children reported disliking expressed clear behaviourally restrictive physical qualities (e.g., buses and subways, hallways, elevators), while 'liked spaces' tended to be less physically constrained (e.g. parks, playgrounds and their apartments). In their review of potentially stressful effects of the urban built environment, Kaminoff and Proshansky (1981) also support the notion that high density living conditions are likely to restrict children's normal play activities.

In general it appears that a key factor in young urbanites' density experiences centres around transportation and human motion, as well as the tendency to interpret urban situations on an interpersonal/social basis, rather than being aware of restrictive spatial or design features.

Cognitive style
The urban child's development of cognitive style is likely to be directly influenced by his/her socio-physical surroundings, which in turn are significantly governed by external human-produced controls. Aspects like video-cameras, elevators, doormen, guards, monitoring neighbours and queues are only a few

examples of the extensive external structurization which occurs within the urban context. While participants in this study expressed some understanding of the personal
advantages (in terms of security, safety, speed, etc.) of some of these systems,
it also frequently resulted in the experience of conflict and frustration.

Children from this sample appeared inclined to employ field-dependent strategies
in the sense of having to be vigilant and alert in an environment which contains
diverse levels of human-produced stimuli. However, children's actions on a more
direct interpersonal level with regard to friends also involve clear field-independent
characteristics. While the latter may be merely a reflection of the interaction between the child's focus on personal need satisfaction and the demands of the larger
socio-physical environment, children nevertheless tended to act as if more field-
independent on interpersonal dimensions while a back-drop of field dependency
occurred in their interactions on more extended socio-physical dimensions. This
observation appears to be on par with Saegert's (1980) argument that children's
development of field-independent qualities may be inhibited by frequent exposure to
overloaded conditions.

A developmental speculation in this realm would involve a gradual change in the
child's directedness on both levels as a function of familiarity and the development
of a more autonomous self-concept.

Another aspect involves environmental incentives to take part in formally
structured group activities, which seems to be based as much upon feelings of insecurity
as upon a commitment or need to belong to and identify with a group as such.

While existing theoretical frameworks are too limited in their scope to capture
or integrate the collage of impressions and issues which have been raised by this
preliminary study, it also follows that more questions and speculations than 'clear'
answers have emerged. However, it seems that relevant future directions would include
the following:

An investigation of differential contextual influences on the child's identity formation.

A developmental investigation of changes in the child's socio-cognitive development.

The urban child's conscious experience of behavioural choice in his/her personal
life world.

The impact of crowding experiences on the child's development of coping strategies
and its implications with regard to cognitive/social development.

References

Altman, I. (1978). Historical and contemporary trends in crowding research. In: A. Baum and
    Y. M. Epstein (eds), *Human Response to Crowding*, Hillsdale, New Jersey: Lawrence
    Erlbaum.
Booth, A. and Johnson, D. (1975). The effect of crowding on child health and development.
    American Behavioral Scientist, **18**, 736–49.
Chandler, M. J., Koch, D. and Paget, K. F. (1978). Developmental changes in the
    response of children to conditions of crowding and congestion. In: McGurk, H. (ed.),
    Ecological Factors in Human Development. Amsterdam: North-Holland.
Cohen, S., Glass, D. and Singer, J. E. (1973). Apartment noise, auditory discrimination and
    reading ability in children. *Journal of Experimental Social Psychology*, **9**, 407–27.
Cohen, S., Evans, G. W., Krantz, D. S. and Stokols, D. (1980). Physiological,

motivational and cognitive effects of aircraft noise on children: moving from the laboratory to the field. *American Psychologist*, **35**, 231–43.

Dean, L., Pugh, W. and Gunderson, E. (1978). The behavioral effects of crowding: Definitions and Methods. *Environment and Behavior*, **10**, 419–31.

Elkind, D. (1975). Recent research on cognitive development in adolescence. In Dragastin, S. E. and Elder, G. (eds), *Adolescence in the Life Cycle*, Washington, D.C.: Hemisphere.

Epstein, Y. M. and Baum, A. (1978). Crowding: methods of study. In: A. Baum and Y. M. Epstein (eds), *Human Response to Crowding*. Hillside, New Jersey: Erlbaum.

Epstein, Y. M. (1981). Crowding stress and human behavior. *Journal of Social Issues*, **37**, 126–44.

Fagot, B. (1977). Variations in density: Effect on task and social behaviors of preschool children. *Developmental Psychology*, **13**, 166–7.

Gasparini, A. (1973). Influence of the dwelling on the family. *Ekistics*, **216**, 344–8.

Gillis, A. R. (1974). Population density and social pathology: The case of building type, social allowance and juvenile delinquency. *Social Forces*, **52**, 306–14.

Heft, H. (1979). Background and focal environmental conditions of the home and attention in young children. *Journal of Applied Social Psychology*, **9**, 47–69.

Kaminoff, R. D. and Proshansky, H. M. (1982). Stress as a consequence of the urban built environment. In: L. Goldberger and S. Breznitz (eds), *Handbook of Stress*. New York: Free Press/Macmillan.

Konopka, G. (1973). Requirements for healthy development of adolescent youth. *Adolescence*, Fall, **8**, 2.

Lipsitz, J. (1980). The age-group. In: M. Johnson (ed.), *Toward Adolescence: The Middle School Years*. Chicago: The National Society for the Study of Education, pp. 7–31.

Loo, C. (1972). The effects of spatial density on the social behavior of children. *Journal of Applied Social Psychology*, **2**, 372–81.

Loo, C. and Smetand, J. (1977). The effect of crowding on the behaviors and perceptions of 10-year old boys. *Environmental Psychology and Nonverbal Behavior*, **2**, 226–49.

Loo, C. (1977). The consequences of crowding on children. In: M. Gurbaynak and W. A. LeCompte (eds), *Human Consequences of Crowding*. NATO series no. 3, New York: Plenum, pp. 94–113.

Marshall, J. and Heslin, R. (1975). Boys and girls together: sexual composition and the effect of density and group size on cohesiveness. *Journal of Personality and Social Psychology*, **31**, 952–61.

McGrew, P. L. (1970) Social and spatial density effects on spacing behavior in preschool children. *Journal of Child Psychology and Psychiatry*, **11**, 197–205.

Murray, R. (1974). The influence of crowding on children's behavior. In: Canter, D. and Lee, T. (eds), *Psychology and the Built Environment*, New York: Halstead Press.

Rapoport, A. Toward a redefinition of density. *Environment and Behavior*, **7**, 133–58.

Robinson, P. W. (1981). *Fundamental of experimental psychology*. Englewood Cliffs N.I.: Prentice-Hall.

Rodin, J. (1976). Density, perceived choice and response to controllable and uncontrollable outcomes. *Journal of Experimental Social Psychology*, **12**, 564–78.

Saegert, S. (1980). *The effects of residential density on low-income children*. Paper presented at the APA annual meeting, Montreal, Canada.

van Staden, F. J. (1983). From crowding to density: conceptual developments, methods and research. *South African Journal of Psychology*, **13**, 128–34.

Sundstrom, E. (1978). Crowding as a sequential process: review of research on the effects of population density on humans. In: A. Baum and Y. M. Epstein (eds), *Human Response to Crowding*. Hillsdale New Jersey: Lawrende Erlbaum.

Wolfe, M. (1975). Room size and density: behavior patterns in a children's psychiatric facility. *Environment and Behavior*, **7**, 199–225.

THE EMERGENCE OF ADOLESCENT TERRITORIES IN A LARGE URBAN LEISURE ENVIRONMENT[1]

JOHN L. COTTERELL

Department of Education, University of Queensland,
St Lucia, Brisbane, Australia 4072

Abstract

This study explored person–environment fit relationships between age and gender characteristics of adolescents and their usage of leisure places within an unusual leisure environment, World Expo '88, and the extent to which adolescent territories emerged as a result of concentrated usage of some places by adolescents. Data from a survey of 465 young people and from interviews with a subsample of these, revealed distinct concentrations of adolescents in entertainment-related places, and in pavilions which adopted 'dynamic' rather than static displays. Interview data further suggest that the places preferred by adolescent visitors had a distinctive atmosphere associated with fun, excitement and relaxation, and allowed them room for personal expression. This cognition of particular leisure places as having a distinctive personal 'feel' seems to contribute, along with frequent usage, to the development of adolescent territories.

Introduction

Despite the growth of theme parks and nature parks as major leisure venues, research on the impact of these large-scale leisure environments on visitors is sparse (Pearce & Moscardo, 1985), particularly investigations of distinctive sub-environments within them and corresponding patterns of visitor responses. This study examined the impact of Expo '88, the special world fair held in Australia in its bicentennial year. It specifically sought to identify places within this urban leisure environment where adolescent visits were concentrated, and then to describe the features of these favoured places and how they functioned as adolescent leisure territories. The term 'territory' is used in the sense of familiar places, similar to the sense of neighbourhood possessed by children (Bryant, 1985) and adults (Stanton, 1986), but derived in the present study from group behavior rather than from individual perceptions.

The modern Expo traces its lineage back to the London exposition of 1851, and has always featured the technological achievements of exhibiting countries. Australia's Expo '88, while focusing on technology, took leisure as the special theme which all national and corporate exhibits had to address in their displays, on a site a mere half-mile from the heart of metropolitan Brisbane. The combination of a leisure theme, the national bicentennial euphoria, and an excellent physical location, provided the conditions for creating a modern leisure park which could be easily integrated with the existing urban services and recreational facilities, and readily accessible to the vast majority of metropolitan residents.

This study deals with the natural environments of adolescents and the problem of people-place relationships, a topic pursued in the classic work of the Muchows in the early 1930s (Muchow & Muchow, 1935; Wohlwill, 1985), where site-specific surveys were undertaken of the places frequented by children. The difficulty faced by researchers interested in characterizing the larger-scale environments in which children live and play (on the scale, for example of neighbourhoods) is finding a method of maintaining focus on both environment and behavior phenomena. The primacy of individual concerns among social psychologists has meant that subjects' mental maps and cognitive processes have received more attention than the environmental surround of molar behavior itself (e.g. Evans *et al.*, 1981; Stanton, 1986). The literature on adolescent leisure, for example, provides little information on the environments in which it occurs (see Kruse & Arlt, 1984), the main focus being on *leisure-time activities*, and as Silbereisen *et al.*, (1986, p. 89) have noted, 'no conclusion as to where these activities actually recur can be drawn'.

In recent years, several authors (e.g. Kaminski, 1986; Sime, 1986; Wapner, 1987) have voiced concern about the continuing gulf between psychological and physical or architectural approaches to the study of human responses to the everyday environment. Sime (1986) for example, argues that human behavior in relation to the physical built environment has been neglected as a topic of study by both architecture (which has focused on the physical dimensions of space and form) and psychology (in its concern with inner psychological processes). He advocates the use of the term 'place' as a means of focusing (architects') attention beyond geographical space and upon people's actions in, and their experience with, particular landscapes. In a similar vein, Wapner (1987) urges the adoption of a 'holistic' approach in environmental research, which captures the transactions of people with the environmental system in 'its physical, social, and interpersonal aspects'.

Kaminski (1986) has drawn attention to the 'differences in grain' among studies of behavior–environment relationships, noting that in broad-focused studies of social environments, the grain must necessarily be coarser than in those concerned with the goals and experiences of individual inhabitants. Studies of entire populations of small communities (e.g. Barker & Schoggen, 1973) have employed behavior settings as naturally occurring 'empirical entities of the ecological environment' (Barker, 1983, p. 173). While these units are 'coarse-grained', they are useful on their own terms for describing the texture of the environment; i.e. 'as an eco-behavioral and not as an individual-biased discipline' (Fuhrer, 1986, p. 367). Other units are needed, however, to describe the ecological environments of kindergartens, playgrounds, or leisure parks. Moore (1986) employed 'activity-pockets' as units of the environment of children's kindergartens, and Gump (1974) and Cotterell (1984) used 'activity-formats' to segment the stream of classroom activity. On a broader scale, the geographer Hagerstrand (1975) devised the concept of 'domain' as a spatial unit of no fixed size, in which people perform roles and exercise responsibilities, and from which a complete space-time ecology could be constructed. Studies of children's territories have examined their 'home range' in terms of distances roamed (Hart, 1978); Van Vliet (1983) has employed a similar technique to study adolescents in Toronto.

In the present study 'place' is used in the sense discussed by Canter (1977) and recently revived by Sime (1986) as a unit having 'geographical, architectural, and social connotations' (Canter, 1977, p. 6), containing distinctive events and patterns of activity, and existing (like behavior settings) independently of the observer. Silbereisen *et al.* (1986) examined adolescent leisure behavior in Berlin in terms of selected leisure places such as discos, shopping malls and swimming pools, and observed the flow of behavior in relation to component situations. The present study uses person–environment fit (Kaplan, 1983) as a guiding principle to focus on adolescent leisure patterns within a large-scale leisure environment and on the extent to which territories may emerge as a result of concentrated usage of some places by adolescents.

Method

Expo environment

The Expo '88 site occupied 44 hectares (109 acres) in a long rectangular shape adjoining the Brisbane River. Figure 1 shows its location in relation to the city, and emphasizes the compact nature of the site, with open spaces and display pavilions shaded by giant tentlike polyester sails. The site contained 62 national and corporate pavilions, 30 restaurants, and over 50 shops, as well as a mixture of natural environments, from lagoons and waterfalls to geysers and rainforests. Adjoining the site was the Funpark, a five-hectare futuristic amusement park containing exciting roller coaster rides and amusements. Expo operated a single ticket multiple-entry system comprising three-day passes and season passes in addition to day passes, thus allowing access to all attractions at no charge, with the exception of the Funpark, where access was unrestricted but rides carried an additional cost.

The Expo site contained national displays from every continent except South America, staffed by people from those countries. Many of the national pavilions contained restaurants and eateries serving foods from the country concerned, and in addition performed cultural displays; e.g. the New Zealand pavilion staged regular singing displays by Maoris outside the pavilion, and the Chinese pavilion contained a section where skilled artists and craftspeople worked.

A non-stop free entertainment program was scheduled for each day of the six month duration of Expo, comprising a total of 25,000 events, ranging from formal stage productions to informal ones, some with a larrikin style. Simultaneous programs of entertainment occurred in several distinct locations: on the traditional stages of the Amphitheatre and the Riverstage, in the Piazza's theatre-in-the-round, on the streets (for daily midday and evening parades, and wandering groups of street entertainers), and in the water of the Aquacade's diving pool and on the Brisbane river itself (for daily water-ski shows). In addition, the national pavilions hosted dancing and singing shows. Nightly features included a laser light show which concluded with a fireworks display. All national days of exhibiting countries were celebrated with special events, and the six-month calendar of Expo was divided into theme weeks which tied in as much as possible with these (e.g. 'True Blue' week, which coincided with Australia bicentennial celebrations). Other theme weeks included 'Flights of Fantasy' week, focusing on 'anything aviational'; 'Music Music Music' week, featuring brass bands, youth orchestras, and rock groups; and 'Come Dancing' week, with dancing from ballroom to ballet.

Sample

Subjects were students between 13 and 19 years of age, drawn from two suburban high schools, and from two colleges adjacent to the Expo location—a private girls' school and a technical college. There was a wide socio-economic mix: students came from working-class and middle-class suburbs, ranging in distance from four to 30 kilometres (two to 18 miles) from the site of Expo. Prior to the beginning of Expo the students were surveyed in class groups at their school or college about their general leisure activities

FIGURE 1. Low oblique aerial photograph of the Expo '88 site and adjoining areas of Brisbane city.

and their plans to attend Expo. At this stage, 84% said that they planned to visit Expo, and 74% of the sample had already obtained a pass. Most of these passes (71%) were season passes. Approximately three months later (midway through the Expo period) they were surveyed at school again, concerning their visits to a detailed list of Expo places. Of the 533 students surveyed at follow-up, 51 students had not attended Expo by that time, and some cases had missing data, leaving a sample of 465 students. The ratio of females to males was almost 3:1.

In addition, telephone interviews were conducted with a small subgroup of the sample who were identified from the Expo visits survey as being frequent attenders relative to others in their particular school ($N = 82$). The ratio of girls to boys in this group was 6:1. Interview respondents were asked a set of questions about the timing of their visits to Expo and how they planned these out, the places which they preferred at Expo, and their emotional reactions to these places. Of particular interest in the present paper were youth responses to two broad questions (each of which was further explored): (1) Is each visit different from the previous one or is it much the same? and (2) Are there some parts of Expo that you find you keep going back to over and over again?

This study relied on self-reports of behavior, rather than actual setting counts, to investigate people–place relationships. In the preliminary phase of the study, some behavior tracking of randomly chosen targets was attempted, and setting counts of a few sites were undertaken, but the heavy crowding (average daily attendances were about 90,000), queuing arrangements, and slow rates of movement around the site, frustrated the use of systematic observation, given the limited resources available for the study. Unfortunately, no large class of University students was available for setting counts over the time required, and management authorities controlling Expo were unwilling to supply data on pavilion attendances, despite repeated requests for assistance.

Places
The survey and interviews recognized 108 places within the Expo environment, ranging in size from large and densely populated ones like the Funpark and the Riverstage to small pavilions like Kenya and the Australian Opal Mine. Shops and street stalls were not included in the survey, although restaurants (but not small takeaway stalls) were listed. Students checked whether they had visited the particular place at all, and then wrote an estimate of how many visits they had made.

In order to compare the appeal of different types of pavilions to teenage visitors, pavilion places were classified into eight groups, from categories developed by the author prior to the collection of the data, and based on the observation of Expo settings. Pavilions were classified on the basis of publicly available information which briefly described each pavilion display. The coding had two dimensions: *theme or content emphasized* (tourism, traditional culture, science and technology, physical recreation), and *mode of presentation* (static display of objects and artifacts or dynamic display, using live shows and high-tech presentations). The pavilions featuring traditional culture usually adopted a static mode of display, (for example, traditional carvings in the Nepalese pavilion) but some included live dancing, laser images, robots and multi-media film presentations (for example, the Japanese pavilion). Among the pavilions featuring sport, some displayed sports equipment and sporting achievements (static) while others (like the U.S. pavilion) in addition encouraged skills testing and gave demonstrations of new team sports activities (dynamic). It is worth noting that this static/dynamic distinction in mode of presentation was related to variety and audience participation, but was not necessarily a distinction in terms of vividness; i.e. between novel and unusual displays on the one hand and mundane ones on the other. The national pavilions, for example, displayed those cultural and geographical features which were seen by their governments to reflect their country's national character, and pavilions which differed in display mode were often very similar in their degree of vividness and exoticness.

Coding of pavilion types was undertaken by two coders blind to the goals of the study. The coding procedure was as follows: first, each coder independently examined the official descriptions of each pavilion and coded the details into two categories—theme and mode; second, where either coder felt that the official description had omitted details relevant to the two categories, he or she added these to the listed descriptions and verified the details by conferring with the other coder, and where necessary by visiting the pavilion; third, each coder independently scored the pavilions according to each type and mode (i.e. six scores) for the number of distinctive features of the display listed in the publicized information. For example, the Canadian information included the following: 'The exhibit is an *interactive theatre* (mode: dynamic) designed to appeal to a family audience. During each 17 minute performance, the theatre computer provides the audience with descriptions of *adventures* (content/theme: tourism) awaiting them, and invites them to choose the adventure that interests them most'. Checks on coding procedures showed very few instances of disagreement, and differences where they occurred, were resolved by negotiation. Pavilions were each given six scores (one for each theme and mode) by the coders, and on the basis of their category scores, were assigned to types by the author, using median-split procedures. Sixty of the 62 pavilions at Expo were able to be classified into one of the eight categories.

Results

Attendance patterns

Participation rates of male and female adolescents were calculated for each of the 108 Expo places surveyed. These rates show the proportion of the group which visited a particular place at least once, and are reported for major places in Table 1.

Inspection of the figures in Table 1 shows that the most popular places were the entertainment areas: the Riverstage, Funpark, Piazza, Amphitheatre, Aquacade, and the Fireworks. Other large-area spaces which had wide appeal to young people were the Pacific Lagoon and the Boardwalk. Comparison of the rates for girls and boys reveals that the majority of Expo places were better attended by girls, being particularly marked in respect of historical and religious exhibits (e.g. Magna Carta, Vatican, Pavilion of Promise). Boys achieved higher participation rates where displays focused on sport and motors (e.g. U.S.A., Ford Motors). Technology exhibits held an appeal for both sexes (e.g. the Technoplaza, Fujitsu Electronics).

In addition, the distribution of reported setting attendances of adolescents across the list of 108 Expo places was plotted on a map of the site, yielding the generalized density patterns shown in Figure 2. What is clear from Figure 2 is the concentration of adolescent visits into distinct sectors of the Expo site, particularly those large-area places characteristically associated with fun and entertainment. By contrast, the pavilions achieve only medium density levels, and these are scattered across the whole

TABLE 1

Adolescent participation rates at selected Expo places

Place	Girls (percent) (N = 334)	Boys (percent) (N = 131)
Entertainments		
Amphitheatre	65·3	55·7
Aquacade	68·0	46·6
Boardwalk	73·1	57·3
Fireworks	79·9	67·2
Pacific lagoon	62·6	61·1
Day and night parades	63·2	48·1
Piazza	70·1	55·0
Riverstage	88·0	77·1
Rock Ski water-ski	58·5	48·9
Funpark	86·8	76·3
Waterfalls area	47·0	46·1
National pavilions		
Britain	37·4	38·2
Canada	61·7	49·6
France	53·6	46·1
West Germany	41·0	33·0
Republic of China	54·2	45·0
Japan	74·6	54·2
Kenya	34·5	22·9
Republic of Korea	42·4	35·0
Malaysia	53·0	34·8
New Zealand	36·2	32·1
Papua New Guinea	36·5	22·9
Queensland	75·7	68·7
South Pacific Islands	50·9	40·5
Switzerland	67·1	56·5
U.S.A.	62·0	71·8
U.S.S.R.	49·1	48·1
Corporate pavilions		
Cook Maritime Museum	20·4	20·6
Ford Motors (Australia)	15·9	30·5
Fujitsu Electronics	52·4	45·8
IBM	39·8	38·2
Magna Carta	32·0	11·5
Pavilion of Promise	63·8	19·8
Queensland Newspapers	49·7	37·8
Suncorp Insurance	33·0	34·8
Technoplaza	60·2	43·5
TV0 Network 10	52·2	40·4
Univations (universities)	35·0	26·4
Vatican (Holy See)	23·1	14·5
Restaurants and taverns		
American Village Restaurant	28·7	17·1
Britannia Inn	12·0	8·7
French Pavilion Restaurant	28·2	10·5
Galaxy Restaurant	37·7	19·5
Munich Festhaus	23·2	20·1
Ship Inn Tavern	11·7	12·9

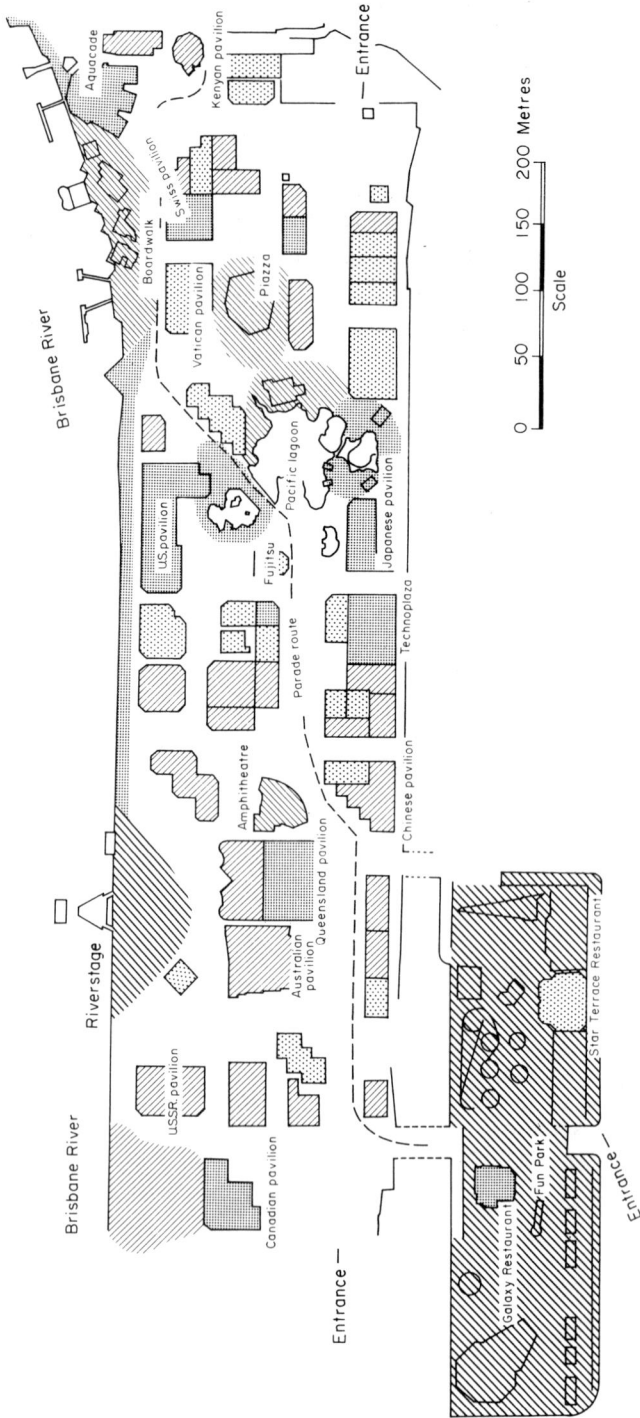

FIGURE 2. Patterns of youth attendance of Expo attractions, May to July 1988. Key of number of visits: ▨ = 1500–3500; ▤ = 750–1499; ▥ = 400–749; ▦ = 200–399; ▦ = 1–199.

TABLE 2
Places listed by interview respondents as ones which were visited 'over and over'

Place	Percent of girls	Percent of boys	Total percent
Funpark	53·6	61·3	54·9
Riverstage	34·8	38·5	33·4
Canadian pavilion	28·9	46·2	31·7
Queensland pavilion	20·3	7·7	18·3
Aquacade diving show	18·8	15·4	18·3
Boardwalk	15·9	7·7	14·6
Piazza	13·0	9·7	12·2
U.S.A. pavilion	5·8	30·8	9·8
Japanese pavilion	10·1	0	8·5
Pacific lagoon	10·1	0	8·5

Expo area, from the Canadian pavilion in the West to the Swiss pavilion and the Cadbury pavilion in the East. While the entertainment areas appear to draw adolescents to the fringes of the Expo site, the attendance data fall into no distinct zonal pattern in respect of pavilion visits. There are no distinct contours of density across the site to suggest the influence of local environmental factors (apart from the zone of open space which linked the Piazza and Lagoon); instead, favored places sit alongside places which appear almost ignored. Since the site was almost level (apart from the Funpark, which was level but elevated) no local geographical factors appeared to be operating to limit the accessibility to some settings.

Concentrations of adolescent attendances at particular Expo places were also estimated from the interview records, which contained comments about the places revisited 'over and over'. These results, shown in Table 2, confirm the findings displayed in Figure 2, and show that the most frequently visited areas at Expo were the Funpark and the Riverstage, and that the pavilion displays of Canada, Queensland, U.S.A. and Japan were the most likely to merit return visits. Thus on empirical grounds there is evidence to suggest that particular sectors of the Expo environment functioned as territories, in the sense that young people assembled in those places in large concentrations and made frequent return visits there.

Sense of place
The use of the term 'territory' in this paper is unrelated to the concept of territoriality, but akin to the sense of 'place' discussed by Canter (1977), whereby familiarity from use of a geographical space generates an image of its distinctive characteristics or atmosphere and a sense of belonging to that place, and thus a desire to return to the familiar 'haunt'. Canter noted that a sense of place emerges from reconciling knowledge about the activities associated with a place and about the physical features of the setting with conceptions people hold of the behavior appropriate for that setting. Recently Korpela (1989) has explored the relationship between children's needs and their resort to specific characteristics of favourite places to meet these needs. The present study examined the comments made by interview respondents as these related to features of places at Expo which met individual needs for freedom, relaxation and pleasure. Thus territory is seen in this section as becoming established through the fit between environmental features and individual feelings and needs.

FIGURE 3. Low oblique aerial photograph of Riverstage and Funpark.

The Riverstage and the Funpark (Figure 3) were places distinctly associated with youth visitors. Respondents clearly recognized a different 'feel' towards these places when attempting to explain why they kept going back to them; one girl observed that 'the Riverstage always gets you in a good mood, whereas pavilions leave you in a quiet mood, and the Funpark makes you more outgoing'. Another respondent contrasted these places with the pavilions in terms of their variety: 'because they change all the time'.

The Riverstage was a place of 'high-tech' excitement, which featured nightly entertainment events ranging from opera and dance to country music and magician shows. Each Friday night, a rock concert or a major overseas musical performance was staged, attended by crowds of up to 12,000, seated on the sloping lawn of the river bank, with the river as background. The appeal of the area lay not only in its programme, which 'offered something out of the ordinary', but also in its physical attributes, 'so open and spacious' which communicated a 'relaxed, no pressure' feeling. It served as a familiar place where people could meet friends, sit in the open air by the river, and enjoy the variety of live concerts.

The appeal of the Funpark could be attributed to its exciting rides, the lighting, and the spacious surrounds, which encouraged young people to go there when they 'were in a lively mood' to take the rides or watch the floorshows at the Galaxies and Star Terrace Restaurants. Those interviewed referred to the Funpark as a place 'to go with friends, go on rides, and get things to eat', as 'a good meeting place', a place 'to muck around, go on rides, watch people'. They described it as 'exciting, action-packed, with a variety of rides', a 'place where I never get bored'. Social contacts were paramount: 'where everyone talks to everyone', 'the place where we mainly hang out because of the thrills of the rides and because most younger people go there', where 'we know everyone, sit down, talk and just hang out'. Observations undertaken in the Funpark by members of the research team found large numbers (up to 1000) adolescents present at the one time, very few of whom were in the company of adults. Apart from those who were seated in eating areas or standing in queues at ride locations, there were continual bursts of movement as groups walked from one ride point to another.

Pavilions

The pavilions were associated with a different tempo from that of the entertainment areas, namely quieter, more individual (less social) and more educational. The contrasts are illustrated in remarks like: 'When I go to the Funpark I'm excited: when I'm visiting pavilions I get more relaxed'; 'during the day we see pavilions and at night it's more exciting—we see concerts and the Funpark'; 'sometimes we go to see pavilions and other days we go to the Funpark and Boardwalk'; 'my family's attitude to Expo (is) it's educational, and so we go and see pavilions; with friends it's more relaxing and we just go to have fun'. These remarks, volunteered by respondents to explain different visiting patterns, reflect the influences of mood, time of day and group membership; but what they also suggest is the extent to which places within Expo were differentiated by young people in terms of atmosphere and function. Moreover, comments made in the interviews suggest that particular features of the pavilions themselves (e.g. the fitness test machines in the Canadian pavilion, the 'people-mover' in the Queensland pavilion, the ski slope in the Swiss pavilion) may have elicited the different rates of patronage.

From the perspective of person–environment fit, differences in adolescent attendance at pavilions should be related to differences in the stimulus characteristics of the pavilion settings. In order to test for differences in person–environment fit according to age-typic and/or gender-typic preferences for particular types of pavilion or kinds of pavilion display, the attendance rates reported[2] by younger and older male and female adolescents were analysed using loglinear/logistic analysis (Fienberg, 1980). Distribution statistics of attendance at each pavilion type, for each age and gender group, are displayed in Table 3, and results of the loglinear analysis follow in Tables 4a and 4b.

TABLE 3

The attendance (percent) across eight categories of mode of display within pavilion type for four populations defined by gender and two age groups

	Tourism		Traditional culture		Science/ technology		Sport	
Pavilion type:	Dynamic	Static	Dynamic	Static	Dynamic	Static	Dynamic	Static
Frequency of attendance								
Male ≤15	178	150	148	162	147	175	175	39
Male >15	302	177	219	225	257	232	252	49
Female ≤15	920	653	898	818	782	663	643	176
Female >15	792	600	763	707	583	630	616	167
Distribution of attendance								
Male ≤15	15·18	12·78	12·61	13·80	12·52	14·91	14·91	3·32
Male >15	17·63	10·33	12·78	13·13	15·00	13·54	14·71	2·86
Female ≤15	16·57	11·76	16·17	14·73	14·08	11·94	11·58	3·17
Female >15	16·30	12·35	13·71	14·55	12·00	12·97	12·68	3·44

TABLE 4(a)

Log-linear/logistic analysis of the distribution of atten-dance for eight mode within type of pavilion categories by age and sex

Source	DF	χ^2	Probability (p)
T[a]	3	451·57	<0·001
M[b](T)	4	532·51	<0·001
M(T1)	1	65·61	<0·001
M(T2)	1	0·10	0·75
M(T3)	1	0·01	92
M(T4)	1	466·79	<0·001
T·S[c]	3	20·20	<0·001
M(T)·S	4	8·37	0·08
T·A[d]	3	0·15	0·98
T·S·A	3	2·87	0·41
M(T)·S·A	4	17·37	0·002
M(T1)·S·A	1	6·85	0·009
M(T2)·S·A	1	0·24	0·62
M(T3)·S·A	1	10·06	0·002
M(T4)·S·A	1	0·23	0·63

Notes: [a]T = Type of pavilion; T1, tourism; T2, traditional culture; T3, science/technology; T4, sport.
[b]M = Mode.
[c]S = Sex.
[d]A = Age group.

Table 3 shows that raw frequencies of attendance by girls greatly exceeded those by boys, reflecting the larger numbers of females in the sample. However, when proportions are considered (Distribution of attendance), the percentages of visits reported by each adolescent population fell within the range of 10 to 18%, with the exception of some pavilions featuring sport, which had lower attendance rates.

Results of the loglinear analysis show main effects for pavilion type and mode of display, and interaction effects between pavilion type and gender, and between display mode, gender and age level. No main effects for age or for gender were found, nor were age and pavilion-type interactions detected. Inspection of Table 4(a) reveals that the display-mode effect is located in two types of pavilion—tourism and sport, while gender × age interactions with display mode occur in tourism and also in science/technology settings.

The nature of the interactions detected in the main analysis (shown in Table 4(a)) may be understood by examining sex and age effects within pavilion type and display mode, and this is shown in Table 4(b). Again, no distinct age effect is apparent. Noteworthy is the siting of significant effects at the $p < 0.002$ level within the dynamic mode of display, in traditional culture and sport pavilions (sex effect) and in science/technology (sex × age effect).

The direction of the various effects reported above can be appreciated from examination of Figure 4, which shows display-mode differences favouring dynamic over static display in the tourism and sport pavilions, while *within* pavilions in the dynamic display mode, girls exceed boys in attendance at traditional culture settings

TABLE 4(b)

Logistic analysis of the significance of age and sex on attendance within each of eight mode within pavilion type categories

Pavilion type:		Tourism		Traditional culture		Science/technology		Sport	
Mode:		Dynamic	Static	Dynamic	Static	Dynamic	Static	Dynamic	Static
Source	DF	χ^2 Probability (p)	χ^2 Probability (p)	χ^2 Probability (p)	χ^2 Probability (p)	χ^2 Probability (p)	χ^2 Probability (p)	χ^2 Probability (p)	χ^2 Probability (p)
Sex	1	0·03 0·85	1·05 0·30	17·68 <0·001	2·65 0·10	2·04 0·15	5·19 0·02	13·47 <0·001	0·48 0·49
Age	1	0·24 0·62	0·01 0·93	0·28 0·60	0·21 0·65	3·54 0·06	0·80 0·37	2·03 0·15	0·14 0·71
Sex·age	1	2·96 0·08	4·97 0·03	0·16 0·69	0·12 0·73	9·80 0·002	2·79 0·09	0·94 0·33	0·95 0·33

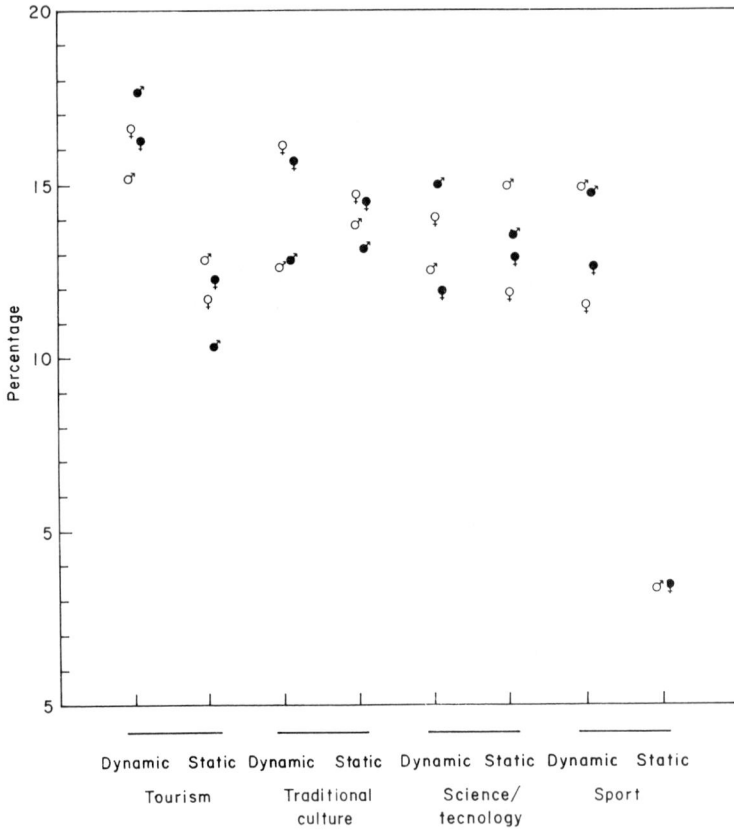

FIGURE 4. The distribution of attendance across eight categories of mode of display within pavilion type for four populations defined by gender and two age groups. (The open symbols refer to the younger age group, and the closed symbols to the older age group.)

while boys exceed girls in attendance at sport. These patterns reflect traditional gender-typic leisure interests, which may also have been expected to show up in the science/technology area. However, the picture here is complicated by an interaction with age, such that younger girls and older boys outscored their same-sex counterparts.

The tendency for adolescents to prefer pavilion displays which featured dynamic rather than static modes of presenting their themes was strongly borne out in the interviews. When interviewees were asked to nominate the places they went back to 'over and over' there was a strong tendency for respondents to choose dynamic forms of display; of the 85 citations made, 75 of these were for pavilions in the dynamic mode. Comments offered spontaneously in interviews capture the appeal of dynamic displays; for example: 'The 3-D movie (Fujitsu)'; 'the skiing (Switzerland); 'the fitness test—we go back to see if we've improved' (Canada); 'I like the competitive games demos' (U.S.A.); 'different things are happening all the time' (Japan); 'the people-mover and the men with video-screen faces' (Queensland).

Person–environment fit was also evident in the tendency for interviewees to express a preference for leisure places which were compatible with their regular leisure patterns. For example, comparison of the places named by respondents as regular local haunts with the pavilions they frequently visited at Expo revealed that those who were sport 'types' made frequent visits to the U.S.A. and Canada pavilions, which were sports-oriented; those who went regularly to nightclubs and restaurants in their leisure time listed the Boardwalk and the Riverstage as their favourite places; and those who liked to go into town to shop, go to the movies, or to hang out in the mall listed the Funpark and the Riverstage as key places. Adolescents also appeared to transfer their leisure activities from their regular 'haunts' to an area in Expo; for example, statements like these occurred in interviews: 'Last Friday night, we were going to see a movie, but decided we'd go to Expo instead'; 'We usually start off in town (but) Expo is easier to hang around at than town'. Despite their impressionistic nature, these findings are consistent with the general pattern of results reported here.

Discussion

This paper investigated the utilization of places within an urban leisure environment from the viewpoint of person–environment fit. It specifically examined whether adolescents who visited particular leisure places were likely to do so in terms of the possible satisfaction of gender- and age-typic needs, and discussed such patterns of setting use as marking adolescent territories. Results from both the survey and interview material suggest that the places most favored by adolescents within the larger areas of Expo, at least in terms of reported frequency of use, were those which allowed youth room to congregate and to interact, while they also enjoyed musical entertainment, the excitement of rides, or displays of skill. Apart from these larger areas, the national and corporate pavilions also contained distinctive sub-environments with their own appeal to adolescents. Differences in visiting rates to pavilions were found to relate to the degree to which displays portrayed information in a dynamic or interactive mode, and to the extent to which displays invited active and participatory forms of audience response.

Moreover, it seems that certain places within a leisure park communicate distinctive atmospheres to adolescents which can form the emotional basis for the establishment of adolescent territories. Given the interconnection of places within a leisure park, a consequence of continual visits to Expo was that young people had an opportunity to sample and evaluate the suitability of a range of environments, incidental to their overt leisure goals at the time. There was evidence that adolescents in this study were conscious of differences in mood and tempo within the Expo environment, and that their choice of a particular place to visit was based on an assumption that its atmosphere was compatible with their needs at the time.

These results should be regarded as exploratory until more work of this kind is carried out, but they indicate that the concentrations of adolescents in particular sectors of the Expo environment can be understood in terms of some kind of fit between the leisure needs and interests of adolescents and the stimulus characteristics of the places involved. Leisure places appear to 'go with' particular needs and leisure styles, allowing adolescents to establish territories within Expo, based upon their own conceptions of a particular place and how it should function, which they had derived from experiences elsewhere. However, no age-typic patterns were identified within the adolescent population. It is likely that a wider age-range, which included young adults and older adults, would highlight specifically adolescent leisure preferences.

The findings of this study leave many questions about behavior–environment relations unanswered, while hopefully pointing to an important group of environments deserving of study by social scientists. A limitation of this study is that, in treating adolescents as a group and focusing on molar activities, systematic examination was not made into the relationships between individual leisure preferences (based on general recreation patterns) and patterns of usage of places in a leisure environment. For example, it is possible that distinctive leisure styles exist among young people, which are related to their rate of utilization of leisure parks (Roberts, 1983).

There are also limitations which should be acknowledged with regard to the characterization of environments. Owing to its coarse-grained viewpoint, the study did not examine in detail the particular distinguishing features of the leisure places themselves as operating environments (e.g. the prevailing activities, the age distribution of inhabitants, and differentiation of roles), which could shed further information on the component situations within them (e.g. Silbereisen *et al.*, 1986). Studies using a more fine-grained approach could examine systematically the behavioral ecology of those leisure places which adolescents prefer, and compare them with places preferred by other age groups but unpopular with adolescents, along the lines developed by Sommer *et al.* (1981) in their study of supermarkets. Such a study would yield valuable insights into the attributes of leisure attractions which draw repeated visits from adolescents, and into the way these places become incorporated into their familiar territory, and contribute to our understanding of the fit between person and environment. To the extent that we have entered the era of the leisured society (Darton, 1986), knowledge of the appeal of leisure places, particularly those contained in capital-intensive leisure parks, is of importance for environmental planners, to the end that physical spaces become *designed places* (Sime, 1986) which incorporate rich insights about the behavior of potential inhabitants.

Acknowledgements

Thanks are extended to Phil Schoggen for helpful comments on an earlier version of this paper. I also wish to thank Dr David Chant for his assistance with statistical analysis, and Hugh Stewart-Killick, cartographer in the Department of Geography, for his preparation of the map in this article. Grateful acknowledgement is also given to the Commanding Officer, Amberley Airforce Base, Australian Department of Defence, for supplying the aerial photographs, and the Editor of the *Brisbane Courier/Mail*.

Notes

(1) This research was supported in part from a grant to the author from the Brisbane City Council.
(2) It should be remembered that the data are based on reported attendances rather than actual setting counts, and are subject to memory distortions, which may favor recall of vivid over less vivid experiences. However, all pavilion settings may be regarded as unusual and to that extent vivid and novel rather than everyday, in that they were staffed by nationals, and contained authentic national artifacts. Moreover the juxtaposition of different national and corporate pavilions increased perceptual contrast.

References

Barker, R. G. (1983). Comments: will the enigma go away? (Discussion of 'The Enigma of Ecological Psychology' by G. Kaminski). *Journal of Environmental Psychology*, **3**, 173–174.

Barker, R. G. & Schoggen, P. (1973). *Qualities of Community Life*. San Francisco: Jossey-Bass.

Bryant, B. K. (1985). The neighborhood walk: sources of support in middle childhood. *Monographs of the Society for Research in Child Development* (Serial No. 210).

Canter, D. (1977). *The Psychology of Place*. London: The Architectural Press.

Cotterell, J. L. (1984). Effects of architectural design on student and teacher anxiety. *Environment and Behavior*, **16**, 455–479.

Darton, J. (1986). The leisured society. *Leisure Management*, **6**, 7–8.

Evans, G. W., Marrero, D. G. & Butler, P. A. (1981). Environmental learning and cognitive mapping. *Environment and Behavior*, **13**, 83–104.

Fienberg, S. E. (1980). *The Analysis of Cross-Classified Categorical Data*, 2nd edit. Cambridge, MA: MIT Press.

Fuhrer, U. (1986). Beyond the behavior setting. *Journal of Environmental Psychology*, **6**, 359–369.

Gump, P. V. (1974). Operating environments in open and traditional schools. *The School Review*, **84**, 575–593.

Hagerstrand, T. (1975). Space, time, and human conditions. In A. Karlqvist, L. Lundqvist & F. Snickars, Eds., *Dynamic Allocation of Urban Space*. Farnborough: Saxon House, pp. 3–12.

Hart, R. A. (1978). *Children's Experience of Place: A Developmental Study*. New York: Irvington Press.

Kaminski, G. (Ed.). (1986). *Ordnung and Variabilitat im Alltagsgeschehen* (Order and variability in everyday happenings). Gottingen, Toronto: C. J. Hogrefe.

Kaplan, S. (1983). A model of person-environment compatibility. *Environment and Behavior*, **15**, 311–332.

Korpela, K. M. (1989). Place-identity as a product of environmental self-regulation. *Journal of Environmental Psychology*, **9**, 241–256.

Kruse, L. & Arlt, R. (1984). *Environment and Behavior*. Munchen: Saur, vols. 1–2.

Moore, G. T. (1986). Effects of the spatial definition of behavior settings on children's behavior: A quasi-experimental field study. *Journal of Environmental Psychology*, **6**, 205–211.

Muchow, M. & Muchow, H. (1935). *Der Lebensraum des Grossstadtkindes*. Hamburg: Reigel.

Pearce, P. L. & Moscardo, G. (1985). Tourist theme parks: research practices and possibilities. *Australian Psychologist*, **20**, 303–312.

Roberts, K. (1983). *Youth and Leisure*. London: Allen and Unwin.

Silbereisen, R. K., Noack, P. & Eyferth, K. (1986). Place for development: adolescents, leisure settings, and developmental tasks. In R. K. Silbereisen & K. Eyferth, Eds., *Development as Action in Context: Problem Behavior and Normal Youth Development*. New York: Springer-Verlag, pp. 87–107.

Sime, J. D. (1986). Creating places or designing spaces? *Journal of Environmental Psychology*, **6**, 49–63.

Sommer, R., Herrick, J. & Sommer, T. R. (1981). The behavioral ecology of supermarkets and farmers' markets. *Journal of Environmental Psychology*, **1**, 13–19.

Stanton, B. H. (1986). The incidence of home grounds and experiential networks. *Environment and Behavior*, **18**, 299–329.

Van Vliet, W. (1983). Exploring the fourth environment: An examination of the home range of city and suburban teenagers. *Environment and Behavior*, **15**, 567–588.

Wapner, S. (1987). 1970–1972: Years of transition. *Journal of Environmental Psychology*, **7**, 389–408.

Wohlwill, J. F. (1985). Martha Muchow and the life space of the urban child. *Human Development*, **28**, 200–209.

CHILDREN'S LANDSCAPE PREFERENCES: FROM REJECTION TO ATTRACTION

FERNANDO G. BERNÁLDEZ*, DOLORES GALLARDO**, and ROSA P. ABELLÓ*

Departamento de Ecología, Universidad Autónoma, Madrid and *Facultad de Bellas Artes. Universidad de La Laguna, Tenerife***

Abstract

Multivariate analysis of the preference responses of 483 children to landscape photographs allowed the identification of three independent preference dimensions: the 1st and 3rd dimensions (illuminated vs shadowed; rough, harsh vs bland, smooth texture or relief) were considered as forms of a more general 'risk, uncertainty factor' often influencing landscape preference. Younger children (11 years old) showed less preference for both shadowed, less illuminated scenes (1st dimension) and harsh, rough scenes with aggressive forms (3rd dimension) than older children (16 years old). There were no significant differences for the 2nd dimension (landscape diversity).

Introduction

Despite its importance in environmental design and education (Bernáldez, Benayas and De Lucio, 1987) the study of children's landscape preferences is a relatively unexplored subject. Zube, Pitt and Evans (1983) highlight the scarcity of research in this topic and report on the importance of water as a scenic component for children's preferences and the differences between children's, and adults' preferences regarding the naturalistic character and physical complexity of the scene. Bernáldez *et al.* (1987) describe changes in landscape preference following participation in environmental education activities and confirm a preference for water, while Bernáldez, Ruiz and Ruiz (1984) agreed with Zube *et al.* on the fact that children tend to prefer less naturalistic, less complex landscapes than adults. In a different context Francès (1968, 1979) identified differences in aesthetic evaluation of themes of varying complexity and congruence according to age and cultural level of children and young adults.

The aim of the study was to analyse the differences in landscape appraisal by children of different ages trying to detect consistent trends in landscape desirability related to age. Former work by Appleton (1978), Bernáldez *et al.* (1984), Abelló and Bernáldez (1986) and Abelló, Bernáldez and Galiano (1986) on insecurity versus security-giving characteristics in natural landscape has been used to elaborate the experiment's hypothesis.

According to the results of these studies the risk and uncertainty connotations of some natural settings are important ingredients of natural landscape preferences. Moreover, the 'alarming, deterring' or 'stimulating, exciting' character of certain landscape features depends on personal capacity for accepting risk or challenge. These are related to sex, age, familiarity with the subject and personality. Specifically, this study addresses the following hypothesis:

1. Some directions of variation in the patterns of landscape preference are related to visual characteristics with both 'deterring, frightening' and 'challenging, stimulating, exciting' effects.

2. The effect of these characteristics on preference can change with age. Younger children perceive them as frightening and tend to reject scenes exhibiting these features while older children can perceive them as stimulating thus evaluating them positively.

Method

The subjects were children from Gran Canaria and Tenerife (Canary Island) interviewed in their schools. They belonged to two age-groups: 191 were 11 years old and 292 were 16 years old.

Pairwise comparison was employed as described by Rodenas, Sancho Royo and González Bernáldez (1975), Bernáldez and Parra (1979) and Bernáldez, Parra and Quintas (1981). Characteristic landscapes of Tenerife were systematically photographed (coasts, mountains, rocks, woodlands, thickets and croplands) to form a collection of representative slides. These were grouped into classes (strata) with comparable themes (rocks compared to rocks, woodlands to woodlands etc.), similar camera distances and framing, then randomly paired within groups. Fifty pairs were obtained and projected by using two coupled slide-projectors to the 483 subjects in sessions of approximately 70 children.

Subjects were asked to mark on a sheet which photo (left or right) they preferred of each pair. These data can be analyzed by various multivariate techniques (Sancho Royo, 1974; Ruiz, 1985). On the ground of previous experience (see for instance Ruiz, 1985) the following methods were used: (1) correspondence analysis, often called reciprocal averaging (Benzecri, 1973); and (2) detrended correspondence analysis (DCA) an improved eigenvector ordination technique based on reciprocal averaging (Hill, 1979).

The dimensions or independent trends of variation obtained could be interpreted by observing the common characteristics within the series of photo pairs with the highest loadings on a given dimension and simultaneously the consistent differences with their counterparts in each pair. During this inspection the pairs are reversed if the sign of the loading is negative. (Rodenas *et al.* 1975; Bernáldez and Parra, 1979; Bernáldez *et al.* 1981). As the interpretations were clear-cut, more sophisticated methods (Abelló *et al.* 1986) were not deemed necessary.

The correct notation with these methods is 0 for preference of one side (right for instance) and 2 for the other side (n and $n + 2$ or n, $n + 1$ and $n + 2$ if $n + 1$ represent ties or indecisions). In this way the response vector is symmetric and centered, allowing a correct relation between preferences for the right or left side of the pairs (Ruiz, 1985). This notation is equivalent to the more straightforward $+1$, 0, -1 but avoids the use of negative values incompatible with correspondence analysis programmes.

Results

Interpretation of the analysis dimensions
Detrended correspondence analysis of the 483 respondents multiplied by 50 photo-pairs matrix provided three readily interpretable dimensions or axes. The most

FIGURE 1. Examples of picture-pairs with the largest contributions to the three first dimensions obtained in the analysis of landscape preference data. Top, 1st axis, interpreted as an opposition of contrasting landscapes: clear, illuminated scenes rich in detail vs darker, shadowed, scenes with less detail definition. Middle, 2nd axis: diverse, contrasted, varied vs more monotonous landscapes. Bottom, 3rd axis: harshness: rough, rasping, with edges and aggressive forms bland, smooth surfaces.

important photopairs for each of the dimensions are shown in Figure 1. Inspection of the pairs of pictures with loadings higher than 0·30 in absolute value (a convenient threshold chosen after former experimenting with DCA analysis) on the first dimension revealed that this trend of variation consisted in the opposition between clear, well-illuminated scenes rich in detail, with high definition and darker scenes with shadows or with gloomy areas that made the perception of detail more difficult. These pairs consistently possessed high negative loadings when the illuminated pictures were at the right side of the pair and high positive loadings when they were at the left side.

The 10 pairs exhibiting loadings higher than 0·30 in absolute value showed these differences with no exception, making the interpretation very clear-cut. In eight cases the corresponding picture was darker in all its surface and in two cases large but partial shadows and less contrast in the illuminated part diminished the definition of details. In one instance mist added to the shadow which concealed the details.

The second dimension can easily be interpreted as a diversity factor, as the six pairs with loadings higher than 0·30 consistently exhibited this characteristic in the right side picture of the pairs with positive loadings (and vice versa) contrasting with the more monotonous character or their counterparts. 'Diversity' was used to describe the more heterogeneous character of the scene: more balanced proportions of sky, forest, rock, water and grass surfaces, versus much greater dominance of one of these components (for instance, plenty of grass-covered surface with no rock and a tiny portion of sky and forest). The diversity-endowed series of photographs showed, in most cases, more contrasting colours, more marked shadow patterns or were more obviously structured than their 'monotonous' counterparts. This diversity factor is identical to the one found by Sancho Royo (1974) with very different picture material and subject populations.

The third dimension was interpreted as a 'harshness' factor as the corresponding pictures of the eight pairs with loadings higher than 0·3 exhibited a rougher texture containing edges and points, often with a broken contour in the horizon line or in mountain slopes forming sharp, acute or pointed features like teeth. These pictures give a general impression of harshness, evoking concepts of rasping or pricking. Their counterparts (in this case at the right side of the pairs with positive loadings and vice versa) were characterized by consistently exhibiting a relatively smooth surface and an absence or relative scarcity of edges, points or sharp forms. The general impression is one of softer, blander and smoother texture or relief.

Similar 'harshness' factors have already been discovered in factor analysis of landscapes used as stimuli. Aggressive forms in plants, pointed leaves, dead branches, etc. are also responsible for this kind of factor (Abelló and Bernáldez, 1986).

Reciprocal averaging (correspondence analysis) yielded very similar results, the correlation between the three dimensions obtained by both methods was 0·999, 0·900 and 0·968 respectively. The interpretation of the three first dimensions is identical, the only noticeable differences being slight changes in the rank of some photopairs when ordered according to their loadings. In the case of the first dimension these changes seem to reinforce the 'definition of prominent details in dark areas' aspect, at the expense of the 'general cast of shadow' aspect but this does not affect the general interpretation. No meaning could be attributed to the small changes in the other axis.

Relation of preference to age
The coordinates on the first three axes of the 11- and 16-year-old subject populations were compared and significant differences detected for the 1st and the 3rd dimensions by means of a Student's t-test (Figure 2).

The 11-year-old children significantly differ from the 16-year-olds ($t = 4·09$, $P < 0·01$) in their tendency to dislike the darker scenes with less detail characterizing the 1st dimensions of the analysis. The same age group significantly differ from their older counterports ($t = 2·92$, $P < 0·01$) in a tendency to dislike the scenes endowed with characteristics giving rise to the 3rd axis of the analysis: harshness, roughness, abruptness, presence of edges, rasping and pricking features. There is no significant difference in the case of the diversity factor represented by the 2nd dimension.

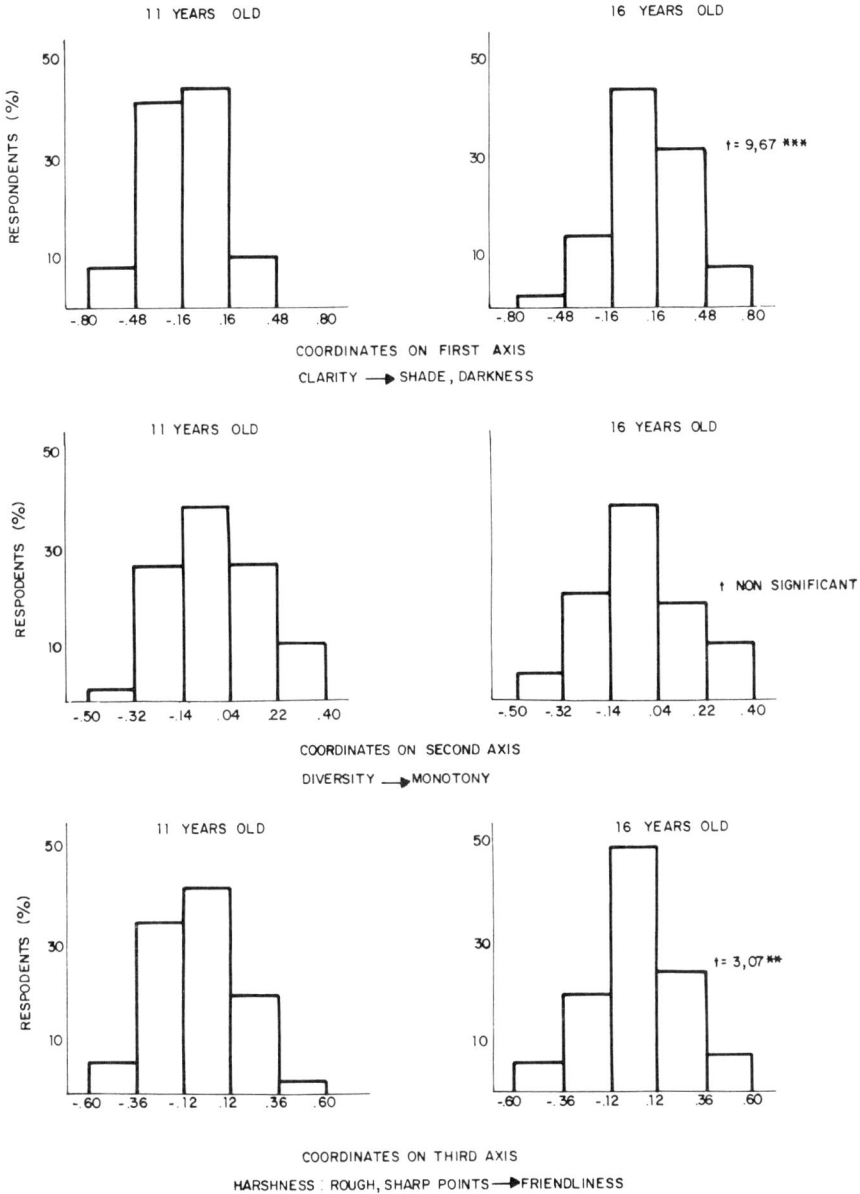

FIGURE 2. Comparison of the position of the 11- and 16-year-old subject populations on the three axes obtained in the analysis of landscape preference data. The 11-year-old children show significantly less preference for both darker, less illuminated scenes (1st axis) and harsh, rough scenes with aggressive forms (3rd axis). There are no significant differences for the 2nd axis (landscape diversity).

Discussion

The 1st and 3rd dimensions (illuminated vs shadowed and harsh, rough vs bland, smooth) relationship to age may be considered as forms of a more general 'risk, uncertainty factor' playing an important role in landscape preference analysis

(Bernáldez and Parra, 1979; Abelló and Bernáldez, 1986). Fear and insecurity have been already considered very important ingredients of landscape aesthetics by Appleton (1978).

The rocky, arid and mountainous landscapes common in the Canary Islands with its lava formations are rich in stirring-up alarming or risk-evoking features but they are by no means unique in this respect. Examples of European landscapes like the pictures from Spain and Germany used by Sancho Royo (1974) contained many features apparently perceived as hostile or risk-evoking by the subjects. For instance, the relief of many alpine and rocky areas and some plant forms were perceived as hostile or aggressive. Moreover, winter landscape with cold signs (snow, mud, defoliated trees, etc.) that rated high as hostile environments are lacking in the Canary collections.

The opposition between illuminated scenes, rich in detail and darker, shadowed scenes with less detail definition is reminiscent of the concepts of both *mystery* and *legibility* of Kaplan and Kaplan (1982). But mystery, as originally defined, implies the consideration of a third dimension, as the observer feels that more information could be obtained if walking deeper into the scene. Legibility as defined by Kaplan and Kaplan (1982) is more closely related to the possibility of the subject finding his/her way easily and making sense of the environment as one wandered farther and farther into the scene.

Darkness and deep shadow, like 'mystery', conceal a part of the scene's information but have a strong risk and uncertainty connotation, related to fears typical of children. Darkness and shadow, like Kaplan and Kaplan's 'mystery', can also stimulate the curiosity of some observers. These circumstances may explain the change from negative influences on preference when the characteristics are perceived as frightening, to positive effect on preference when they are perceived as exciting and stimulating.

After reviewing the effects of challenging characteristic of landscape on preference, Abelló and Bernáldez (1986) concluded that the same visual features may act positively or negatively on preference depending on the subject's personal strategy and his/her capacity to accept challenge. The transition from an alarm or insecurity factor (for younger children) to a stimulating role or ingredient of artistic quality seems to be reflected here. Fear of darkness has often been described as one of the most frequently confirmed phobias related to age, whilst shade, light and shade effects, silhouettes, etc. are some of the most frequently employed resources in paintings, drawings and photography to achieve 'artistic' or 'dramatic' effects.

The change of appraisal with age of the characteristics represented in the 3rd dimension: 'Harshness: rough, rasping; with edges, points and aggressive forms vs bland, smooth, blunt' is very similar. These visual characteristics have challenging connotations that can both alarm and cause anxiety, resulting in a negative influence on the preference of younger children and young adults.

The detection of risk-evoking, hostile characteristics associated with wild, untamed disordered landscapes as factors influencing preference is not new. In some cases they have been reported as influencing appraisal differently depending on age and sociocultural peculiarities (Bernáldez and Parra, 1979). Other work on arousal-eliciting environments are also of interest in this connection. Francès (1979) noted a relationship between a scene's complexity and incongruence and the degree of preference, depending on age and education. Zube *et al.* (1983) found that higher relative relief strongly enhances scenic quality for young adults, has a moderate effect for the elderly, but is irrelevant as a beautifying factor for young children. They comment that developmental work suggests that optimal arousal levels in children in comparison to adults are elicited by less physically complex stimuli (Wohlwill, 1974). Both Zube *et al.* (1983) and Bernáldez, *et al.* (1984) noticed the different preference response to naturalistic landscape of children and the elderly as compared to young and middle-aged adults. In that study it was concluded that the disordered, complex, 'wild' environment was less appreciated by both those under 15 and over 35.

The common factor of these experiments is the risk and/or a factor of uncertainty revealed in the information content (complexity, disorder), mystery or incongruence, which produces rejection (by children or the elderly) or stimulation (youth or adults).

Gender has also been related to the preference for wild, spontaneous, disordered landscapes preferred by male university students vs more ordered humanized landscapes preferred by female students (Bernáldez and Parra, 1979).

The change from the role in preference (from rejection to attraction) of challenging or stimulating may be compared with the observations of Baldwin and Baldwin (1981) in non-human primates. Maturing males increasingly accept environmental challenge and develop an explorative, 'centrifugal' behaviour, tending to explore more remote areas and 'risky' places within the territory hunted by the group. These areas, which are usually avoided by females and younger males, can be accurately mapped (like very dense thickets, wasp nest areas, etc).

References

Abelló, R. P. and Bernáldez, F. G. (1986). Landscape preference and personality. *Landscape and Urban Planning*, **13**, 19–28.

Abello, R. P., Bernáldez, F. G. and Galiano, E. F. (1986). Consensus and contrast in landscape preference. *Environment and Behavior*, **18**, 155–178.

Appleton, J. (1978). *The Experience of Landscape*. J. Wiley & Sons: London.

Baldwin, J. D. and Baldwin, J. I. (1981). *Beyond Sociobiology*. Elsevier: New York.

Benzecri, J. P. (1973). *L'analyse de données*. Vol. I and II. Dumond: Paris.

Bernáldez, F. G. and Parra, F. (1979). Dimensions of landscape preferences from pairwise comparisons. *National Conference on Applied Techniques for Analysis of the Visual Resources*. USDA. Nevada.

Bernáldez, F. G., Benayas, J. and De Lucio, J. V. (1987). Changes in environmental attitudes as revealed by activity preferences and landscape tastes. *The Environmentalist*, **7**, (1), 21–30.

Bernáldez, F. G., Ruiz, J. P. and Ruiz, M. (1984). Landscape perception and appraisal: ethics, aesthetics and utility. *8th International Conference on Environment and Human Action*. Berlin. IAPS 8.

Bernáldez, F. G., Parra, F. and Quintas, G. M. (1981). Environmental preferences in outdoor recreation areas in Madrid (Spain) *J Environmental Management* **13**, 13–26.

Francès, R. (1968). *Psychologie de l'esthétique*. PUF: Paris.

Francès, R. (1979). *Psychologie de l'art et de l'esthétique*. PUF: Paris.

Hill, M. O. (1979). *DECORANA- A FORTRAN Program for Detrended Correspondance Analysis and Reciprocal Averaging*. Ithaca, N.Y: Cornell University.

Kaplan, S. and Kaplan, R. (1982). *Cognition and environment*. Praeger: New York.

Rodenas, M., Sancho Royo, F. and González Bernáldez, F. (1975). Structure of landscape preferences: a study based on large dams viewed in their landscape setting. *Landscape Planning* **2**, 159–178.

Ruiz, J. P. (1985). *Percepción y Gestión del Ecosistema Pastoral de los Ganaderos de la Sierra de Madrid.* Doctoral Thesis. Universidad Autónoma, Madrid.

Sancho Royo, F. (1974). *Actitudes ante el Paisaje Experimental.* Publicaciones de la Universidad de Sevilla: Sevilla.

Wohlwill, J. F. (1974). Human response to levels of environmental stimulation. *Human Ecology*, **2**, 127–147.

Zube, E. H., Pitt, D. G. and Evans, G. W. (1983). A lifespan developmental study of landscape assessment. *Journal of Environmental Psychology*, **3**, 115–128.

PRIVACY-SEEKING BEHAVIOR IN AN ELEMENTARY CLASSROOM

CAROL S. WEINSTEIN*

Rutgers, The State University of New Jersey, U.S.A.

Abstract

This study observed privacy-seeking behavior in an elementary classroom, investigated individual differences in privacy seeking, and compared preferences for private spaces varying in degree of enclosure. Four privacy booths were placed in a fourth-grade classroom. A ticket system was used to assess booth use. Information on personality and background variables was obtained with self-report, peer, teacher, and parent questionnaires. After an initial period of enthusiasm, overall booth use declined sharply. However, analysis revealed substantial individual variation in booth use that remained consistent throughout the study. For boys, booth use was significantly correlated with teachers' ratings of sociability, aggressiveness, and distractibility. For girls, a significant positive relationship was found between privacy seeking at home and in school. Self-reported desire for privacy was uncorrelated with actual privacy-seeking behavior in the classroom. No significant differences in the use of the various booths were found, although self-reported preferences clearly favored the booth that allowed visual access to the rest of the classroom when desired.

Introduction

In recent years, numerous researchers have attempted to assess desire for privacy and privacy behaviors (Lawton and Bader, 1970; Brunetti, 1972; Marshall, 1972; Smith, 1977; Mazeika, 1978; Parke and Sawin, 1979; Shumaker, 1980). Most of this research has relied on self-report; only a few studies have attempted to observe actual privacy-seeking behavior. One such investigation by Golan (1978) examined privacy seeking, self-esteem, and social interaction among adolescents living in two residential psychiatric centers. Golan found that those subjects who frequently sought out privacy (defined as chosen physical aloneness) tended to engage in isolated active behavior when in social situations. Contrary to expectations, no relationship was found between privacy seeking and self-esteem. Moffitt (1974) provided third- and fourth-graders with the opportunity to use 'privacy booths' during study and free times. One of her most interesting results was a significant negative correlation between booth use and the extent of privacy opportunities at home.

The general goal of the present research was to observe privacy-seeking behavior in an elementary classroom where opportunities for privacy had been minimal. The study adopted Golan's (1978) definition of privacy, chosen physical aloneness. As in Moffitt's (1974) work, desire for privacy was operationalized as the voluntary use of a privacy booth. The specific objectives of this study were: (1) to determine if children would take advantage of opportunities for privacy if they were made available; (2) to investigate individual differences in desire for privacy; (3) to

---

* Requests for reprints should be sent to Carol S. Weinstein, Department of Science and Humanities Education, Rutgers Graduate School of Education, New Brunswick, N.J. 08903.

examine the relationship between privacy behavior at home and at school; (4) to determine the relationship between self-reported desire for privacy and observed privacy-seeking behavior; and (5) to compare preferences for private spaces varying in degree of enclosure.

A number of these objectives warrant elaboration. Given the fact that the typical classroom is a high density environment with little available privacy, considerable interest in using the privacy booths was anticipated. Yet, it was also expected that this interest would vary substantially from person to person. A study of seating preferences in a university library (Sommer, 1970), for example, discovered that almost half of the students perferred to sit in a public area rather than a more private area. Sommer concluded that some students feel the need to have activity going on around them in order to maintain their concentration. A major aim of the present study was to assess the extent of such individual differences in desire for privacy and to explain these differences in terms of personality and home background variables.

The first of the personality variables chosen for study was self-esteem. Although Golan (1978) found no relationship between self-esteem and privacy seeking, she did find that subjects who had their own bedrooms scored higher on a measure of self-esteem than those who shared a bedroom. Similarly, a study with institutionalized and noninstitutionalized elderly people (Aloia, 1973) found a correlation between perceived privacy opportunities and self-esteem. These findings are compatible with theoretical work that views privacy as critical to the development and maintenance of self-identity and personal autonomy (Plant, 1930; Schwartz, 1968; Westin, 1970, Simmel, 1971; Wolfe and Laufer, 1975). It has also been argued that highly anxious individuals with a poor self-image need privacy as a form of protection and restoration (Bates, 1964; Schwartz, 1968). If personal space is viewed as a mechanism to achieve privacy (Altman, 1975), these ideas find further support in the personal space literature (Smith, 1954; Luft, 1966; Dosey and Meisels, 1969; Karabenick and Meisels, 1972; Patterson, 1973). Thus, it was hypothesized that self-esteem would be negatively correlated with privacy-seeking behavior.

The second personality variable selected for study was sociability. Plant (1930) has suggested that crowded environments can lead to mental strain because of the constant pressure to get along with other people. It is likely that this strain is even more severe for those individuals who find social interaction difficult. Indeed, Berscheid (1977) has hypothesized that individual differences in the need for privacy may parallel individual differences in the dimension extroversion–introversion. This notion is supported by Marshall (1970) who found a significant relationship between privacy preference scores and scores on an extroversion–introversion measure. In addition, several studies have demonstrated a relationship between extroversion–introversion and personal space, with introverts preferring greater distances from others (Patterson and Holmes, 1966; Cook, 1970; Mehrabian and Diamond, 1971; Williams, 1971). In order to limit the amount of self-report data in the present research, peer and teacher ratings of sociability were used instead of a self-report measure of extroversion–introversion. It was expected that individuals who were rated low in sociability would be more likely to take advantage of opportunities for privacy than those who were high in sociability.

The third personality variable to be examined was aggressiveness. Golan (1978) has reported that in a psychiatric hospital adolescents who had more privacy were less aggressive. Murray (1974) found that high levels of home density were associated with more aggressiveness in school settings. Furthermore, numerous crowding studies also find aggressiveness associated with high levels of density (for a review, see Sundstrom, 1978). These studies suggest a link between the availability of privacy and aggressive behavior, but they do not demonstrate a clear relationship between aggressiveness as a trait and desire for privacy. In accord with the previous hypothesis, it was predicted that aggressive individuals would seek out privacy more than others, since they experience particular difficulty in social situations.

The final personality characteristic chosen for study was distractibility. Although several studies have examined the impact of study carrels on the performance of emotionally disturbed (Shores and Haubrich, 1969) and hyperactive children (Cruickshank et al., 1961; Rost and Charles, 1967), there appears to be no information on whether or not distractible individuals will *choose* to work alone in a privacy carrel. It was hypothesized that students who were more distractible would use the privacy booths more than those who had no difficulty concentrating.

In addition to studying the relationship between desire for privacy and personality variables, another objective of the present research was to examine the impact of home environment variables on privacy seeking in school. Both drive reduction theory and adaptation level theory have been put forth to explain variations in privacy preferences. From a drive reduction perspective (Altman, 1975) one would hypothesize that a lack of privacy opportunities at home results in greater privacy seeking at school, while sufficient privacy opportunities at home are reflected in less privacy seeking in school. Adaptation level theory (Helson, 1964) predicts that students who have little available privacy at home will demonstrate little interest in achieving privacy at school. There are apparently only two studies (Marshall, 1972; Moffitt, 1974) that have empirically tested these conflicting approaches, and their findings are contradictory. The present research was designed to examine these two positions by relating privacy seeking in school with information provided by parents on home density and opportunities for privacy. Parents were also asked about privacy seeking behavior at home, in order to assess the extent to which such behavior is consistent across situations.

A different thrust of the present study concerned the attractiveness of various designs for providing privacy. Previous research has indicated that both degree of enclosure (Gramza, 1970) and opportunities for observing others (Curtis and Smith, 1974) affect the appeal of private spaces. In the present study, all privacy booths were identical except for the presence or absence of a window. In one booth, moreover, the window could be covered from the inside with an opaque panel, producing a state of greater enclosure. It was expected that this booth would be preferred over all the others. This prediction also follows from Altman's (1975) theoretical analysis of privacy, which emphasizes the importance of control over access to self rather than mere seclusion.

Method

Subjects

Two classes of fourth-grade students (ages 9–10) in a middle class, suburban school served as subjects. There were 10 boys and 14 girls in Class A and 10 boys and 13 girls in Class B (total $N = 47$). The two groups of children exchanged classrooms for

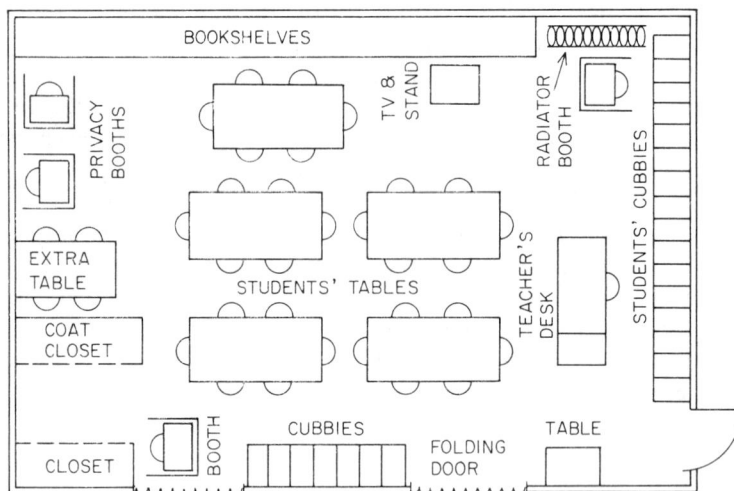

FIGURE 1. Floor Plan of Classroom A. (Drawing not to scale.)

one hour a day, so that Teacher A could teach language arts to both groups, and Teacher B could teach mathematics to both. The study was conducted in Teacher A's classroom. The children had access to the booths only during the one hour of language arts.

Materials

Privacy booths. Four three-sided cubicles (each side 0·9 meters wide × 1.5 m high) were constructed from double-wall cardboard (see Figure 1). Each cubicle enclosed a desk and chair. On the outside of each booth was hung a box for tickets and a sign that read 'vacant' on one side and 'occupied' on the other. Two of the booths had no window; one booth had a Plexiglass window (0·3 m × 0·3 m) on a side wall, at eye level for a child seated at the desk. The fourth booth had an identical window plus a cardboard panel that could be lowered from inside to cover it completely.

Piers-Harris Children's self concept scale. This is an 80-item self-report instrument that contains statements such as 'I am an important member of my class' and 'I am often mean to other people'. The scale has been used frequently and is well-regarded for its reliability and validity (see Crandall, 1973, for a critical review; also Piers, 1977).

Peer rating questionnaire. A simple peer rating questionnaire called 'What Are the Children In Your Class Like?' was developed for use in this study. The questionnaire contained nine statements, each beginning with a child's name (e.g. 'Linda is one of the first to be chosen for games' and 'Mark picks on other children'). Students responded by circling 'usually', 'sometimes', or 'hardly ever'. Each child in the class was rated by five other children on each of the nine items, and these five ratings were pooled to obtain an average score for each item. The questionnaire was designed to tap peer perceptions of sociability, aggressiveness and distractibility, each represented by three items. The internal consistency (Cronbach, 1951) of these three measures was 0·84, 0·59 and 0·79, respectively. A final question asked students to name the three children with whom they would most like to play and the three with whom they would least like to play.

Desire for privacy scale. This is a 51-item questionnaire (Weinstein and Aiello, 1980) containing such statements as 'I would rather share a bedroom than have my own room' and 'I like to have time alone to read or think'. Children respond by circling 'yes' or 'no'. The questionnaire was designed to represent three key dimensions of privacy reported by Wolfe and Laufer (1975): aloneness, information management, and access to spaces. The reliability of this scale was 0.84.

Home environment inventory. This is a brief questionnaire filled out by parents that asks questions about home density (number of people in the house, number of rooms), opportunities for privacy (e.g. 'Does your child have his/her own bedroom?'), and privacy-seeking behavior at home (e.g., "Does your child put up privacy markers like signs that read 'keep out' or 'do not disturb?' ").

Teacher questionnaire. This instrument required the teachers to rate each child in the class on five dimensions: sociability, self-confidence, aggressiveness, distractibility, and achievement. Teachers used a five-point rating scale with one equal to 'below average' and five equal to 'above average'. The correlations between the two teachers' ratings were 0.74 (sociability), 0.60 (self-confidence), 0.71 (aggressiveness), 0.52 (distractibility), and 0.77 (achievement).

Procedure

Two privacy booths (one with no window and one with a closable window) were introduced into Teacher A's classroom. Already present in the room were five rectangular tables that could seat six students each. The two groups of fourth-graders who used the room for language arts followed an individualized program. Students normally worked independently at the tables, lining up by the teacher's desk for help and instruction when needed. The introduction of the booths meant that children could choose to do their assigned work while seated at a table with other children or while seated alone in a privacy booth.

One week was allowed for the expected excitement over the booths to diminish; during this time, children's reactions were anecdotally recorded. It soon became apparent that competition for the booths was so extreme that many children were unable to use a booth when they wished. Thus, at the end of one week, a third booth (the booth with the open window) was placed in the classroom. Although formal data collection was begun at this time, it was clear that demand for the booths was still extremely high. A fourth booth (with no window) was therefore added seven days later. In this way, it was finally assured that children would have the opportunity to work alone whenever they desired. All four booths were positioned so that the open sides were next to a wall, creating the effect of a four-sided cubicle. Three times during the study, the locations of the booths were switched.

Teacher A introduced the booths to the students following a brief script prepared by the investigator. She explained that the booths were gifts from a person interested in learning about the kinds of classroom spaces fourth-grade students preferred. (The term 'privacy booth' was used in this explanation and was subsequently adopted by the children.) The teacher then outlined the procedure for use. A child wishing to use a booth got a ticket from a box on the teacher's desk, filled out his/her name, the date, and the time. The child then went to the booth, set a timer for ten minutes, turned the sign to 'occupied', and entered the booth. When the time was up, he/she filled out the time on the ticket, put it in the box attached to the booth, and turned the sign to 'vacant'.

At the end of each day, the used tickets were collected. These constituted the record of that day's booth use. Data were collected Monday through Thursday, since the instructional program did not allow the booths to be used on Fridays. At the close of the study, all questionnaires were administered. In addition, students were asked to indicate their booth preferences and to list their reasons for using the booths.

Results

Booth use over time
The principal dependent variable was the number of times a student used a privacy booth in a given day. When averaged over 25 consecutive days of the study and over subjects, the mean value of this variable was 0·351. However, there was considerable variation in total booth use from day to day, ranging from a maximum of 0·822 (37 visits) to a minimum of 0·021 (one visit). Figure 2 shows that this variation was not random; there was a strong fall off in booth use as the study progressed.

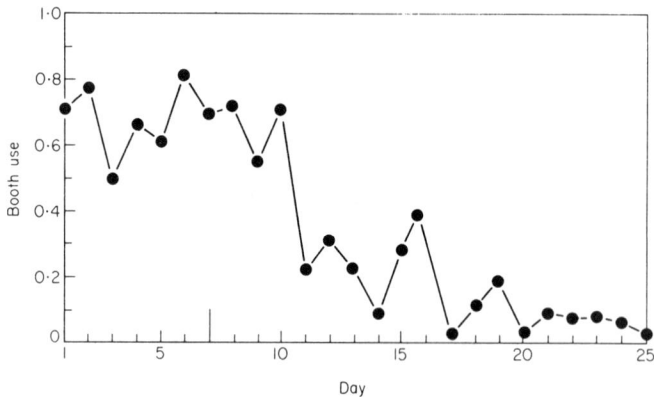

FIGURE 2. Mean booth use (no. of visits/no. of students) for successive days of the study. The heavy tick mark at Day 7 indicates the introduction of the fourth privacy booth.

Individual variation in booth use
The first seven days of formal data collection do not provide an accurate picture of an individual's desire for privacy; the three booths available during that period were so heavily used that often a student wishing to use a booth did not find one vacant. Therefore, the measure of individual privacy seeking was calculated as the average booth use over days eight through 25, rather than over the entire 25-day period. Individual differences in booth use were substantial, ranging from a low of zero to a high of 0·895. (See Table 1.)

Analysis revealed that children from the homeroom in which the booths were located used them less frequently than children from the second room, means of 0·168 and 0·291, respectively, $t(45) = 2·11$, $P < 0.05$. (This and all following tests of significance are two-tailed.) Since these group differences might be misleading when assessing individual differences in desire for privacy, a corrected individual booth use

TABLE 1

Distribution of booth use across children
(days 8–25)

No. of visits to a privacy booth	N	Percentage of students
0	9	19
1–5	22	47
6–10	12	26
11–15	3	6
16–20	1	2
	47	100

score was obtained by subtracting the room mean from the individual students' scores. This measure was then used in correlational analyses with personality and background variables.

The internal consistency of the measure of individual privacy seeking was examined by calculating separate scores based on the odd days and the even days during the period Day 8–Day 25. The odd–even correlation was 0·60. According to the Spearman Brown formula (Ghiselli, 1964), this suggests that the overall reliability of the 18-day average was 0·75. Thus, although total booth use declined over time, differences among individuals tended to be consistent.

Correlations between booth use and individual variables
The Pearson Product–Moment Correlation Coefficient was used to relate individual booth use with personality and home background variables. The results are presented in Table 2, along with the intercorrelations among the independent variables.* As can be seen, no relationship was found between booth use and self-esteem as measured by Piers-Harris. For boys, booth use was significantly correlated with the teachers' ratings of sociability, distractibility, and aggressiveness. (These three teacher ratings were highly intercorrelated; thus, the associations with booth use may not represent independent relationships, but may indicate a relationship between booth use and some general behavior dimension.) For girls, booth use was significantly correlated only with privacy-seeking behavior at home, a composite variable that included items asking about closing of doors, use of privacy markers, and an overall assessment of privacy seeking.

The analysis revealed no relationship between booth use and home density (total number of rooms/number of people living in home), not between booth use and having one's own bedroom. Finally, there was no relationship between booth use and scores on the Desire for Privacy Scale.

Booth preferences
There were no significant differences in the use of the four booths. However, students' self-reported preferences indicated a marginally significant preference for the

---

* Other analyses used the number of *days* on which a student used a booth, rather than the number of separate visits. None of the conclusions were altered.

TABLE 2
Correlation matrix of study variables

	Variables	(1)	(2)	(3)	(4)	(5)	(6)
	(1) Booth use		0·04	−0·10	0·36	0·36	−0·12
	(2) Self-esteem	−0·01		0·66†	−0·40	−0·29	0·60†
	(3) Peer rating of popularity	0·17	−0·04		−0·57†	−0·45*	0·78†
	(4) Peer rating of distractibility	−0·27	0·24	−0·81†		0·72	−0·50*
	(5) Peer rating of aggressiveness	−0·12	0·27	−0·73†	0·80†		−0·39
	(6) Peer nominations—popularity	0·01	−0·09	0·39*	−0·34	−0·19	
	(7) Peer nominations—unpopularity	−0·16	−0·24	−0·47†	0·33	0·28	−0·43*
G I	(8) Teacher ratings of sociability	0·19	−0·11	0·74†	−0·81†	−0·62†	0·45*
R L	(9) Teacher ratings of distractibility	−0·06	0·25	−0·47†	0·54†	0·25	−0·46*
S	(10) Teacher ratings of aggressiveness	−0·11	0·07	−0·73†	0·76†	0·68†	−0·39*
	(11) Teacher ratings of self-confidence	0·23	0·19	0·06	0·12	0·36	0·41*
	(12) Teacher ratings of achievement	0·11	0·15	0·10	−0·10	0·25	0·36
	(13) Self-reported desire for privacy	−0·02	−0·42*	−0·08	−0·06	0·27	−0·50†
	(14) Home density	−0·01	0·03	−0·25	0·22	0·25	0·06
	(15) Own bedroom	−0·09	0·02	0·36	−0·38	−0·36	0·01
	(16) Privacy seeking at home	0·50†	0·04	0·06	−0·03	0·12	−0·19

* < 0·05, † < 0·01.

booths with windows, $t(42) = 1{\cdot}73$, $P = 0{\cdot}09$. Of these two, the booth with the closable window was highly favored over the booth with the open window, mean ranks of 1·86 and 2·70, respectively, $t(42) = 4{\cdot}45$, $P < 0{\cdot}001$.

Discussion

One of the most surprising findings of the study was the low level of interest in using the privacy booths once the novelty had abated. While initial enthusiasm for the booths had been overwhelming, delaying the start of the study several times and eventually leading to the construction of two additional booths, overall booth use declined sharply after about three weeks. There are several possible explanations for the low level of interest in using the booths. First, the students generally came from home environments that provided abundant opportunities for privacy: average home density was 2·28 rooms per person; 70% of the subjects had their own bedrooms; and 75% of the homes contained dens or family rooms. From a drive reduction perspective, one can argue that the children had little need to seek privacy in school.

A second explanation is suggested by a study (Smith, 1977) in which individuals

BOYS

(7)	(8)	(9)	(10)	(11)	(12)	(13)	(14)	(15)	(16)
0·14	−0·58†	0·46*	0·62†	−0·22	−0·24	−0·24	0·29	−0·01	0·21
−0·28	0·46*	−0·13	−0·02	0·39	0·37	−0·19	0·36	−0·31	−0·10
−0·57†	0·57†	0·08	−0·10	0·46*	0·43	0·07	0·06	−0·40	−0·46
0·35	−0·66†	0·38	0·29	−0·69†	−0·75†	0·14	−0·16	0·41	0·05
0·28	−0·44	0·39	0·48*	−0·42	−0·38	−0·05	−0·13	0·08	−0·09
−0·38	0·69†	−0·07	−0·21	0·37	0·44	−0·29	−0·11	−0·23	−0·32
	−0·20	−0·16	−0·16	−0·21	−0·28	0·12	−0·20	0·21	0·35
−0·45*		−0·56*	−0·59†	0·60†	0·60†	0·01	−0·27	−0·13	−0·18
0·36	−0·70†		0·56*	−0·55*	−0·49*	−0·16	0·29	−0·29	−0·26
0·58†	−0·86†	0·53†		−0·14	−0·13	−0·37	0·54*	−0·34	−0·23
−0·31	0·09	−0·44*	0·06		0·91†	0·11	−0·02	−0·18	−0·15
−0·28	0·30	−0·68†	−0·19	0·80†		−0·20	−0·02	−0·09	−0·17
0·30	−0·25	0·23	0·12	−0·46*	−0·35		−0·35	0·05	0·00
0·18	−0·21	0·28	0·20	0·10	0·02	−0·07		−0·59†	−0·14
−0·17	0·23	0·11	−0·27	−0·48*	−0·48*	−0·06	−0·33		0·08
−0·13	−0·12	0·16	−0·01	0·03	−0·03	0·01	0·01	−0·02	

were moved from an antiquated prison, where privacy opportunities were minimal, to a new building which offered a high degree of privacy. Questionnaires administered before and after the move showed that inmates' expectations for achieving privacy increased, while the value they placed on privacy decreased. The same phenomenon may have been occurring in the present study: once the previously unattainable goal of privacy was obtained, it was no longer so attractive. Moreover, the students were at an age when the peer group is especially significant. Since moving to a booth meant giving up the opportunity to socialize and to co-operate on learning activities, it may have entailed more of a 'sacrifice' than students had envisioned. This, too, would serve to diminish the value placed on achieving privacy.

Third, it is important to keep in mind that desire for privacy depends on the context. As Wolfe (1980) points out, concepts and patterns of privacy behavior are defined by cultural norms, developmental stage, and the social–physical properties of the setting. Although classrooms are generally high density situations, our culture does not view the educational activities that take place in elementary school as requiring physical privacy. Instead of being physically secluded, children are taught to minimize interaction and limit the sharing of information. It seems likely that having experienced this situation since entry into school, students do not label learning activities as tasks requiring privacy in the form of chosen aloneness.

A fourth explanation for the low level of interest may be the design of the booths themselves. Perhaps students would prefer privacy booths in which two people could work together or areas that provided both privacy and 'environmental softness' (Jones & Prescott, 1978)—rockers, pillows, stuffed chairs, and rugs.

Despite the overall decline in booth use, individual differences in privacy seeking remained consistent throughout the study. The personality and home environment variables examined were only partially helpful in explaining these differences. No relationship was found for either sex between self-esteem and booth use. Interestingly, however, students with low self-esteem had higher scores on the Desire for Privacy Scale ($r = -0.31$, $P < 0.05$). Perhaps such students would have liked to have used the booths, but lacked the assertiveness or confidence to do so.

Sociability, aggressiveness, and distractibility (as rated by the teachers) were significantly related to boys' booth use, while privacy-seeking behavior at home was the only variable significantly correlated with girls' booth use. The different pattern of results for boys and girls is difficult to explain. However, comparison of the mean teacher ratings for boys and girls on the three personality variables demonstrates significant differences in ratings for aggressiveness and distractibility, with boys rated significantly higher on both dimensions, $P < 0.05$ and $P < 0.005$, respectively. There were no significant sex differences in ratings of sociability, nor any differences in the variability of the ratings for these three dimensions. It is possible that the different correlations for boys and girls are due to a threshold effect, whereby only high levels of distractibility and aggressiveness increase privacy-seeking behavior.

No relationship was found between home density and booth use, a finding that is inconsistent with the negative correlation found by Moffitt (1974). The lack of a correlation also provides no support for either drive reduction or adaptation level theory. It may be necessary to have either higher levels of density or greater variability to observe the effects predicted from either perspective.

Although the near-zero correlation between booth use and scores on the Desire for Privacy Scale was unexpected, in retrospect, it is not entirely surprising. While the scale was designed to assess a broadly-defined desire for privacy, encompassing control over access to spaces, information management, and aloneness, the behavioral measure focused on the specific act of choosing to work in a privacy booth. The lack of a significant relationship points out the difficulty in using multiple-act self-reports as predictors of behavior in single-act situations.

The discrepancy between self-reported desire for privacy and booth use also raises an important question about the validity of the dependent variable. Clearly, the booths provided a degree of seclusion, but was booth use actually measuring desire for privacy? Several potential problems exist. For example, once a child left a table to use a booth, the remaining children had more space and, one might argue, less need to seek privacy in one of the booths. Furthermore, within the classroom context, booth use may have taken on socio-psychological meanings unrelated to privacy seeking; going to a booth might have been an excuse for getting up and walking around or even a way of attracting attention.

It is impossible to say with certainty that such confounding effects did not occur. There is reason to believe, however, that booth use did constitute a valid measure of privacy seeking. Sitting at the large tables was hardly a private situation, even when someone had left to use a booth. There were almost always four children remaining at each table, sitting in full view of their classmates. In addition, children's reasons for using the booths—to be alone, to not be bothered, to have privacy, and to get work done—indicate that they did, in fact, view the booths as providing the opportunity to have privacy.

It is not easy to find a pure behavioral exemplar of a psychological construct like privacy. Unless we are willing to use 'less than pure' behavioral measures, we will be forced to rely solely on self-report data. The use of privacy booths appears to be a promising paradigm for the study of privacy needs.

Conclusion

The present study is a first step in examining the significance of privacy opportunities in the classroom. As such, it has generated as many questions as it has answered: What developmental, social, and contextual variables influence the level of overall privacy seeking? What factors in addition to those examined in the present study account for individual differences in privacy behavior? What factors are responsible for the different correlational patterns obtained for boys and girls? Does the opportunity for privacy increase positive social interaction in the classroom? What is the impact of privacy on school performance and attitudes about school? Clearly, further research is needed before we can decide how much privacy should be available to children in school and how important such privacy opportunities are.

Acknowledgements

I am grateful to Fred Cohen, Carol Bertrand, Tom Brown and the fourth-grade students at Moss School, Metuchen, New Jersey, for allowing me to conduct this study in their school. I also wish to thank Susan Kindervatter for her expert assistance in data gathering and Judy Mezzatesta for her accurate coding.

References

Aloia, A. (1973). Relationships between perceived privacy options, self-esteem and internal control among aged people (Unpublished Doctoral dissertation, California School of Professional Psychology). *Dissertation Abstracts International*, **34**, 5180B.

Altman, I. (1975). *The Environment and Social Behavior*. Monterey, CA: Brooks/Cole Publishing Co.

Bates, A. (1964). Privacy—a useful concept? *Social Forces*, **42**, 432–4.

Berscheid, E. (1977). Privacy: A hidden variable in experimental social psychology. *Journal of Social Issues*, **33** (3), 85–101.

Brunetti, F. A. (1972). Noise, distraction and privacy in conventional and open school environments. *Proceedings of the EDRA Conference*. Los Angeles: University of California.

Cook, M. (1970). Experiments on orientation and proxemics. *Human Relations*, **23** (1), 61–76.

Crandall, R. (1973). Measures of self esteem. In J. P. Robinson and P. R. Shaver (eds), *Measures of Social Psychological Attitudes. Revised Edition*. Ann Arbor: University of Michigan Institute for Social Research.

Cronbach, L. J. (1951). Coefficient alpha and the internal structure of tests. *Psychometrika*, **16**, 297–334.

Cruikshank, W. M., Bentzen, F. A., Ratzburg, F. H. and Tannhauser, M. T. (1961). *A Teaching Method for Brain-Injured and Hyperactive Children*. Syracuse: Syracuse University Press.

Curtis, P. and Smith, R. (1974). A child's exploration of space. *School Review*, **82** (4), 671–9.

Dosey, M. A., and Meisels, M. (1969). Personal space and self-protection. *Journal of Personality and Social Psychology*, **11**, 93–7.

Ghiselli, E. E. (1964). *Theory of Psychological Measurement*. New York: McGraw Hill, Inc.

Golan, M. B. (1978). Privacy, interaction and self-esteem (Unpublished Doctoral dissertation, City University of New York). *Dissertation Abstracts International*, **39** (3–B), 1541B.

Gramza, A. F. (1970). Preferences of preschool children for enterable play boxes. *Perceptual and Motor Skills*, **31**, 177–8.

Helson, H. (1964). *Adaptation Level Theory*. New York: Harper & Row.

Jones, E. and Prescott, E. (1978). *Dimensions of Teaching–Learning Environments. II. Focus on Day Care*. Pasadena, CA: Pacific Oaks College.

Karabenick, S. and Meisels, M. (1972). Effects of performance evaluation on interpersonal distance. *Journal of Personality*, **40** (2), 275–86.

Lawton, P. and Bader, J. (1970). Wish for privacy in young and old. *Journal of Gerontology*, **25**, 48–54.

Luft, J. (1966). On nonverbal interaction. *Journal of Psychology*, **63**, 261–8.

Marshall, N. J. (1970). Personality correlates of orientations toward privacy. In J. Archea and C. Eastman (eds), *EDRA 2: Proceedings of the Second Annual EDRA Conference*. Pittsburgh: Carnegie Mellon University.

Marshall, N. J. (1972). Privacy and environment. *Human Ecology*, **2**, 93–110.

Mazeika, E. R. The relationship between perceived privacy, life satisfaction, and selected environmental dimensions in homes for the aged (Unpublished Doctoral dissertation, University of Southern California). *Dissertation Abstracts International*, **39** (5–B), 2509.

Mehrabian, A. and Diamond, S. G. (1971). Effects of furniture arrangement, props, and personality on social interaction. *Journal of Personality and Social Psychology*, **20**, 18–30.

Moffitt, R. A. (1974). The effects of privacy and noise attenuation alternatives on the art-related problem solving performance of third and fourth grade children (Unpublished Doctoral dissertation, Arizona State University). *Dissertation Abstracts International*, **35** (3–A), 1506–7.

Murray, R. (1974). The influence of crowding on children's behavior. In D. Canter and T. Lee (eds), *Psychology and the Built Environment*. London: Architectural Press.

Parke, R. D. and Sawin, D. B. (1979). Children's privacy in the home: Developmental, ecological, and child-rearing determinants. *Environment and Behavior*, **11** (1), 87–104.

Patterson, M. L. (1973). Stability of nonverbal immediacy behaviors. *Journal of Experimental Social Psychology*, **9**, 97–109.

Patterson, M. L. and Holmes, D. S. (1966). Social interaction correlates of MMPI extroversion–introversion scale. *American Psychologist*, **21**, 724–5.

Piers, E. V. (1977). *The Piers-Harris Children's Self Concept Scale. Research Monograph #1*. Nashville, Tenn.: Counselor Recordings and Tests.

Plant, J. S. (1930). Some psychiatric aspects of crowded living conditions. *American Journal of Psychiatry*, **9**, 849–60.

Rost, L. J. and Charles, D. C. (1967). Academic achievement of brain-injured and hyperactive children in isolation. *Exceptional Children*, **34**, 125–6.

Schwartz, B. (1968). The social psychology of privacy. *American Journal of Sociology*, **73**, 741–52.

Shores, R. E. and Haubrich, P. A. (1969). Effect of cubicles in educating emotionally disturbed children. *Exceptional Children*, **36** (1), 21–24.

Shumaker, S. A. (1980). Adjusting the physical environment to the discomforts of inadequate privacy. Paper presented at the 1980 meeting of the American Psychological Association, Montreal, Canada.

Simmel, A. (1971). Privacy is not an isolated freedom. In J. Pennock and J. Chapman (eds), *Privacy*. New York: Atherton Press.

Smith, D. E. (1977). Privacy and environment: A field experiment (Unpublished Doctoral dissertation, The University of Florida). *Dissertation Abstracts International*, **38** (7–B), 3475.

Smith, G. H. (1954). Personality scores and personal distance effect. *Journal of Social Psychology*, **39**, 57–62.

Sommer, R. (1970). The ecology of privacy. In H. M. Proshanky, W. H. Ittelson and L. G. Rivlin (eds), *Environmental Psychology: Man and His Physical Setting*. New York: Holt, Rinehart, & Winston.

Sundstrom, E. (1978). Crowding as a sequential process: review of research on the effects of population density on humans. In A. Baum and Y. M. Epstein (eds), *Human Response to Crowding*. Hillsdale, N. J.: Lawrence Erlbaum Associates.

Weinstein, C. S. and Aiello, J. (1980). Desire for Privacy Scale. Unpublished questionnaire, Rutgers, The State University of New Jersey.

Westin, A. (1970). *Privacy and Freedom*. New York: Atheneum.

Williams, J. L. (1971). Personal space and its relation to extroversion–introversion. *Canadian Journal of Behavioral Science*, **3** (2), 156–60.

Wolfe, M. (1980). Childhood and privacy. In I. Altman and J. Wohlwill (eds), *Children and the Environment. Human Behavior and the Environment, Advances in Theory and Research*, Vol. III. New York: Plenum Press.

Wolfe, M. and Laufer, R. S. (1975). The concept of privacy in childhood and adolescence. In D. H. Carson (ed.), *Man–Environment Interactions: Evaluations and Applications Part II*. (EDRA V). Stroudsburg, Pa: Dowden, Hutchinson & Ross.

EDUCATIONAL ISSUES, SCHOOL SETTINGS, AND ENVIRONMENTAL PSYCHOLOGY*

LEANNE G. RIVLIN

City University of New York

and CAROL S. WEINSTEIN†

Rutgers University

Abstract

This paper reviews selected research on classroom and school environments, using a framework that views schools from three perspectives—as places for learning, as places for socialization and as places for psychological development. Studies are included that deal with the impact of noise and classroom design on learning; the relationship between seating position, achievement and status; spatial cognition; the classroom environment and sex role stereotyping; privacy; and density. The need for classrooms to enhance children's feelings of competence, security and self-esteem is also stressed. The goal of the paper is to point out ways in which environmental psychologists can contribute to the improvement of the educational system and to the quality of life in schools.

Introduction

One of the strengths of work in environmental psychology has been the grounding of research in daily life experiences, as one can see from the range of topics covered by this and other journals dealing with person/environment relationships. It is thus surprising that educational issues and school environments have not received more attention. Research in what we now identify as environmental psychology began in the late 1950s, with the sixties and seventies its most active period of growth. Simultaneously, the educational critiques of the sixties and seventies (e.g. Holt, 1964; Kozol, 1967; Silberman, 1970), the school building boom and the experimentation with alternative school models stimulated assessment of the design and functioning of schools and classrooms. With a few notable exceptions, however, most school building research was done by educational specialists. Environmental psychologists, both in the U.S.A. and abroad, largely focused their attention on housing, offices and health facilities, psychiatric hospitals in particular (Ittelson *et al.*, 1970; Moos, 1972; Sommer and Ross, 1958). They generally turned to schools less out of an interest in educational issues than because they were convenient sites for research, with their

*A version of this paper was presented as part of a symposium, Environmental Psychology in Health, Work, Schools, Prisons, Crime, Housing and Communities, American Psychological Association, Anaheim, California, August, 1983.

†To whom requests for reprints should be sent at Rutgers Graduate School of Education, New Brunswick, NJ 08903, U.S.A.

captive occupants and relatively stable physical qualities. There are some real questions, therefore, about the contribution of environmental psychology to our understanding of school settings.

In a period when our educational system is under intense scrutiny, if not attack, and numerous national reports have recommended strategies for school improvement (e.g. Boyer, 1983; National Commission on Excellence in Education, 1983), a greater contribution by environmental psychology would be welcome. For the most part, educators and educational critics tend to ignore two facts: first, that schools and classrooms are *physical entities* as well as organizational units; and second, that the physical characteristics of a setting can influence both the behavior of its users and the educational program. Environmental psychologists can stimulate a recognition and awareness of these facts and ensure that the efforts to create high quality educational experiences for our children include consideration of the physical milieu. Towards this effort, the present paper will review representative research and thinking on school environment/student behavior relationships and suggest ways in which environmental psychologists can contribute to current attempts to improve public education. (For a more complete review of the literature on the physical environment of the school, see Weinstein, 1979.)

When examining the work on school settings, it is important to keep in mind that schools, like all physical and organizational structures, serve a variety of functions. Most obvious is the school's responsibility to *educate*—to foster cognitive development, to transmit information about subjects of the curriculum, and to communicate the joy and excitement of learning. Indeed, it is this function that is at the heart of the current furor over education. Calls to improve students' basic skills, to upgrade science and mathematics instruction, to require more homework, and to lengthen the school day (National Commission on Excellence in Education, 1983)— whether reasonable or not—are concerned with this most fundamental role of the school. In addition to education, however, schools historically have been expected to assume a major responsibility for *socialization* (Ornstein and Miller, 1980). From this perspective, a primary function of schools is to transmit the ideas and values of society and to prepare children for their adult roles and responsibilities. A third and final function is the school's obligation to support *individual psychological development* (Armstrong *et al.*, 1983). American society has traditionally placed a high premium on 'rugged individualism' and has valued individual dignity and achievement. Proponents of this orientation argue that schools must provide for students' individual differences and should foster youngsters' creativity, independence, feelings of self-esteem, and self-fulfillment.

If schooling is examined in an historical context, it becomes clear that one or another of these functions has been emphasized at different points in time, while the others have receded in importance. Nonetheless, all three of these interrelated functions are recognized as legitimate and necessary, and thus, they will form the framework used here for reviewing research and thinking on school environments. Despite the fact that the physical setting is generally neglected in discussions of education and educational problems, schooling takes place in *buildings*. How these buildings affect the processes of learning, socialization, and individual psychological development is the concern of the present paper.

Schools as Places for Learning

Any review of history quickly reveals that what is perceived as appropriate school content has varied greatly over the years and is itself socially based. As one looks at the changes both in who should be educated as well as what should be taught, it becomes clear that schooling cannot be considered independently of its social context (Bowles and Gintes, 1976; Butts and Cremin, 1953; Nasaw, 1979). The school curriculum has focused at different times on literacy, practical tasks, classics, rote learning, independent learning, the arts, the 'basics', health, manual skills and habit training, among other subjects. Whatever the focus, however, one must ask how the physical environment impacts on the transmittal of information, on achievement, and on children's cognitive development. Studies on the effects of noise are relevant to this question.

Noise and learning

Although extremely limited, the research on school noise can be divided into two categories—studies that have looked at the impact of external noise, generated by airplanes and surface traffic, and studies of internal noise, generated by the daily activities of teachers and students. A provocative study by Bronzaft and McCarthy (1975) falls in the first category. These researchers looked at the reading and achievement scores of children attending an elementary school located within 220 feet of an elevated train (with 80 trains passing daily). Although classes on the noisy side and the quiet side were initially of comparable intellectual and achievement levels and were taught with similar methods, there were significant differences between the two sides by the end of the school year. Students on the noisy side had lower overall achievement test scores. Nine of the ten classes had reading scores that lagged three to four months behind the quiet side classes, and in one instance there was an 11 month lag. Bronzaft and McCarthy suggest that students' performance might have suffered because of a tendency for people in noisy environments to block out sound, whether relevant or not (Deutsch, 1964; Cohen et al., 1973). The result would be impaired auditory discrimination, which might, in turn, lead to difficulty in reading. Another, more direct, explanation is the loss of instructional time due to noise. The authors cite Stempler's (1973) finding that passing trains can result in an 11% loss of teaching time.

The study by Bronzaft and McCarthy is particularly germane to a discussion of the contribution of environmental psychology because of its real-world consequences. After the publication of the article, the New York City Transit Authority installed rubber padding on the train tracks closest to the schools, and the Board of Education placed sound absorbent ceilings in three of the noisiest rooms of the schools. Subsequent comparisons of reading scores showed children on the noisy side reading as well as those on the quiet side (Bronzaft, 1981).

Research by Cohen et al. (1980) has underscored the need to look at more than achievement scores when assessing the impact of noise. In this study, third and fourth grade students from four elementary schools in the Los Angeles Airport air corridor were matched with students from quiet schools. Although noise did not affect math or reading achievement scores, the noisy school students performed less well on a puzzle task, were less persistent and had higher blood pressure measures. There was also evidence of increasing distractibility with longer exposure.

One of the few existing studies of internally generated school noise was conducted by Slater (1968), who compared seventh-graders' performance on a reading test under quiet, average, and noisy school conditions. Data analysis revealed no noise effects on speed or accuracy of performance. A more recent study by Weinstein and Weinstein (1979) looked at the impact of noise in an open space school on fourth-graders' reading comprehension. Consistent with Slater's study, there was no difference in performance under conditions of quiet and normal background noise.

Despite this apparent lack of direct, internal noise effects on achievement, numerous surveys conducted in open space schools indicate that teachers find the noise from adjacent groups to be a problem. With few sound barriers to obscure noise, teachers report the sound level to be grating, to interfere with communication and to reduce the number of instructional options at their disposal (see, for example, Fitzpatrick and Angus, 1975). To remedy such problems, environmental researchers have occasionally provided assistance, either in the form of environmental workshops with school personnel or by way of direct design recommendations (Brunetti, 1972; Rivlin and Rothenberg, 1976; Rothenberg and Rivlin, 1975). A good example is a project by Evans and Lovell (1979) that sought to alleviate problems of student distraction, class interruptions, high noise levels and poor traffic flow patterns. Variable height, sound absorbent partitions were added to redirect traffic away from class areas and to define class boundaries. The modifications significantly reduced classroom interruptions and increased substantive, content questioning.

Classroom design and learning

A few studies have looked specifically at the effect of classroom design on achievement or achievement-related variables. Unfortunately, they provide no consistent picture of the relationship. For example, Horowitz and Otto (1973) compared the achievement of college students in a tradtional room and a colorful 'alternative learning facility', complete with a complex lighting system and flexible, comfortable seating. Grades on two term papers and a final examination showed no difference in learning between the two groups. More recently, however, Wollen and Montagne (1981) found that achievement *was* related to the aesthetic appearance of a college classroom. In this study, two sections of a college course were taught in two rooms of identical size and shape. One room was decorated in an attractive, colorful fashion, while the other was plain and monochromatic. Each class spent five weeks in each room. In the more attractive room, student achievement was superior, teachers were rated more positively, and attitudes towards the room were more positive. Unfortunately, no measures were taken of instructors' behavior under the two conditions, other than a single observation of student–teacher interaction that showed no room differences.

Another college study conducted by Sommer and Olsen (1980) looked at the effects of design on participation, a variable that may be related to achievement. In this project, modifications were made to a rather unattractive, traditional college classroom, creating a 'soft' space with tiered, upholstered benches, carpeting, and adjustable lighting. Observations indicated that the participation of students in this soft classroom (both in terms of the percentage of students voluntarily speaking and the number of comments per student) was substantially higher than in traditional classrooms that the researchers had studied previously.

Another indication that settings may indirectly influence learning comes from Santrock's (1976) study which found that the 'affective quality' of a setting influences the persistence of students at a task. Three environmental conditions were created for first- and second-graders—happy, sad and neutral. This was done by decorating a private room in the children's school with happy, sad, or neutral pictures. The experimenter also told the children happy or sad stories on the way to the room or acted in a neutral manner. During the motor task given in the room, the children were asked to describe happy, sad or neutral thoughts. Findings revealed that the environmental intervention had the most powerful effect on persistence: children worked longer in the happy room. An additive effect was also found, with children working longest in the happy room with the happy experimenter and with happy thoughts. These results underline the powerful impact of situational variables on task involvement, although one might question the ecological validity of the research approach.

The relationship between classroom design and learning has also been of particular interest to researchers involved in special education, where there has been a tendency to prescribe simple, stimulus-free environments for children with learning difficulties (e.g. Cruickshank *et al.*, 1961). However, a recent study by Weiland (1984) had results that run counter to conventional and professional wisdom. Weiland compared the performance of retarded, neurologically impaired, learning disabled, and normal children working in a 'stimulus-reduced' classroom and a 'stimulus-enriched' one. In the 'reduced' condition, the classroom was functional but spartan. The 'enriched' setting looked like a well-appointed classroom with more color, more equipment and more moveable and manipulable elements. All of the students worked on problems more effectively in the enriched setting. There was no indication that children were distracted by the elements of the enriched environment; indeed, they appeared to be more attentive, more persistent and more creative. Weiland's study is a good example of the ways in which environmental researchers can fight against the stereotypic thinking behind much environmental design and contribute to educational practice.

While much of the research on achievement has been conducted on the college level, studies looking at the effect of classroom design on cognitive development have focused, not surprisingly, on the preschool level. A recent study by Nash (1981), for example, compared the behavior and cognitive development of children in 19 'randomly arranged' preschool classrooms with that of children in 19 'spatially planned' rooms. In the randomly arranged rooms, equipment and materials were set up in either a haphazard fashion or according to pragmatic criteria (e.g. the water table should be near a sink). In the spatially planned rooms, the same equipment and materials were thoughtfully and intentionally organized to promote specific learning outcomes. Scheduling, activity choices, and interaction patterns were similar in all rooms. Yet, not only did children in the spatially planned rooms engage in more manipulative activities, they also produced more complex shape, color and number patterns using those materials (beads, pegboards, unit blocks, etc.). The most striking finding was that conservation was achieved earlier and by a greater number of children in the spatially planned rooms. In a similar vein, Moore (in press) found that the spatial definition of behavior settings is related to cognitive development behavior such as degree of engagement in developmental activities and exploratory behavior.

Seating position and achievement

As far back as 1932, Waller observed that the choice of a classroom seat did not appear to be random.

> 'In the front row is a plentiful sprinkling of over-dependent types, mixed perhaps with a number of extraordinarily zealous students. In the back row are persons in rebellion ... Quantitative investigation of these phenomena would be long and difficult, but not impossible.' (p. 161).

Fifty years later, a small body of empirical data exists that supports Waller's astute observation. Some of the research has examined personality variables and attitudes associated with various seats within the classroom. These studies generally demonstrate that students with more positive attitudes toward school and higher self-esteem sit toward the front of the classroom (see Weinstein, in press). Other research has focused specifically on achievement. Among this latter work is a college self-report study conducted by Becker *et al.* (1973). This investigation found that students sitting toward the front and the center of the room had higher course grades than those sitting toward the rear and the sides.

In an effort to discover whether this front-center phenomenon is due to self-selection or environment, Stires (1980) compared the test scores of two sections of a college course taught by the same instructor. Students in the 'choice' condition selected permanent seats on the second day of class; students in the 'no-choice' condition were seated alphabetically. In both sections, those in the middle of the room received higher grades than students at the sides, results that clearly support the environmental explanation.

Findings that bolster the self-selection explanation come from a very similar study by Wulf (1977). She found a front-center effect only when students had chosen their seats, and even then, only for participation. Grades were unrelated to seating position in both the choice and no-choice conditions. Studies by Millard and Stimpson (1980) and Kinarthy (1976) were also unable to find any evidence that assigned seating position in college classrooms affects grades.

Spatial cognition

There is one additional component of cognitive development and environmental research in schools that needs to be addressed: that is, the issue of spatial cognition. Several studies have looked at children's ability to represent classroom space and to translate their images of classrooms to scale models. For example, Rivlin, *et al.* (1974) gave second- to fourth-graders an unfurnished, scaled model of their classroom and asked them to set it up to look like their real room using scaled furnishings. The children were able to translate their images of what their room was like to the model. Moreover, they were able to use the model to answer detailed questions about classroom functions, suggesting that they had rather well-defined images of the spaces.

While children's ability to represent space changes developmentally (Piaget and Inhelder, 1956), features of the environment may serve to enhance developing representations. Siegel and Schadler (1977) showed that the accuracy of kindergarteners' models of their classrooms was improved when the model contained cues or pre-placed landmarks. Similarly, Liben *et al.* (1982) demonstrated that in the actual class-room, preschoolers remembered the location of furniture placed near a wall more accurately than comparable items located in the middle of the room. These findings suggest that representation of the spatial environment may be enhanced through the systematic use of bounding features (rugs, colored floor tiles, room dividers, etc.) and salient, distinctive landmarks. Within the classroom, such landmarks may be created with colorful signs and pictures if they do not exist as fixed features in the space.

There is also evidence that children's spatial representation is facilitated by the organization of the environment. Golbeck (1984), for example, asked preschoolers to arrange a small-scale model of their classroom under two condtions. One emphasized the functional relationships between items in the classroom (e.g. art activities, housekeeping, library), while the second lacked this emphasis. Children's constructions were more accurate when their attention was directed to the functional organization of the room, although some preliminary experience with the model task seems to be required with preschool aged children.

Hart's (1979) comprehensive study, *Children's Experiences of Place*, is also relevant to a discussion of schools and spatial cognition. Hart found that it was rare for children under eight years of age to be able to integrate accurately their school bus trip into the models of maps that they produced. School journeys and the school location had to be experienced by walking through the setting before children could represent the spatial relationships involved. The need for direct experience with places as a basis for spatial cognition is largely ignored in schools where neighborhood trips and journeys into the world are infrequent. Movement through the school itself is generally supervised, leaving few day-by-day opportunities for this important area of learning.

In evaluating work on spatial cognition and its implications for schools, it is important to remember that environmental psychologists studying cognition have typically been interested in children's knowledge of *specific places*, while researchers in spatial cognition have been primarily concerned with *abstract spatial concepts* (see Golbeck, in press; Liben, 1981 for further discussion of this distinction). Hart's work illustrates a concern with the former, while that of Siegel and Schadler (1977) and Golbeck (1984) illustrates a concern with the latter. However, schools and classrooms are particular physical places which continually present spatial representational problems to children. These experiences in the everyday world with specific places serve to nourish the development of abstract spatial concepts.

Implications for future work

One of the problems in interpreting the significance of the work on the school environment, achievement and cognitive development is the realization that we are dealing with a range of ages, a variety of research strategies, different physical con-ditions, and different independent variables and dependent measures. Most of the research described here has investigated the impact of 'microlevel' variables (room decor, seating position, etc.). Other research, however, has examined highly abstract, 'macrolevel' factors. A recent example of this type is a study by Wiatrowski *et al.*

(1983) that categorized the physical, social, and organizational characteristics of schools along two major dimensions of school differences—urban social disorganization and academic suburbanism. The authors suggest that this diagnosis could enable intervention. They conclude that 'small, decentralized, autonomous schools may be conducive to the propagation of orderly, crime-free environments' (p. 72), qualities associated with suburban schools. The specific features of the suburban schools—superior facilities and resources and teachers who helped students after regular school hours—are the ingredients of a more positive school career. In other words, such characteristics would enhance opportunities for positive learning experiences and, presumably, achievement.

The rich and varied opportunities offered by the school environment as stimuli for learning need close attention and increased documentation. We think that we know a great deal about the impacts of poorly designed or impoverished schools. It is time for environmental researchers to address the positive contributions that schools can make in children's intellectual development through increasing the environmental consciousness of responsible school personnel and evaluating and documenting innovative programs.

Schools as Places for Socialization

The powerful socializing role of the educational system cannot be ignored. Schools act as a major system for integrating children into society and designating their places within it. Schools become the arena for communicating to children the value system of our culture, one that is largely middle class and white. They concretize the norms by which behavior is to be judged, identify status, separate children from each other, and continue a system that will be perpetuated throughout most children's lives. In considering the social organization of a classroom, it is possible to examine this process and its environmental components. Seating position is one aspect that stands out as a communicator of status.

Seating position and status

A longitudinal study by Rist (1970), conducted over two and one-half years, traced the experiences of black ghetto children entering the school system. Rist vividly describes the caste system that developed from the differential treatment of the children by the teacher (who also was black), based on her varying expectations of their ability. What is most troubling about Rist's findings is the fact that these expectations were formed about the kindergartners even before they entered the classroom. Moreover, the expectations were based on information about the children that bore little or no relationship to learning potential—information from pre-registration forms (e.g. the presence or absence of a phone in the home), from lists of families on public welfare, from interviews with mothers and children during registration and from the experiences of the teacher or other teachers with the children's older siblings.

The teacher's expectations helped to determine where the children would sit and how much attention they would receive from her. On the eighth day of school, she assigned permanent seats, and from that point on, the class was divided into 'those expected to learn and those expected not to' learn (p. 423). The 'best' group sat at table 1, immediately in front of the teacher, with the best view of the blackboard. The children at this table differed from the children at tables 2 and 3 in several noticeable ways. They dressed better; few had very dark skin; they did not have matted hair and none of the children smelled of urine. In terms of behavior, the table 1 group interacted more easily with the teacher and used Standard American English more than their peers at the tables farther away. Finally, children at table 1 came from smaller, better educated families with higher incomes.

This early attribution of the children's abilities had long term effects on their school careers: Rist found that the tracks defined for them at age five continued in almost every case through second grade. This study clearly identifies the powerful impact that teacher expectations can have on the academic futures of children, with the institutional system and use of the physical environment partners in this process. Sitting at table 1, 2, or 3 in kindergarten defined much of these children's school futures.

Classroom environments and sex role stereotyping

If there is evidence that schools reinforce the class structure in our own society, they have also been indicted as potent social agents in instilling sex stereotypes. At the very least, they are settings in which prevailing stereotypic behaviors are played out and reinforced. Rothenberg's (1977) observations of children in a series of classrooms provided an opportunity to examine the distribution of class members over various sectors of rooms. She found that location and activities were often segregated by sex. For example, block areas of rooms were male-dominated, although the females valued these areas as well, reifying the gender-based view of appropriate activities and the options available depending on the child's sex.

A number of writers have suggested that resources in the environment are responsible for much of this sex-typed behavior on the part of very young children. Hartley *et al.* (1952) asked a group of pre-school teachers why they thought boys did not play at being fathers in the housekeeping areas. The teachers' subsequent examination of these areas revealed that the materials were generally female-oriented.

> 'As soon as items of men's clothing, such as shoes, jackets, and caps were included in the doll corner, as well as tool boxes, lunch boxes, pipes, and other masculine paraphernalia, the boys participated far more frequently. When, in addition, water-play was introduced in relation to household equipment such as egg-beaters, pots, muffin tins, mops, and dishes, the attraction of domestic play for boys increased still further.' (p. 49)

Osmon (1971) further recommends equipping the housekeeping area with a variety of 'junk materials'—bottles, boxes, flower pots, gears, wheels—that can serve as a stimulus for all kinds of dramatic play, lessening dependence on traditional sex-typed props.

Kinsman and Berk (1979) joined the block and housekeeping areas in a preschool and a kindergarten in an effort to encourage children to enter the opposite-sex setting. Although the results were not entirely consistent with the hypothesis, the intervention did lead to an increase in mixed-sex groups in both areas. Interestingly, the younger children seemed more amenable to the changes than the kindergarteners, who tried to rebuild the wall between the areas using trucks, a mirror, an ironing board, and other equipment. Kinsman and Berk conclude that:

'the separate environments promoted interaction among children of the same sex, separation of boys from girls, and less constructive and involved play by children who entered an "opposite-sex" area. Girls, for whom sex-related play alternatives are less rigid and prescribed, adapted more readily to the removal of shelves than boys. Girls showed an especially impressive increase in relevant, constructive use of the block area after the settings were joined.' (p. 73)

This study by Kinsman and Berk is a particularly interesting illustration of how environmental manipulations can bring about significant changes in children's social behavior.

Implications for the future

Environmental psychologists have focused on the role of the home in the formation of our sense of ourselves as productive, worthy persons, as persons capable in particular areas and able to pursue them. The available research suggests that it would be well to turn attention to schools as carriers of symbolic meanings of personhood, as potent social forces that shape people's identities.

Whether it is the assignment of a seating position or the arrangement of a room, the social significance of the action must be considered in light of its impact on children's images of themselves as individuals and as members of a group. Much of this has individual psychological significance for children, an issue that will be addressed in the next section. But a good deal defines the social world of children, the limits and opportunities that they see for themselves. As evaluators of social institutions, schools among them, environmental psychologists can describe these impacts and outline some of the directions for change.

Schools as Places for Psychological Development: School as Haven

Adapting Rainwater's (1966, 1970) view of housing in his description of house-as-haven, one can view school as more than a place to learn and more than a context for socialization. It is a place with familiar scenes and people that can protect children from the unknown, the unfriendly, the things to be feared. It is an arena for psychological development. It is a place with meaning both for children's images of themselves and for their image of learning the domain of ideas, facts, and other information. Bettelheim (1955), writing about his residential school for severely disturbed children, the Orthogenic School, presents a similar view.

'As the sheltering function of the School demonstrates to the child that his basic needs of physical protection, for warmth, food, and rest, will always be satisfied, the strange buildings slowly become "his home". This is particularly true as meaningful relations to protecting adults are added to the satisfaction of all his other needs ... Only after a child is able to feel "at home" in the physical environment of the School does he begin to recognize and accept other children, to establish contact with his counselors, and gradually to form close relations with them.' (p. 37)

There is a real difference, however, between a home and a school, even a residential school, and for this reason the term 'haven' has been chosen to connote a safe place, a refuge, a sanctuary. The degree to which schools reach the level of haven will obviously vary. Current feelings of general dissatisfaction with the schools, as well as problems with truancy and vandalism (Zeisel, 1974, 1976) suggest that many children find no haven in the educational system; indeed, that they are running away from it.

Yet, the school remains, for most children, the setting in which they spend the greatest portion of their time, after the home. Increasing numbers of children are introduced into educational settings earlier and earlier. Even some infant day care programs take school as their model. With Head Start programs, nursery schools, and now suggestions that school begin at age four and that the school year be lengthened, the educational system becomes part of children's life space and social world in even more intense ways.

Environmental autobiographies (Cooper Marcus, 1978; Hester, 1978; Horwitz and Klein, 1978) testify to the powerful effects of early living environments on people's adult worlds. With school a major, early environmental experience, the first continuous one involving departure from the home nest, the protective qualities that schools offer, albeit varying for different children, must be considered.

What is formed in school is a large part of people's sense of themselves, their sense of competence, their ability to relate to peers and adults, equals and authorities. A good deal is abstracted from the teacher's responses to the child as an individual and the quality of the setting as haven. Whether children are able to project and develop their individual interests and skills, whether there are opportunities for privacy, whether there are places in the room that the child can personalize and with which they identify are issues salient to environmental psychologists and their colleagues. These have been examined to some extent, although not sufficiently.

Fostering satisfaction, competence and security
The work of Roger Barker's team is relevant to a consideration of school as haven. This classic research showed that the experiences of students in big schools and small schools are vastly different (Barker and Gump, 1964). The numbers of students available for positions of responsibility and work can affect the level of participation and feelings of satisfaction of any one student. Small schools that are 'under-manned' allow the participation of those not necessarily expert in an area, while large schools need draw on only those who excel. The school atmosphere, morale, and activities will be greatly affected by the kinds of skills necessary for participation, and these relate to other aspects of school functions. Certainly, size will affect the nature of the experience for students, defining the degree of bureaucracy, access to services and resources, and the ability to understand the environment. Yet classes and schools are increasing the numbers of children accommodated. As school populations drop there is a tendency to close them and consolidate with other schools.

The recent work by Moore and his colleagues (Moore *et al.*, 1979) on child care settings also speaks to the issue of school as haven. They have compiled an extensive list of recommendations to ensure that the child care environment enhances children's feelings of competence, individuality, and self-esteem. For example, they urge that environments be 'child-scaled' so that fountains, sinks and doorknobs are accessible to children; storage should be low, open and well-organized; the first view of the classroom should be inviting, familiar and friendly, so that children can be reassured that 'good things happen in this place'.

In a similar fashion, Olds (1979) argues that classrooms for young children should be designed to promote feelings of comfort and competence. She writes the following.

'Teachers often complain that preschool-age children ... have short attention spans and are unable to attend to activities long enough to get beyond superficial levels of involvement with materials ... Yet young children will not engage in genuine exploratory and discovery behaviors unless they first feel comfortable and secure in their physical surroundings. In-depth inquiry requires an ambience in which children can "lose" themselves, not in the sense of being confused and disoriented, but in the sense of feeling safe from attack, intrusion, and exposure, which makes them self-conscious about their performance. Comfortable surroundings foster playful attitudes toward events and materials that help lower anxiety, promote understanding, and enable children to be more open in divulging their personal responses to events ... To support the development of attentional processes, memory, and mastery, it is quite important that classrooms be inviting, comfortable places that entice learners to pause, play, and stay for a while.' (p. 94)

The need for privacy

A limited amount of research has focused on children's need for privacy, although once again it is difficult to abstract generalizations because of the variety of approaches used. What is most apparent is the fact that privacy, 'chosen physical aloneness' (Golan, 1978), is important to children in schools and does not vanish as a domain of life in the group atmosphere of the classroom. In fact, some of the alternative teaching methods associated with open education may intensify the need to withdraw from the stimulation of the surround (Rivlin and Rothenberg, 1976). The degree to which this is possible in the classroom may help to define some of the aspects of school as haven.

Weinstein's experimental study (1982) introduced privacy booths, places where students could be alone, into a fourth-grade classroom. Some interesting sex differences appeared. Teachers' ratings of sociability, aggressiveness and distractibility were related to use of the booths by boys—with less sociable, more aggressive and more distractible individuals using the booths more—while for girls the only variable that correlated significantly with booth use was privacy-seeking behavior at home. Overall, there was substantial individual variation in booth use that remained consistent throughout the study.

The impact of density

Density within the classroom also has been the subject of research. Interestingly, much of the work has focused on either preschool or college-level classrooms. Preschool studies have looked at children's level of involvement and their social behavior under different density conditions. For example, Shapiro (1975) found that noninvolved behavior was most frequent in classrooms where space was less than 30 square feet per pupil and in those with more than 50 ft² per pupil. In this study, optimal conditions for involvement fell between the extremes.

A number of studies have pointed to a link between density and preschool aggression (Bates, 1972; Hutt and Vaizey, 1966; Loo and Kennelly, 1978). However, there also is evidence that the amount of space is not independent of the resources available within the space. Studies by Rohe and Patterson (1974) and Smith and Connolly (1972) have observed no increase in aggression with greater density when resources also are increased.

Smith and Connolly (1980) set up a preschool in which to vary and study many of the density attributes, especially the interrelationship of the number of children in a group, amounts of space and play equipment, preschool curricula, and staff: child ratio. Sex differences were also examined. The findings revealed that smaller classes (as few as ten children) tended to have more cross-sex friendships and fantasy play. When classroom size was varied (25, 50 or 75 ft^2 per child), the larger spaces had more space for running, chasing, and 'vigorous or unusual uses of apparatus' but no substantial change in social or aggressive behavior (p. 309).

On the college level, Schettino and Bordon (1976) obtained self-reports from students indicating that males experienced greater aggressiveness with increasing density, while females experienced increased feelings of nervousness and crowdedness.

The subtle interaction between density of the home environment, opportunities for privacy in the home and school, and school performance is another area of interest to the environmental researcher. Saegert (1981) has reviewed this literature citing evidence that home density has some measured negative effects on school performance, both behavioral and academic, although there is a clear need for controlled research in this area. Her own study (Saegert, 1981) of the relationship between residential density and children's well-being revealed that children from high density apartments were rated by teachers as more anxious and more hyperactive-distractible. In addition, higher density was related to lower vocabulary scores and to lower than average scores on reading comprehension, especially in this case, for boys.

Despite the considerable interest in density by environmental specialists, there is a limited amount of this research in school settings. In addition, researchers have used different levels of density and different indicators of effects. When combined with major gaps in the age groups studied, the ability to generalize about the impacts of density is sorely constrained.

Implications for the future

We are far from understanding the complex ramifications of school density, prior density experiences, or the importance of opportunities for privacy and personalization of school settings—that is, the importance of school as haven. The studies such as the ones described here have begun to articulate the important parameters on which to focus and set directions for future work.

If places serve to provide a sense of individuality to those within them—'place-identity' in Proshansky's terms (Proshansky *et al.*, 1983) certainly the school must make a significant contribution toward children's personal development. One's sense of identity may begin to develop at home, but the school is a major social and psychological force that adds much to a child's sense of self and the interests, skills, and personal qualities that define identity.

Concluding Thoughts

Environmental change

Schools have often been used as a context for assessing the effects of environmental manipulations. Indeed, the educational setting is an especially good place for pre-post studies designed to evaluate the process of environmental change. There has been research on changes intended to increase opportunities for privacy (Weinstein, 1982), changes to increase cooperative play and use of specific areas of a room (Sutfin, 1982), and changes to distribute children over the room, broaden their range of behavior and change the frequency of specific behaviors (Phyfe-Perkins, 1982; Weinstein, 1977). Environmental researchers need to direct more attention to this area, however, along with feedback to the participants.

One other point concerns the actors involved in this change effort. Usually the researcher focuses on the school administration and teaching staff, rather than the children. The work of a number of environmental researchers (e.g. Van Wagenberg *et al.*, 1981) suggests that children can be involved in this process, especially when they have been exposed to some environmental education. Children can address problems in their classrooms with great realism and sensitivity, come up with creative ideas for change, build scaled models and then implement the changes. These skills need to be developed, evaluated, and recognized as an important part of children's education. It is one component of a broad environmental education program to which environmental researchers can contribute.

Environmental psychology and education

In summary, the work accomplished by environmental researchers provides a skeleton, a framework that needs to be filled out and developed. We are not starting at point zero, but neither can we be complacent about the relatively little that we know. We need more information about the subtle effects of exclusion, whether by caste or sex. We need more attention to the enduring effects of educational experiences by focusing on influences that can be traced over time. We need attention to the school as *place*, as a physical entity and continuing experience in children's lives. Its appearance, the comfort and safety it affords, are important to children's personal and intellectual development. Gump (1978) has suggested that in environmental research no other variable has been shown to have the power of size. But there are many important variables imperfectly or insufficiently studied that need attention in the future by a wide range of professionals, including environmental psychologists.

This is not a good time for education. Slowly all but the basics are being eliminated. Schools are being closed and children consolidated in unreflective ways. The enthusiasm of environmental psychologists and their collaborative work with students, teachers, and educational researchers may offer some relief from increasing program and economic restrictions. Such a participatory research model could serve, at the very least, as one way of raising environmental consciousness among school personnel. Perhaps it might even serve to improve the quality of life in schools.

References

Adams, R. S. (1969). Location as a feature of instructional interaction. *Merrill–Palmer Quarterly of Behavior and Development*, **15**, 309–22.

Adams, R. S. and Biddle, B. J. (1970). *Realities of Teaching: Explorations With Videotape*. New York: Holt, Rinehart & Winston.

Armstrong, D. G., Henson, K. T., Savage, T. V. (1981). *Education: An Introduction*. New York: Macmillan Publishing Co.

Barker, R. G. and Gump, P. V. (1964). *Big School, Small School*. Stanford, California: Stanford University Press.

Bates, B. C. (1972). *Effects of Social Density on the Behavior of Nursery School Children*. University of Oregon, Center for Environmental Research.

Becker, F. D., Sommer, R., Bee, J. and Oxley, B. (1973). College classroom ecology. *Sociometry*, **36**, 514–25.

Bettelheim, B. (1955). *Truants From Life*. Glencoe, Illinois: The Free Press.

Bowles, S., and Gintes, H. (1976). *Schooling in Capitalist America: Educational Reform and the Contradictions of Economic Life*. New York: Basic Books.

Boyer, E. L. (1984). *Panel on the Preparation of Beginning Teachers*. New Jersey: State Department of Education.

Bronzaft, A. (1981). The effects of a noise abatement program on reading ability. *Journal of Educational Psychology*, **1**, 215–22.

Bronzaft, A. and McCarthy, D. P. (1975). The effect of elevated train noise on reading ability. *Environment and Behavior*, **7**, 517–27.

Brunetti, F. A. (1972). Noise, distraction, and privacy in conventional and open school environments. In W. J. Mitchell (ed.), *Environmental Design: Research and Practice*. Proceedings of EDRA 3/AR 8 Conference, Los Angeles, University of California Press.

Butts, F. R. and Cremin, L. A. (1953). *A History of Education in American Culture*. New York: Holt, Rinehart & Winston.

Cohen, S., Glass, D. C. and Singer, J. E. (1973). Apartment noise, auditory discrimination, and reading ability in children. *Journal of Experimental Social Psychology*, **9**, 407–422.

Cohen, S., Krantz, D., Evans, G. W. and Stokols, D. (1980). Community noise and children: cognitive, motivational and physiological effects. In J. Tobias (ed.), *The Proceedings of the Third International Congress on Noise as a Public Health Problem*. Washington D.C.: American Speech and Hearing Association.

Cooper Marcus, C. (1978). Remembrance of landscapes past. *Landscape*, **22**, 34–6.

Cruikshank, W. M., Bentzen, F. A., Ratzeburg, F. H. and Tannhauser, M. T. (1961). *A Teaching Method for Brain-injured and Hyperactive Children*. Syracuse: Syracuse University Press.

Deutsch, C. P. (1964). Auditory discrimination and learning: social factors. *The Merill-Palmer Quarterly of Behavior and Development*, **10**, 277–96.

Evans, G. W. and Lovell, B. (1979). Design modification in an open plan school. *Journal of Educational Psychology*, **71**, 41–9.

Fitzpatrick, G. S., and Angus, M. J. (1974). *Through Teachers' Eyes: Teaching in an Open Space Primary School. Technical Report No. 1*. Education Department of Western Australia, West Perth, ED118 530.

Golan, M. B. (1978). *Privacy, Interaction and Self-esteem*. Unpublished Doctoral Dissertation, City University of New York.

Golbeck, S. (1984). Constructing a model of large-scale space with the space in view: effects of task structure and cognitive style. Paper presented at the Southeastern Conference on Human Development, Athens, GA.

Golbeck, S. (in press). Spatial cognition as a function of environmental characteristics. In R. Cohen (ed.), *The Development of Spatial Cognition*. Hillsdale, New Jersey, Lawrence Erlbaum Associates.

Gump, P. V. (1978). School environments. In I. Altman and J. F. Wohlwill (eds), *Children and the Environment*. New York: Plenum Press.

Hart, R. (1979). *Children's Experience of Place*. New York: Halsted Press.

Hartley, R. E., Frank, L. K. and Goldenson, R. M. (1952). *Understanding Children's Play*. New York: Columbia University Press.

Hester, R. (1978). Favorite spaces. *Childhood City Newsletter*, **14**, 15–17.

Holt, J. (1964). *How Children Fail*. New York: Pitman Publishing Corporation.

Horowitz, P. and Otto, D. (1973). *The Teaching Effectiveness of an Alternative Teaching Facility. (ERIC Document Reproduction Service No. ED 083 242.)* Alberta, Canada: University of Alberta.

Horwitz, J. and Klein, S. (1978). An exercise in the use of environmental autobiography for programming and design of a day care center. *Childhood City Newsletter*, **14**, 18–19.

Hutt, C. and Vaizey, M. J. (1966). Differential effects of group density on social behavior. *Nature*, **209**, 1371–1372.

Ittelson, W. H., Proshansky, H. M. and Rivlin, L. G. (1970). The environmental psychology of the psychiatric ward. In H. M. Proshansky, W. H. Ittelson and L. G. Rivlin (eds) *Environmental Psychology*, 1st Edit. New York: Holt, Rinehart & Winston.

Kinarthy, L. (1976). The effects of seating position on performance and personality in a college classroom. (Doctoral dissertation, University of Southern California). *Dissertation Abstracts International*, **37**, 2078A.

Kinsman, C. A. and Berk, L. E. (1979). Joining the block and housekeeping areas: changes in play and social behavior. *Young Children*, **35**, 66–75.

Kozol, J. (1967). *Death at an Early Age*. Boston, Massachusetts: Houghton Mifflin Co.

Liben, L. (1981). Spatial representation and behavior: Multiple perspectives. In L. S. Liben, A. H. Patterson and N. Newcombe (eds), *Spatial Representation and Behavior Across the Life-Span*. New York: Academic Press.

Liben, L., Moore, M. and Golbeck, S. (1982). Preschoolers' knowledge of their classroom environment: evidence from small-scale and life-sized spatial tasks. *Child Development*, **53**, 1275–84.

Loo, C. and Kennelly, D. (1978). Social density: its effects on behaviors and perceptions of preschoolers. Paper presented at the American Psychological Association Meeting, Toronto, Cánada.

Millard, R. J. and Stimpson, D. V. (1980). Enjoyment and productivity as a function of classroom seating location. *Perceptual and Motor Skills*, **50**, 439–44.

Moore, G. T. (in press). The role of the socio-physical environment in cognitive development. In T. G. David and C. S. Weinstein, *Spaces for Children: The Built Environment and Child Development*. New York: Plenum.

Moore, G. T., Lane, C. G., Hill, A. B., Cohen, U. and McGinty, T. (1979). *Recommendations for Child Care Centers*. Milwaukee, Wisconsin: Center for Architecture and Urban Planning.

Moos, R. H. (1972). Assessment of the psychosocial environments of community-oriented psychiatric treatment programs. *Journal of Abnormal Psychology*, **79**, 9–18.

Nasaw, D. (1979). *Schooled to Order: A Social History of Public Schooling in the United States*. New York: Oxford University Press.

Nash, B. C. (1981). The effects of classroom spatial organization on four- and five-year-old children's learning. *British Journal of Educational Psychology*, **51**, 144–55.

National Commission on Excellence in Education (1983). *A Nation at Risk: The Imperative for Educational Reform*. Washington, D.C.: U.S. Government Printing Office.

Olds, A. R. (1979). Designing developmentally optimal classrooms for children with special needs. In S. J. Meisels (ed.), *Special Education and Development: Perspectives on Young Children with Special Needs*. Baltimore, Maryland: University Park Press.

Ornstein, A. C. and Miller, H. L. (1980). *Looking into Teaching*. Chicago: Rand McNally College Publishing Co.

Osmon, F. L. (1971). *Patterns for Designing Children's Centers*. New York: Educational Facilities Laboratories.

Phyfe-Perkins, E. (1982). The preschool setting and children's behaviour: an environmental intervention. *Journal of Man–Environment Relations*, **1**, 10–29.

Piaget, J. and Inhelder, B. (1956). *The Child's Conception of Space*. New York: W. W. Norton, & Co.

Proshansky, H. M., Fabian, A. K. and Kaminoff, R. (1983). Place-identity: Physical world socialization of the self. *Journal of Environmental Psychology*, **3**, 57–83.

Rainwater, L. (1966). Fear and the house-as-haven in the lower class. *Journal of the American Institute of Planners*, **32**, 23–31.

Rainwater, L. (1970). *Behind Ghetto Walls*. Chicago: Aldine.

Rist, R. C. (1970). Student social class and teacher expectations: The self-fulfilling prophesy in ghetto education. *Harvard Educational Review*, **40**, 411–51.

Rivlin, L. G. and Rothenberg, M. (1976). The use of space in open classrooms. In H. M.

Proshansky, W. H. Ittelson and L. G. Rivlin (eds), *Environmental Psychology: People and Their Physical Settings* 2nd Edit. New York: Holt, Rinehart & Winston.

Rivlin, L. G., Rothenberg, M., Justa, F., Wallis, A. and Wheeler, F. G., Jr. (1974). Children's conceptions of open classrooms through use of scaled models. In D. H. Carson (ed.) *Man–Environment Interactions: Evaluations and Applications.* Washington, D.C.: Environmental Design Research Association.

Rohe, W. and Patterson, A. H. (1974). The effects of varied levels of resources and density of behavior in a day care center. In D. H. Carson, (ed.) *Man–Environment Interactions: Evaluation and Applications.* Washington, D.C.: Environmental Design Research Association.

Rothenberg, M. S. and Rivlin, L. G. (1975). An ecological approach to the study of open classrooms. *Conference on Ecological Factors in Human Development,* International Society for the Study of Behavioral Development, University of Surrey, England.

Rothenberg, M. S. (1977). *The Social and Spatial Organization of Boys and Girls in Open Classrooms.* Unpublished Doctoral Dissertation, City University of New York.

Saegert, S. (1981). Environment and children's mental health: residential density and low income children. In A. Baum and J. Singer (eds) *Handbook of Psychology and Health* (Vol. 2). Hillside, New Jersey: Erlbaum Associates.

Santrock, J. W. (1976). Affect and facilitative self-control: influence of ecological setting, cognition, and social agent. *Journal of Educational Psychology,* **68**, 529–35.

Schettino, A. P. and Borden, R. J. (1976). Sex differences in response to naturalistic crowding: affective reactions to group size and group density. *Personality and Social Psychology Bulletin,* **2**, 67–70.

Shapiro, S. (1975). Preschool ecology: A study of three environmental variables. *Reading Improvement,* **12**, 236–41.

Siegel, A. W. and Schadler, M. (1977). The development of young children's spatial representations of their classrooms. *Child Development,* **48**, 388–94.

Silberman, C. E. (1970). *Crisis in the Classroom.* New York: Random House.

Silverstein, B. and Krate, R. (1975). *Children of the Dark Ghetto: A Developmental Psychology.* New York: Praeger.

Slater, B. (1968). Effects of noise on pupil performance. *Journal of Educational Psychology,* **59**, 239–43.

Smith, P. L. and Connolly, K. J. (1972). Patterns of play and social interaction in preschool children. In N. Blurton Jones (ed.), *Ethological Studies of Child Behavior.* Cambridge, England: Cambridge University Press.

Smith, P. K. and Connolly, K. J. (1980). *The Ecology of Preschool Behavior.* Cambridge, England: Cambridge University Press.

Sommer, R. and Olsen, H. (1980). The soft classroom. *Environment and Behavior,* **12**, 3–16.

Sommer, R. and Ross, H. (1958). Social interaction on a geriatrics ward. *International Journal of Social Psychiatry,* **4**, 128–33.

Stempler, S. (1973). *Subway Noise Impact on P.S. 98. Report BNA No. 31130.* New York: Department of Air Resources.

Stires, L. (1980). The effect of classroom seating location on student grades and attitudes: environment or self-selection? *Environment and Behavior,* **12**, 241–54.

Sutfin, H. D. (1982). The effect on children's behavior of a change in the physical design of a kindergarten classroom. *Journal of Man–Environment Relations,* **1**, 30–41.

Van Wagenberg, D., Krasner, M. and Krasner, L. (1981). Children planning an ideal classroom: environmental design in an elementary school. *Environment and Behavior,* **13**, 349–59.

Walberg, H. (1969). Physical and psychological distance in the classroom. *School Review,* **77**, 64–70.

Waller, W. (1932, 1965). *The Sociology of Teaching.* New York: John Wiley.

Weiland, G. (1984). *A Psychological Examination of Learning Environments for Handicapped Children.* Unpublished Doctoral Dissertation, City University of New York.

Weinstein, C. S. (1977). Modifying student behavior in an open classroom through changes in the physical design. *American Educational Research Journal,* **14**, 249–62.

Weinstein, C. S. (1979). The physical environment of the school: a review of research. *Review of Educational Research*, **49**, 577–610.

Weinstein, C. S. (1981). Classroom design as an external condition for learning. *Educational Technology*, **21**, 12–19.

Weinstein, C. S. (1982). Privacy-seeking behavior in an elementary classroom. *Journal of Environmental Psychology*, **2**, 23–35.

Weinstein, C. S. (in press). Seating arrangements in the classroom. *International Encyclopedia of Education*. Oxford: Pergamon Press.

Weinstein, C. S. and Weinstein, N. D. (1979). Noise and reading performance in an open space school. *Journal of Educational Research*, **72**, 210–13.

Wiatrowski, M. D., Gottfredson, G. and Roberts, M. (1983). Understanding school behavior disruption: classifying school environments. *Environment and Behavior*, **15**, 53–76.

Wolfe, M. and Rivlin, L. G. (in press). The institutions in children's lives. In T. G. David and C. S. Weinstein, *Spaces for Children: The Built Environment and Child Development*. New York: Plenum.

Wollin, D. D. and Montagne, M. (1981). College classroom environment: effects of sterility vs. amiability on student and teacher performance. *Environment and Behavior*, **13**, 707–16.

Wulf, K. M. (1977). Relationship of assigned classroom seating area to achievement variables. *Educational Research Quarterly*, **21**, 56–62.

Zeisel, J. (1974). Designing out unintentional school property damage: a checklist. In D. H. Carson (ed.) *Man–Environment Interactions: Evaluations and Applications*. Washington, D.C.: Environmental Design Research Association.

Zeisel, J. (1976). *Stopping School Property Damage: Design and Administrative Guidelines to Reduce School Vandalism*. Arlington, Virginia: American Association of School Administrators, and New York: Educational Facilities Laboratories.

THE TRANSITION FROM HOME TO BOARDING SCHOOL: A DIARY-STYLE ANALYSIS OF THE PROBLEMS AND WORRIES OF BOARDING SCHOOL PUPILS

SHIRLEY FISHER, NORMAN FRAZER and KEITH MURRAY

Department of Psychology, University of Dundee

Abstract

The investigation concerns the impact of the new school environment on a group of 50 male and female children aged 11–16 years, who leave home to reside temporarily at boarding school, in terms of the characteristics of problems and worries reported and the incidence of spontaneous reports of homesickness. More problems relating to the school than to the home environment were reported but proportionally more worry units were reported associated with home problems for both males and females. There was no sex differences in this respect. The reported level of spontaneously reported homesickness was 16% and there were no sex differences. Factors such as age, geographical distance of move and decision to go away to school were not influential in determining the level of reported problems or incidence of spontaneous reports of homesickness. A relationship was found with level of problems reported and recent life history but the result proved difficult to interpret.

Introduction

One feature of the British education system is the child who leaves home as early as age seven to reside (board) at school (Note 1). The majority of these schools, generally termed 'boarding schools', fall largely within the private education sector, and therefore tend to cater for a high socioeconomic catchment.

There is no research literature addressed to the problem of leaving home to attend school as a boarder for the first time, therefore the extent to which the transition to boarding school is accomplished smoothly with little stress remains unknown. For the children concerned it involves a change in residence with possible accompanying loss of the security of family and friends as well as the need to adjust and conform to a new life style, make new acquaintances, learn new rules and regulations and begin new academic subjects. All these factors are possible sources of pressure on the child. Research aimed at locating the sources of pressure will provide useful insights into aspects of disruption and adaptation. This study is concerned with such an approach based on diary-style analysis of the problems experienced by boarding school pupils in their first two weeks at boarding school.

There is an existing research literature on the effects of relocation in adults, either enforced as with slum clearance schemes (see Fried, 1963) or voluntary as when a move is required because of a job transfer (see Brett, 1982). In general, the research evidence is equivocal with respect to whether the results for physical and mental health are negative (see Syme *et al.*, 1965; Cruze-Coke *et al.*, 1964) or positive, unless there

The work reported here was supported by a grant from the Economic and Social Science Research Council.

are already pressures on the individual created by handicap, poverty or racial integration problems (see Fischer and Steuve, 1977; Newman and Owen, 1982). The possibility remains that some relocations may be desirable with few adverse consequences whereas others are undesired with adverse effects. This raises the question of the meaning given to moves by the individual. Stokols *et al.* (1983) distinguished the *spatial context* of the move (factors relating to home, work and commuting domains) from the *temporal context* (factors relating to previous, current and anticipated events within life history). The authors argue that the *congruence* of the domains (the extent to which they support personal goals) and the temporal context of the move are critical factors.

One of the difficulties with studying the effects of relocation is that the nature of relocations vary greatly. For example Wapner *et al.* (1981) recognize transitions which do not disrupt equilibrium from those which do and are referred to as 'critical transitions'. Examples of critical transitions provided by the authors are 'forced migration', 'obliging one to leave the locus of one's habitual action', 'the loss through death or separation of someone very intimate', 'the abandonment by another', 'the obligation to leave a well-known and familiar environment for one that is alien', 'the obligation to reitre from one's job or vocation' and 'the sudden upheavals in familiar scenes of action occasioned by natural catastrophes'. The authors further point out that it is the experience of the agent that determines whether a transition is critical. They further argue that although not all critical transitions lead to disruption of self–world relationships and the sense of overwhelming loss, there are at least attenuated manifestations of this discernible in all of the above phenomena. They illustrate the point with regard to children obliged to go to school—the school setting is seen as an alien setting or threatening agent from which they must flee; 'they may search for regions there as instrumentalities of refuge or sanctuary' (p. 265).

The relocation a child undergoes when leaving home to attend boarding school is very different in many important respects from the typical forms of relocation experienced by adults, and on Stokol's argument concerning the importance of circumstantial factors, it would be undesirable to generalize from work on other forms of relocation. The relocation the boarding school child undergoes could be described as 'reversible relocation' in the sense that the move is not permanent and the previous 'home' environment continues to exist in a different geographical location and can be visited or contacted. In addition the child leaves school at regular time intervals and undertakes 'holidays' at home which are usually greater than two weeks but in the Summer may last as long as six to seven weeks.

Although, as already emphasised it is possible that each relocation has unique properties and that idiosyncratic features further qualify the circumstance of individual moves, there is good reason to suppose that the transition from primary to secondary education will impose pressures on children at an age which is particularly vulnerable. Mechanic (1983) argues that *any* discontinuity such as a geographical move or a change of school may encourage feelings of isolation and self-consciousness in adolescents. Pupils who attend boarding school for secondary education are usually aged between 11 and 13 years and are on the threshold of adolescence.

The continued existence of home could be argued to create peculiar circumstances for the child. If, because of bereavement, the child had no home or parents a very strong adverse reaction would be expected. In the case of a 'reversible relocation', the child is in a sense temporarily bereaved in that much of the contact, support and affection parents and family provide is less directly obtainable. Moreover, the history of living at home may create a source of daily worries which may continue to feature although the child is no longer physically present and able to influence the outcome.

Interruption Theory as formulated by Mandler (1975) suggests that interruption of on-going activities may produce release of tension and provide a basis of further cognitive activity as the individual seeks to restore equilibrium or to find substitute activity. This model may be very circumscribed, based as it is on laboratory studies of involuntary interruption of tasks, but provides some further basis for arguing that the influence of the 'home' may be expected to be powerful in the beginning of a move to a new environment.

Adjustment to the new school may be affected by the fact that the boarding school child has less direct sources of support from parents and family. He or she is more responsible for a greater proportion of daily decisions. At boarding school a child's 'school life' is effectively extended until bed-time because routines and disciplines are still in existence and the child remains responsible for uniform, clothes, personal possessions, etc. In comparison with a day-school child there are more potential sources of problems and less access to parental help in solving them. The child has to cope with a new school as well as learn self-reliance and independence.

Gump (1980) summarizes some interesting points about the role of the class teacher: Parsons (1955) argued that the teacher is not seen as a mother surrogate but as one who presses pupils away from exclusive familial ties. The teacher for example is less likely to respond to a child's emotional needs and more likely to require activity from him. Schoggen (1975) found that interactions between parents are more individualized and frequent, although sometimes negative, whereas teachers were more likely to make demands and influence the childrens' behaviour. Taken collectively these findings suggest that a world dominated by teachers rather than parents might be rather different from a world dominated by parents in many important respects, and this in itself requires adjustment.

Brown and Armstrong (1982) investigated the reported worries of day-school children in transition from primary to secondary school by means of a quantitative analysis of worries reported in essays written in the first English lesson. Out of 22 categories of worries summarized by the authors the most commonly reported worry (in terms of the number of subjects reporting it) was that of 'general fear of school routines' (62%), followed by 'feeling lost' (35%) and 'strict teachers' (28%). When compared with worries reported in the second term, there was a change towards increased concern with 'tests' (81%), 'detentions' (37%) and 'being late' (35%). Certainly the change in emphasis towards concern with tests may mean that other worries are attenuated (implying a limited 'worry resource') or that attitudes have changed towards on-going concerns at school rather than the problems of new encounters.

The current study is concerned with examining quantitative differences in reported *problems* from 'home' and 'school' during the first two weeks of the school year. The data on reported problems was obtained by means of diary-style techniques. As part of the same diary study additional data was obtained on the number of times during the day that a pupil could remember *worrying* about a particular problem reported. This provided a second dependent variable and the main issue of interest concerns the proportionate amount of worrying associated with home or school oriented problems. Because home problems may have a longer potential problem history and are less controllable because of difficulties produced by physical distance, it is hypothesized that home-oriented problems might be associated with greater levels of worrying. To investigate this a technique is introduced of calculating the problem/ worry ratio. A large value indicates low worry levels and a low value indicates high worry levels for a particular problem.

As part of the same design a second focus concerned obtaining evidence of spontaneous use of the term 'homesickness', which is an anecdotally observed reaction to leaving home, known to those who deal with students in institutions such as boarding schools and universities. It is a term defined in dictionaries as 'pining for home' (*Chambers Dictionary*, 1954) and 'depressed by absence from home' (*Pocket Oxford Dictionary*, 1969). In spite of the common acceptance of the term and its meaning, there is little information about homesickness: what conditions create, modify or ameliorate it or how serious it is for the pupil. Fisher (1984a, 1984b) reported the results of a pilot study in which 64% of a sample of university students reported homesickness in their first term at university. However, in this study, the term was provided and students were asked to indicate whether they had or had not experienced it. There is an initial need to obtain data in conditions where the term is not used and no indication is provided to indicate that it is an issue of interest.

An extra feature of the study depended on whether the balance of homesick/non-homesick subjects provided a division adequate for an experimental comparison. The issue of interest was whether those who report homesickness have a higher number of reported problems at school or at home and whether the problem/worry ratios are lower than for those who do not report homesickness.

In advance of the investigation the following hypotheses were formulated.

(1) There would be a greater number of reported problems which were 'school-orientated' rather than 'home-orientated' because a new school presents many potential demands due to the need to find out about rules and regulations, meet new people, find out about locations and experience new lessons and games activities. In addition boarding school children need to find out about rules, regulations and locations in their place of residence.

(2) Home-orientated problems would be associated with a greater number of reported worries because the physical distance and separation would ensure that they were less easily solved.

In addition, the spontaneous incidence of reports of 'homesickness' was quantified and expressed as a percentage of total number of participants involved in the study to provide a measure of incidence. No predictions were made in advance as to what the incidence would be.

A number of demographic variables were examined with respect to the number of reported problems and the incidence of homesickness. The following hypotheses were formulated.

(3) It was argued that involvement in the decision to go to school could be a factor affecting the number of problems reported and the spontaneous incidence of home-sickness. Those who made the decision to come would be committed to the decision and less likely to incur adverse effects.

(4) It was hypothesized that previous experience of boarding school should have an ameliorating effect on reported homesickness because the child would have been away from home before.

(5) It was hypothesized that the experience of severe events in life history would be more likely to increase the tendency to be vulnerable to stressful life events (Brown and Harris, 1978) and therefore to report more problems, lower problem/worry ratios and to be associated with homesickness.

In addition to the hypotheses formulated, demographic details such as geographical distance of location of home from school, the number of siblings in the same school and the age and sex of respondent, were examined. No directional hypotheses were formulated.

Method

Subjects and school
The subjects were all children between the ages of 13 and 16 years and were resident at school (boarders). Children co-operated on a voluntary basis.

The boarders who took part in the study were in their first year of attendance at a mixed-sex boarding school situated in the centre of a city. The school was a fee-paying school which catered for the educational needs of children between the ages of 11 and 18 years. The total number of pupils attending the school was 461. The entire new intake of 63 boarders (29 females and 34 males) was issued with diaries and 54 diaries were returned (26 female and 28 male). Of the diaries returned four, received from males, had to be discarded because they were incomplete.

Design and measures
A booklet was designed (Note 2) so that each pupil could record information about problems on a daily basis. The booklet was A5 size, cardboard-backed, easily carried and contained complete explanations for provision of data on life history events, personal details and daily worries. The booklet was headed 'School Record Book' and was kept by each subject for two weeks.

The first section of the booklet was concerned with personal details such as age, sex, address, whether this was the first move away from home and the degree to which the pupil was personally responsible for the decision to go to the school. In addition, various other details of background, including the number of siblings and whether the siblings were at the same school, were ascertained.

The method chosen for obtaining life history data was to provide a straight vertical line marked off at yearly intervals on which subjects were asked to enter negative events on the left in the space headed 'unpleasant events' and positive events on the right in the space headed 'pleasant events'. The same procedure was followed for obtaining data on life events in the six months prior to the beginning of the first school term.

A record sheet was provided for each of the days of the study. Each sheet consisted of columns which were clearly separated and marked against a time scale. Respondents were instructed to report any problem by describing it at the bottom of one of the columns and to indicate by means of endorsements on a time scale marked in hourly units when any associated worry occurred. Each record therefore provides data concerning the types and number of problems experienced each day, and the number of 'worry units' associated with each problem.

This method of reporting provided two dependent variables; one was the number of problems, and the other was the 'worry units' associated with each problem.

In order that problem reporting incidence would be analysed over the two weeks of the study, problems were considered on a daily basis. For example, if a pupil reported being worried about teachers on two days, these were considered to be two problems.

Procedure
The study was first instigated by means of a personal visit to the Headmaster, with whom the study was discussed and co-operation sought. The diaries were then delivered to the school ready for the start of the new term at the beginning of September 1983. The diaries were issued by school staff to boarders during their first days at the school. No instructions were given by the staff, since comprehensive instructions were included in the diary. On completion, the diaries were returned via the school staff members.

Results

The number of completed diaries used in the analyses was 50 (24 female, 26 male). Table 1 shows the main results of interest. Problems identified are distinguished from the proportion of reported worry units associated. As shown by the table, the incidence of reported problems averages out at about seven to eight per person during the two weeks and the incidence of reported worry units averages out at 17–18 over the same period.

There was a trend towards reporting more 'school orientated' than 'home orientated' problems. The trend was significant statistically on a Wilcoxon Test ($N = 50$, $z = -4.922$, $P < 0.001$). The levels of reported school problems correlated significantly with reported home problems, although the correlation was not high (Spearman Rank: $r_s = +0.245$; $P < 0.05$).

Calculation of a ratio of problem/worry (P/W) units provides an index of the number of worry units devoted to each problem. A low value indicates higher reported worry units per problem. Calculation of P/W ratios on an individual subject basis, as illustrated in Table 1, showed that school-orientated problems had higher P/W values than home-oriented problems. The result was significant overall ($N = 50$, $z = -4.533$, $P < 0.001$). There were no sex differences in this respect.

TABLE 1

Frequencies of school orientated and home orientated problems and associated worry units (in parentheses): N = 50

	School orientated	Home orientated	Unclassified	Total
Frequency	304 (641)	75 (233)	9 (12)	388 (886)
$\bar{x}$	6·1 (12.8)	1·5 (4·7)	—	7·8 (17·72)
S.D.	7·3	3·6	—	8·9
Percentage	78·4	19·3	2·3	100·0
P/W Ratio	0·58	0·26	—	0·56

Ratio of number worry units per reported problem calculated for each individual subject.

Overall, four males and four females reported feelings of homesickness, in conditions where there was no prior or concurrent use of the term. This represents 16% of the sample.

Factors such as the degree of decision involvement (control), previous experience at boarding school and distance of home from school, showed no association with school problem incidence, home problem or homesickness incidence. However, the presence of siblings at school was positively associated with both level of school problems ($r_s = 0.314$, $P < 0.001$) and home problems ($r_s = 0.23$, $P < 0.05$).

The reported incidence of pleasant life events within the last six months was associated with school problem incidence ($r_s = 0.298$, $P < 0.01$). The reported incidence of unpleasant life events within the last six months was associated with home problem incidence ($r_s = 0.240$, $P < 0.05$). Further analyses revealed that these effects were largely due to correlations for males. The reports were subsequently adjusted to include only those events used in the Life Experience Survey (Sarason *et al.*, 1978), in order to avoid the inclusion of trivial events, e.g. 'birthday party at age 6'. No significant correlations were obtained with the adjusted reports.

There was a tendency to report more problems at the beginning of the diary study. A series of Wilcoxon Tests between first week and second week data showed that the number of school-orientated problems and worry units decreased in the second week ($P < 0.01$) but that for home problems, although the number of problems decreased ($P < 0.05$), the number of worry units did not.

Chi-squared tests on reported worry units across time of day showed that problems were reported non-randomly across the available units 0700 to 2400 h ($\chi^2 = 72.5$, df $= 17$, $P < 0.01$). Consideration of the data showed that 2300 and 2400 h contained no entries. When these hours were omitted the result was not significant statistically ($\chi^2 = 17.4$, df $= 15$). However, the distribution remained significantly non-random for males ($\chi^2 = 30.2$, df $= 15$, $P < 0.01$) in that more worries were reported in the early part of the day.

Table 2 lists a content analysis of the problems reported. The three experimenters reached consensus on the categorization of the problems reported. Thirty-two categories were employed.

Each category was considered to be either school orientated (S), home orientated (H) or unclassified (U), depending upon the immediate focus of the category. In all, twenty-five categories are classified as school orientated, four as home orientated and three are unclassified.

The most common problem categories were 'school work', reported by 58% of the pupils, and 'homework', 'social', 'sports' and 'family at home', each of which were reported by over 30% of the pupils. With the exception of 'family at home', all the other most common categories were school-orientated. Six categories were reported by only one pupil. At least one school-orientated problem was reported by 94% of the children whereas only 38% reported at least one home-orientated problem. Ten per cent of children reported an unclassified problem.

Discussion

The main result of interest was that school-orientated problems are more frequent than home-orientated problems during the first two weeks of experience at the new

TABLE 2

Frequency of children reporting problems during the first two weeks of their first year in boarding school

Type of problem	N	%[a]	Classification[b]
School work	29	58	S
Homework	17	34	S
Family at home	16	32	H
Social	16	32	S
Sports	16	32	S
Security of personal possessions	13	26	S
Being late	12	24	S
Discipline	11	22	S
Going to school	11	22	S
Health	11	22	S
Finding locations	10	20	S
School routines	10	20	S
Tests	10	20	S
Opportunity to contact relatives	8	16	S
School societies and clubs	6	12	S
Bullying	5	10	S
Fagging	5	10	S
Homesickness	8	16	H
Pets at home	4	8	H
Sleeping habits	4	8	S
Feeling lonely	3	6	S
Food	3	6	S
Forgetting things	3	6	U
Opportunities outside school	3	6	S
New subjects	3	6	S
School uniform	3	6	S
Chapel	1	2	S
Current affairs	1	2	U
Leaving home	1	2	H
Personal appearance	1	2	S
New teachers	1	2	S
New language	1	2	U

[a]Percentages are calculated on the total number of children ($N = 50$) experiencing each kind of problem, hence the total will be greater than 100%.

[b]S = School orientated problem; H = home orientated problem; U = Unclassified problem.

school. However, home problems have a more compelling influence in that proportionately more worry units are attached to home problems than to school problems. The lower problem/worry ratio for home-orientated problems can be explained by assuming that problems referring to home are less easily solved than school problems because of the physical distance factor. This would be consistent with the idea that loss of control over a problem raises the tendency to worry (see Fisher, 1984a). An alternative explanation is that there is a larger history of problems concerned with home and these problems thus continue to exert an influence on the individual. It should be emphasised that since there are markedly fewer home problems reported than school problems, the argument based on 'Interruption Theory' which assumes that the interrupted 'home life' continues to be the major source of problems cannot be supported.

It was not possible to make any formal comparison with the content analysis of school problems in transition from primary to secondary school provided by Brown and Armstrong (1982) because they requested information specifically addressed to the new school. Also their categorizing of reported problems may have been different. They reported 22 different categories of school problems, whereas in the current study 25 categories of school problems were identified and there were four home-orientated categories and three unclassified categories. It is perhaps worth noting however that at the top of the Brown and Armstrong list, worries about school routines attracted the greatest percentage of subject endorsements (62%) followed by feeling lost (35%). By comparison, in the current study with boarding school children, school work (58%) and homework (34%) predominated. The pupils studied in this school were more concerned about work than about regulations or being lost than in the day-school transitions studied by Brown and Armstrong. It seems that the difference between the two groups of pupils is perhaps less easily explained by residential status than by the academic attitudes of the two schools. This underlines the need to begin an analysis of the differences between day-school pupils and boarders in the same schools with respect to problems reported in the first two weeks.

The reported incidence of homesickness was very low (16%) and with such a small number of subjects ($N = 8$), it was impossible to make comparisons with regard to the role of demographic variables with any degree of reliability. 'Concern with family at home' was a frequently reported home-orientated problem and it is possible that there is a 'labelling' difficulty. Children may miss home but not be alerted to using the term homesickness to describe the feelings. The spontaneous report data provides useful base data for comparisons with reports when the term is used.

None of the demographic factors examined predicted level of home or school problems. Therefore there is no evidence that age, sex, distance from home, decision to come to school or previous experience away from home are factors that will influence the difficulties encountered in adjustment.

Only the presence of siblings in the same school was found to have a relationship with school and home problem reporting, but the relationship was not in the expected direction. Far from having an ameliorating effect, the presence of siblings increases problem reporting. It is possible that having to live up to the standards set by a sibling in the same school presents problems of adjustment. In the school studied, by far the greatest frequency of reported problems was concerned with either school work or with homework and it is possible that the presence of a sibling in the school serves to create competitive pressures.

As reported in the results, there was no association found between negative (exit and loss) events in life history and adverse reaction to relocation, either in terms of homesickness or in terms of problem reporting. This may have been due to the means of obtaining the data; it was not feasible to administer scales such as the Life Event Scale or Social Readjustment Scale to the pupils. Free recall was the only strategy acceptable to schools. The effects which were unpredicted were the associations between problem reporting and reported life events in the last six months. As these associations were limited to males and only concerned recent events, they are difficult to interpret and may reflect a reporting bias perhaps best explained in terms of a 'positive hedonic set' in that when encountering a new school environment with all its problems, previous life events are seen as more pleasant. That females do not show this effect is interesting and needs further explanation because the attitude of males and females with regard to the past may be differentially influenced by current attitudes to school.

The problem of reporting bias in the provision of details about life history events is not new. For example a study of 76 neurotically-depressed individuals by Schless et al. (1974), using a 43-item list of events on the Holmes and Rahe scale (Holmes and Rahe, 1967), showed that uniformly higher ratings were given to the life events by the depressed group. Equally, normal subjects have been shown to recall pleasant events readily (Beeb-Center, 1932) whereas depressed subjects show some evidence of increased speed and intensity in the recall of unpleasant memories (Lloyd and Lishman, 1975). Such findings do raise a problem for analysis of congruence between expected and encountered conditions in the new environments because the response to encountered events may produce retrospective bias of all kinds, especially with regard to conditions at home.

The diary methodology used in this study has advantages over a retrospective study in that there is at least some probability that problems will be recorded at the time they are experienced. One of the greatest dilemmas is how long a person facing a number of new school events should be additionally burdened with the task of keeping a diary. The results showed a decrease in entries after the first week. Since it seems implausible that problems will have diminished so rapidly, it does seem to suggest that diary keeping is tedious and that the time period should be reduced. This would favour a diary-style methodology with short sampling periods across the school term.

In conclusion the main result of interest in this study is that although the major source of problem is the school, home still continues to exert an effect on the thinking of the boarders. In particular the problems relating to home have a more pervasive effect in terms of associated worry episodes. From the descriptions given in the diaries it is not possible to know much detail about the content of the problems. Of particular interest is whether the move from home creates problems (e.g. no-one to look after a favourite pet) or whether existing problems at the time (e.g. relative ill) may continue to be influential.

Notes

(1) In the United Kingdom the term 'school' refers to primary and secondary education up to the age of 18 years.
(2) A copy of the booklet is available from the authors.
(3) 'Home-orientated' problems were defined as those problems whose content was concerned with home (previous environment). 'School-orientated' problems were those whose content was concerned with the new environment. The partitioning of problems on the basis of subject descriptions was done by consensus of all three experimenters. (Details available from the authors.)

References

Beeb-Center, J. G. (1932). *The Psychology of Pleasantness and Unpleasantness*. New York: Russell and Russell (Reprinted 1965).

Brett, J. M. (1982). Job transfer and well-being. *Journal of Applied Psychology*, **67**, 450–463.

Brown, J. M. and Armstrong, R. (1982). The structure of pupils' worries during transition from junior to secondary school. *British Educational Research Journal*, **8(2)** 123–131.

Brown, G. W. and Harris, T. H. (1978). *Social Origins of Depression: A Study of Psychiatric Disorders in Women*. Great Britain: Tavistock.

Cruze-Coke, R., Etcheverry, R. and Nagel, R. (1964). Influences of migration on blood pressure of Easter Islanders. *Lancet*, 28 March, 697–699.

Fischer, C. S. and Steuve, C. A. (1977). Authentic Community?: The Role of Place in Modern Life. In C. S. Fischer, R. M. Jackson, C. A. Steuve, K. Gerson, L. M. Jones and M. Baldassare (eds), *Networks and Places: Social Relocations in the Urban Setting*. New York: The Free Press, pp. 163–186.

Fisher, S. (1984a). *Stress and the Perception of Control*. Hillsdale, New Jersey: Lawrence Erlbaum Associates.

Fisher, S. (1984b). Stress and geographical mobility. *Architectural Psychology Newsletter*, XII (2 and 3), 1–4.

Fried, M. (1963). Grieving for a lost home. In L. J. Duhl (ed.), *The Urban Condition*. New York: Simon and Schuster, pp. 151–171.

Gump, P. (1980). School environments. In I. Altman and J. Wohlwill (eds), *Children and Environment*. New York and London: Plenum Press pp. 131–174.

Holmes, T. H. and Rahe, R. H. (1967). The social readjustment rating scale. *Journal of Psychosomatic Research*, **11**, 213–218.

Lloyd, G. G. and Lishman, W. A. (1975). Effect of depression on the speed of recall of pleasant and unpleasant experiences. *Psychological Medicine*, **5**, 173–180.

Mandler, G. (1975). *Mind and Emotion*. New York: John Wiley.

Mechanic, D. (1983). Adolescent health and illness behaviour: review of the literature and a new hypothesis for the study of stress. *Journal of Human Stress*, **9**, 4–14.

Newman, S. J. and Owen, M. S. (1982). Residential displacement: extent, nature and effects. *Journal of Social Issues*, **38**, 135–148.

Parsons, T. (1955). The school class as a social system: some of its functions in society. *Harvard Educational Review*, **29**, 297–318.

Sarason, I. G., Johnson, J. H. and Siegel, J. M. (1978). Assessing the impact of life changes: development of the Life Experiences Survey. *Journal of Consulting and Clinical Psychology*, **46**, 932–946.

Schless, A. P., Schwartz, L., Goetz, C. and Mendels, J. (1974). How depressives view the significance of life events. *British Journal of Psychiatry*, **125**, 406–410.

Schoggen, P. (1975). An ecological study of children with physical disabilities in school and at home. In R. Weinberg and F. Wood (eds), *Observation of Pupils and Teachers and Special Educational Settings: Alternative Strategies*. Minneapolis: Leadership Training Institute, University of Minnesota, U.S.A. pp. 123–147.

Stokols, D., Shumaker, S. A. and Martinez, J. (1983). Residential mobility and personal well being. *Journal of Environmental Psychology*, **3**, 5–19.

Syme, S. L., Hyman, M. M. and Enterline, P. E. (1965). Cultural mobility and the occurrence of coronary heart disease. *Journal of Chronic Diseases*, **26**, 13–30.

Wapner, S., Kaplan, B. and Ciottone, R. (1981). Self-world relationships in critical environmental transitions: childhood and beyond. In L. S. Liben, A. H. Patterson, N. Newcombe (eds), *Spatial Representation and Behaviour Across the Life Span*. New York: Academic Press. pp. 251–280.

HOMESICKNESS AND HEALTH IN
BOARDING SCHOOL CHILDREN*

SHIRLEY FISHER, NORMAN FRAZER and KEITH MURRAY

*Stress Research Unit, Department of Psychology, University of
Dundee, Dundee, Scotland*

Abstract

Three studies are reported concerning homesickness in children attending a new boarding school. Homesickness was found to be a complex cognitive/motivational/emotional state. The first study concerned retrospective reports of 115 pupils at the end of the first year. Seventy-one per cent of the group reported having experienced homesickness during the school year. This same group also reported a higher incidence of non-traumatic ailments during the year and more days off school. Previous boarding school experience was found to have an ameliorating effect on reports of homesickness. Two further studies are described which involve a diary style of methodology. The first confirmed incidence levels of 76% and an ameliorating effect of previous boarding school experience was found. The second study devoted exclusively to homesickness reporting showed incidence levels of 71%. Homesickness reporting generally decreased during the two-week period of the diary studies; males showed a different daily reporting patterns from females; 'very homesick' respondents had different daily and weekly reporting pattern from other respondents. The findings are elaborated in terms of the risk model developed by Fisher *et al.* (1985, *J. Environ. Psychol.*, 5, 181–195): a geographical move is a necessary but not a sufficient condition for a homesickness experience; circumstantial and life situations act as 'gate devices' influencing which variables have a moderating effect.

Introduction

Geographical relocations result in both an interruption of daily life and an encounter with a new environment. For a person who leaves home to reside in a new place for educational or vocational reasons both aspects may be influential: there is distancing from the support normally given by family and friends, loss of the familiar 'routine' patten of life and exposure to an implosion of new environmental and psychosocial factors. University students, college students and children who leave home to board at school experience these circumstances.

This article is concerned with the effect of leaving home to reside at school. Boarding schools are a feature of the British primary and secondary educational system. The schools are generally part of private sector education and attract a high socio-economic catchment group. In the primary school sector, children as young as 7 years may be involved. The age for commencing secondary school is 11 to 13 years.

There are relatively few data on the effects of leaving home to reside at school for the first time. For the children involved there is simultaneously a transition to a new school environment which itself may produce new problems (see Brown and Arm-

*This work was supported by a grant from the Economic and Social Research Council.

strong, 1982) and the need to adjust to a change in residence resulting in loss of direct contact with family and with the home environment.

A phenomenon noted by teachers and officers of residential institutions, for those who leave home, is 'homesickness'. It is commonly seen as an adverse reaction with features in common with depression. Dictionary definitions of the term include 'pining for home' (*Chambers Twentieth Century Dictionary*, 1954): 'depressed by absence from home' (*Pocket Oxford Dictionary*, 1969). In spite of the existence of the term in the English language (and other languages) and in spite of common understanding of the term, there is little information available from research about the state of homesickness and its effect on health and welfare.

An early treatise on the subject of homesickness by Harder (1678)* referred to the adverse experience of adults experiencing a relocation in terms of *'maladie du pays'*. He defined a new term *'nostalgia'* from the Greek *'nostos'*, meaning 'a return home', and *'algos'*, meaning pain or sorrow. Harder stated that there are a number of remote, internal or predisposing causes which implant in the mind strong thoughts about returning home. These include differences in climate, customs, habits and food.

Studies by Fisher *et al.* (1985) have suggested that about 60–70% of students report homesickness in their first weeks at university. Results suggested that a move is a necessary but not a sufficient condition for a homesickness experience; factors concerning the new environment (satisfaction with work, social relationships, residence) combine with personality factors such as cognitive failure levels and personal control over the move, in influencing the likelihood of the experience. Equally, factors relating to the geographical transition itself are important; increased distance is positively associated with homesickness reporting.

At least one reason for adverse reaction following the move to a new school may be the interruption of old plans which were dominant in life at home. Interruption has been reported by Mandler (1975) as providing a source of tension. Perhaps homesickness, or yearning for aspects of previous life style, occurs because of the potent effects of interruption. Perhaps homesickness is a prerequisite for later successful adjustment; old plans are revised and attenuated, to be replaced with plans suitable for the new environment (see Fisher, 1984). Homesickness seems to have features in common with the grief reaction produced by the bereaved in which there is high mental preoccupation with the deceased and with previous life shared with the deceased.

It is also possible that the exposure to the new school environment under conditions of reduced contact with family and friends, is itself threatening and that the homesickness response represents a desire to leave the environment. A child at school is exposed to a structured, disciplined world dominated by teachers. Parsons (1955) emphasized that the teacher-dominated world is likely to be different from the parent-dominated world in that the teacher requires activities and competence and is less likely to provide a child with emotional support. This argument is supported by Schoggen (1975) who argues that interactions with parents are sometimes negative but individualized and frequent, whereas teachers are more likely to make demands and to emphasize competence in the activities of the child.

Exposure to a new environment is also likely to be associated with low control. A school environment where discipline is imposed and regulations must be adhered to would give the child very little control over his immediate world. The desire for the

---

*From Rather's translation of *18th Century Medical Essays* (1965).

old 'home' environment may additionally reflect the effort of perceived control loss (see Fisher, 1984).

The Purpose of the Present Research

The purpose of the investigations in this study was to obtain data on the incidence and main features of homesickness as self-reported by groups of boarding school pupils in their first term at a new school and to identify some of the factors which differentiate those who report the experience from those who do not.

There are particular difficulties associated with attempting to undertake this kind of study. Most boarding schools are concerned to reduce the possibility of homesickness in new pupils. It was felt by some school staff that the provision of information by new pupils could help to sensitize them towards the experience. Therefore, the studies reported represent a compromise between ideals and practicalities in this respect.

Three studies are reported: One study was conducted at the end of the school year to avoid any problems which might se from homesickness reporting. This involved two large schools and is termed the 'retrospective study'. Two different diary studies involving smaller subject numbers were conducted in the first few weeks of the new term for new pupils. These are termed the 'diary studies' and provide some further check on incidence as well as providing more data about some of the main features of homesickness. The diary studies vary in the emphasis given to the reporting of homesickness and in the questions asked about it.

Retrospective study (Study 1)

The initial study was retrospective and involved a sample of 117 pupils drawn from two comparable Scottish boarding schools situated 20 miles apart. The study took place at the end of the school year when there was relatively less need to be concerned about whether reporting the experience might sensitize pupils to it. A number of questions concerning environmental and personal antecedents were provided and the following hypotheses were formulated.

Environmental factors

(i) It was hypothesized that increased geographical distance between home and school would be a factor likely to increase homesickness. Fisher et al. (1985) found that, in university students, increased geographical distance was associated with increased homesickness reporting. The distance factor was thought to operate to increase the 'cost' of home contact thus increasing a sense of isolation and loss of control.

(ii) It was hypothesized that features of the new psychosocial environment which helped to reduce the sense of isolation and loss of contact with home might reduced the likelihood of homesickness. Therefore, the presence of one or more siblings at the school was hypothesized to reduce the experience of homesickness.

Control over the decision and personal involvement

(iii) It was hypothesized that a positive attitude towards the move in advance ('want to come to boarding school') and increased responsibility for the decision to attend school would be associated with decreased homesickness reporting. There are two possible reasons: first, a positive attitude could indicate that the transition was desired;

second, decreased control might intensify threat (see Fisher, 1984) and exacerbate negative feelings on arrival.

Previous experience

(iv) It was hypothesized that previous boarding school experience might help to reduce the effects of homesickness because the pupil would be used to leaving home; thus some of the stress produced by the distancing from the support and contact with parents would be reduced. This was given anecdotal support by observations made by some of the staff interviewed at schools during the course of the studies.

(v) It was hypothesized that some individuals might be vulnerable to life stresses because of unstable background factors associated with loss of one or both parents by death or divorce at an early age (see Fisher, 1984). Previous studies by Fisher *et al.* (1984) and Fisher *et al.* (1985) produced no evidence that this was a factor in homesickness but the issue is such an important one that it was felt worthwhile asking the questions again.

Homesickness and physical health

(vi) It was hypothesized that general unhappiness, likely to be a feature of home-sickness, might affect physical health increasing the proportion of non-traumatic ailments during the year. This is in keeping with the view that social and psychological change are preconditions for illness (Totman, 1979; Fisher, 1985).

Diary of daily problems and homesickness (Study 2)

The second investigation was conducted during the first two weeks at a new boarding school. A diary study by Fisher *et al.* (1985) conducted along similar lines showed that when the term 'homesickness' was not introduced as a heading for daily reports in a diary, only 16% of boarding school pupils spontaneously introduced the term. However, although the number of reported problems concerning the 'school' environment exceeded the number of problems concerning the 'home' environment, in terms of the periods spent worried and preoccupied, the latter exerted a greater intrusive effect on pupils. It was conceivable that homesickness was experienced by pupils but not labelled as such.

The diary study reported here involved presentation of the term 'homesickness' on a check sheet for endorsement on a daily basis. This provided a means of checking the incidence of homesickness when reporting was coincident with the experience and the term was provided for possible endorsement.

A second purpose of the diary was to examine some of the environmental and personal factors likely to be associated with homesickness reporting.

Diary study of the 'anatomy' of homesickness (Study 3)

The second diary study had the same aims as the first but concentrated exclusively on homesickness and the conditions reported as associated with it. This provided a further check on incidence in conditions where the study was conducted concurrently with the new term experience but when information about homesickness was specifically re-quested.

In addition, the study enabled details of the phenomenology of the homesickness experience to be obtained. Research questions included details on circumstance at the time a homesickness experience was reported.

From informal observations provided by school staff there was some suggestion that people are more likely to feel homesick if left alone and involved in passive behaviour such as lying on the bed, waiting for meals or lessons. There was some feeling that sharing sleeping accommodation and keeping a high level of daily activity would help ward off homesickness because the pupils would have 'less time to think'.

Method

Retrospective study (Study 1)

Two boarding schools were involved in this study. One was an all female school; one was predominately male with a small proportion of girls. The number involved in the study was 117. The subjects were children between the ages of 11 and 14 years. There were 58 girls and 57 boys in the sample, the total of first year pupils in both schools.

The pupils were asked by the investigators by complete two questionnaires which were randomly presented. Approximate testing time was 20 minutes. The questionnaires were distributed to assembled classes, completed in privacy and then collected by the investigators.

The first questionnaire elicited the following information: (i) distance of school from home; (ii) responsibility for the decision to go to boarding school; (iii) whether the pupil had wanted to attend boarding school; (iv) whether the pupil had brothers or sisters at the school; and (v) whether the pupil had previous boarding school experience. In addition each pupil was asked to provide a definition of the term 'homesickness' and to indicate whether it had been experienced since attending school.

The second (health) questionnaire required the following information to be provided: (i) any ailments or illness experienced during the year; (ii) number of days affected by it; (iii) number of days on which school activities were missed as a result; and (iv) number of occasions a doctor was consulted. The information was freely recalled by pupils.

Diary investigation of problems/worries and homesickness (Study 2)

The investigation involved the cooperation of the school staff and pupils in a mixed sex boarding school. The subjects were children between the ages of 13 and 17 years, in the first term of their first year. The children cooperated on a voluntary basis.

The project was first instigated by a personal visit to the Headmaster. It was decided that only half of the first year intake would be given diaries, to minimize any adverse effects or increased demands on teaching staff. A selection of subjects was chosen by random selection of names from a list of the first-year entries. A total of 18 subjects (8 females and 10 males) were then issued with diaries to be kept anonymously and returned to a collection file in the school after two weeks. Out of those diaries given to subjects, one diary was returned without completion.

A booklet was designed so that pupils could record the details of problems on a daily basis. It was A5 size, cardboard backed and contained complete explanations for the provision of data on life history events, personal details, daily problems and associated worry periods.

Diary study of 'the anatomy of homesickness' (Study 3)

The investigation was the same in all respects as Study 2, with the exception that the diary had a different format.

The format chosen for daily recording of the presence of absence of homesickness involved a grid with four headed columns with cells represented against a time scale. The subject was required to indicate at the top of the page whether or not he had felt homesick. He was then required to indicate by means of printed crosses in the cells on the first column, how often during that day he had felt homesick and then to report whether he was alone or with friends, what activity he was engaged in and where he was at the time by means of crosses in the remaining columns.

The subjects were 21 male children between the ages of 12 and 16 years, who were newly resident at an all male school. Children cooperated on a voluntary basis and record books were anonymous. The subjects were selected by means of random procedure from a school list. The booklets were given out by staff at the school and were returned via a central collection point. All diaries were completed anonymously.

Two dependent variables can be identified. The first is the decision that homesickness has been experienced that day (homesickness problem). The second is the frequency of entries for any day in which a homesickness problem is reported (worry units or 'homesickness' episodes).

Results

Study 1: the retrospective study

The first study showed the incidence of homesickness to be 71%; there were 83 homesick and 34 non-homesick individuals. In order to establish that the two groups did not differ in the meaning attributed to the term 'homesickness', each definition provided was analysed into each of its constituent features. Consensus was reached between all three investigators in this respect. Some definitions provided had only one feature such as for example 'homesickness means missing home'. Other definitions contained multiple features such as for example 'homesickness means being lonely in a new place, crying at night and missing home'. In the latter case three constituents would be extracted 'being lonely', 'crying at night' and 'missing home'. It was also necessary to ensure that the meaning provided by homesick and nonhomesick individuals did not differ in the number of constituent features involved.

Table 1 shows the main features from the definitions provided by 115 respondents; two subjects failed to provide definitions. It indicates the percentage of times that each listed feature was reported. As illustrated by the Table, missing parents, family and people at home, was reported by over 65% in both groups. In addition, 34–46% of respondents in each group included the features of 'missing home environment' and 'missing house, home and area'.

There were no differences between those who reported homesickness and those who did not in terms of either the number of component features provided or the principal component features identified.

Table 2 lists the main results of interest. Neither distance from home nor decisional involvement influenced whether or not homesickness was reported. Those who reported that they wanted to come to school were less likely to report homesickness

TABLE 1

Features utilized in definitions of homesickness for homesick and non-homesick school pupils

Feature categories from definitions provided	Frequency of reported features and percentage of subjects reporting each feature	
	Homesick ($n = 82$) F (%)	Non-homesick ($n = 33$) F (%)
'Missing parent family'; 'longing for people at home'	54 (65·9)	25 (75·8)
'Missing home environment'; 'missing house, home, area, etc.'	28 (34·1)	12 (36·4)
'Wanting to go home'; 'feeling a need to return home'	21 (25·6)	10 (30·3)
'Missing friends'; 'longing for friends'	12 (14·6)	1 (3·0)
'Feeling of loneliness'	10 (12·2)	1 (3·0)
'Crying'	3 (3·7)	3 (9·1)
'Unsettled'	4 (4·9)	1 (3·0)
'Hating the present place'	4 (4·9)	1 (3·0)
'Feeling unhappy'	4 (4·9)	1 (3·0)
'Not getting on with people'	3 (3·7)	0 (0·0)
'Dissatisfaction with present situation'	3 (3·7)	0 (0·0)
'Feeling depressed'	2 (2·4)	1 (3·0)
'Disorientation'; 'feeling lost in new environment'	2 (2·4)	0 (0·0)
'Regret that life had changed', 'A feeling of regret'	2 (2·4)	0 (0·0)
'Never been away from home before'	2 (2·4)	0 (0·0)
'Feeling ill'	2 (2·4)	0 (0·0)
'Unable to do anything'	2 (2·4)	0 (0·0)
'Feeling unloved'	0 (0·0)	2 (6·1)

The following features were endorsed by only one person in the following groups. *Homesick:* 'problem at school'; 'missing someone close to talk to'; 'obsession with thoughts of home'; 'looking for familiar company and faces'; 'feeling isolated'. *Non-homesick:* 'feeling uneasy'; 'unable to cope'; 'feeling full and weary'; 'thinking home is better than here'.

($P<0.001$). Whilst the presence of one or more siblings at school did not influence the reporting of homesickness, previous boarding school experience had an ameliorating effect ($P<0.05$).

The results in Table 2 illustrate that homesick subjects reported more non-traumatic ailments such as colds, headaches and feeling sick ($P<0.05$). The days affected are significantly greater for the homesick group ($P<0.02$). However, the difference is not significant when the ratio of number of days per ailment provides the basis of comparison. The homesick group also report having more visits to a doctor than the non-homesick group ($P<0.05$). By contrast there are no differences in reported number of traumatic disorders (sprains, breakages, muscle damage, etc.), in days affected or visits to a doctor.

Study 2: Diary study of problems, worries and homesickness

Overall, 13 out of 17 subjects reported at least one period of homesickness. This represents 76·4% incidence overall (males, 80%; females, 71·4%). Homesickness

TABLE 2

Variables associated with homesickness reports for boarding school pupils

Variable	Homesick (n = 83)		Non-homesick n = 34		t	P (One-tailed)
	$\bar{x}$	S.D.	$\bar{x}$	S.D.		
Age (years)	12·58	1·06	12·84	0·91	1·28	NS[a]
Distance from home (miles)	374·61	394·66	453·50	407·48	0·89	NS
Responsibility for decision	0·84	0·57	0·97	0·63	1·06	NS
Wanted to come	0·77	0·43	0·97	0·18	3·56	<0·001
Siblings at school	0·37	0·49	0·35	0·49	0·21	NS
Boarding experience	0·36	0·48	0·56	0·50	1·95	<0·05
Non-traumatic ailments	2·95	1·62	2·38	1·28	1·83	<0·05
Days affected	27·28	45·72	14·53	12·71	2·30	<0·01
Activities affected	3·94	11·59	6·09	10·16	0·07	NS
Number of times doctor seen	1·88	1·49	1·32	1·17	1·94	<0·05
Traumatic ailments	0·06	0·24	0·09	0·29	0·34	NS
Days affected	1·95	10·99	1·53	5·23	0·28	NS
Activities affected	1·95	11·11	0·73	3·68	0·88	NS
Number of times doctor seen	0·06	0·24	0·09	0·29	0·54	NS

Larger means indicate higher scores on the attributes listed.
[a]NS = Not significant.

reports were greater than any other problem category. There were a total of 90 reports of homesickness across all subjects for the 14 days, and 201 worry units associated. Males had 27 reports of homesickness with 87 associated worry units. Females had 63 reports of homesickness with 114 associated worry units. The difference was not significant statistically.

Because of the relatively small sample of subjects, within group comparison between the homesick ($N = 13$) and non-homesick group ($N = 4$) must be treated cautiously. Geographical distance, sex, age, control over the decision to go away to school and the presence of siblings at the same school did not differentiate between the homesick and non-homesick groups. The only significant results were that for males, previous boarding school experience was negatively correlated with homesickness ($r_s = -0.58$, $P < 0.01$) and the presence of one or more siblings at the school increased the number of problems reported ($P < 0.01$).

A distinction was made by the investigators as to whether according to the subject's own description of a problem, it was 'school' or 'home' orientated. School orientated problems were then further divided into those which were concerned with *work* (e.g. 'worried about Maths lesson'), *social* (e.g. 'worried about whether my class teacher likes me') or *general* (e.g. 'worried about whether I will find the dining room for lunch'). Home orientated problems (e.g. 'worried about my dog at home') are listed separately from reports of homesickness. A small percentage of personal or health problems which did not fit with the above categories were assigned to an extra *'other'* category. The number of problems was divided by the number of worries on an individual subject basis. This resulted in a derived ratio, the problem/worry ratio. There was no overall difference between school oriented and home oriented problems in terms of problem/worry ratio. Pupils with at least one sibling at school had higher problem reporting levels overall ($P<0.01$) and lower problem/worry ratios for school problems ($P<0.05$).

Spearman rank correlations showed that there was a negative relationship between time of day and worrying for males but generally a positive relationship for females. However, the correlations for males were only significant in the case of the 'work' category ($r_s = -0.48$, $P<0.05$) and only significant for females in the case of 'home' categories ($r_s = 0.70$, $P<0.01$) and homesickness ($r_s - 0.71$, $P<0.01$). Chi-square test results showed that part of the overall significance is attributable to the last two cells available on the time scale for recording worries: 2300 and 2400 hours were not much used. Without these categories, the analysis for hours 0700–2200 is significant only for females ($P<0.01$).

In terms of homesickness there was also a different pattern for females as compared with males. Females reported more homesickness as the day progresses ($P<0.01$), but the pattern was bimodal in character for males.

Analysis of the distribution of reported problems across the 14 days of the diary study in terms of Spearman rank correlations showed that in general there was lowered problem reporting as the duration of the diary study increased. This was true for males: school problems ($r_s = -0.59$, $P<0.05$), home problems ($r_s = -0.72$, $P<0.01$). For females the correlations were negative (school $r_s = -0.33$; home $r_s = -0.34$) but not significant. With regard to reports of homesickness specifically, the correlation was negative both for males ($r_s = 0.78$, $P<0.01$) and females $r_s = 0.45$, $P<0.05$).

There were few significant correlations between life history events and the incidence of problems in different categories. There was however a significant negative correlation between the reporting of unpleasant events and the reported incidence of homesickness but only for females ($r_s = -0.722$, $P<0.05$).

On the basis of reported worry levels, a highly homesick group of three subjects (scores greater than 1.5 times the average worry unit level for the homesick group) was further identified. In view of the small number of subjects in this sample, some caution must be attached to the findings: The subjects reported more problems overall (significant on a Mann Whitney comparison: $P<0.01$). They reported more home orientated problems ($P<0.05$) and more problems in the 'social' category of school orientated problems ($P<0.05$). They reported higher worry levels for home problems and more worries attached to a 'social' and 'general' categories of school orientated problems ($P<0.05$).

Study 3: the anatomy of homesickness

Of the 21 subjects, only six reported no homesickness. The incidence was thus 71%. A total of 58 reported problems and 322 episodes of worry units occurred across subjects. Two subjects accounted for 82% of all episodes. These were identified by means of a criterion of greater than ($1.5 \times$ mean) episodes for the group as a whole.

For all subjects, there was a significant negative Spearman Rank Correlation between reported homesickness *problems* and time within the 14 day study period ($r_s = -0.732$; $P<0.001$). However, the correlation between *episodes* and time was positive but not significant statistically for the group overall ($r_s = 0.494$). When the two very homesick subjects' results were removed, the correlation for the remainder of the subjects was negative and significant ($r_s = -0.673$; $P<0.004$). The two subjects (S1 and S7) reporting high homesick episode levels showed significant positive correlations (S1 $r_s = +852$, $P<0.001$; S7, $r_s = +756$, $p<0.001$). Therefore, two subjects appeared to be getting worse as the diary study progressed, the remainder were improving.

The number of worry units showed a negative relationship with time of day for the hours 0700–2400 ($r_s = -0.728$; $P<0.001$); 0700–2200 ($r_s = -0.610$, $P<0.01$). However, removal of the two very homesick subjects lowered the correlation for the remaining subjects and it was no longer significant ($r_s = 0.286$). Examination of the two very homesick subjects showed that the more severe case (defined by number of episodes) had a correlation which was not significant ($r_s = -0.313$) whereas the less severe case had a positive correlation which was significant ($r_s = 0.719$, $P<0.001$).

None of the demographic variables predicted the homesick response on this occasion. The only finding that was significant was that less reporting of positive life events in the last six months was associated with increased homesickness episodes ($r_s = -0.505$, $P<0.01$).

Out of a total of 322 reported episodes of homesickness, 145 were given descriptive details which enabled further investigation concerning associated circumstances. The low proportion was due to the reporting characteristics of one of the two very homesick individuals who reported so many episodes that there was a natural inclination to use ditto marks or not report circumstances.

A consensus was reached by the investigators as to whether each reported activity was mentally 'active' or 'passive' and physically 'active' or 'passive'. Statistical comparisons are not possible because there is no way of generating an expected distribution of activities. Over 70% of homesickness episodes were associated with periods of mental and physical passivity.

Of the 145 descriptions of social conditions prevailing at the time of a homesickness episode, 78% were reported as occurring when respondents were in the presence of one or more people.

The distribution of reported episodes across reported venues suggested two prominent modes are 'the dormitory' (46% of cases) or 'other school buildings' (31% of cases). Again, it is impossible to generate an expected distribution of where pupils spend their time. In the particular school studied there are four or more pupils to a dormitory. A high proportion of time during the day is in the classrooms for teaching purposes or on game fields or in communal eating and leisure rooms.

Discussion

Homesickness appears to be a complex 'umbrella' term embodying a large number of cognitive, motivational and emotional features centred largely on missing and yearning for family and home. Some of the features reported are *symptoms* (such as 'crying'); others are antecedents (such as 'not wanting to leave home').

The results of the retrospective study confirm the results of a study with university students by Fisher *et al.* (1985) that homesick and non-homesick respondents do not differ in personal meanings attributed to the term. However, informal comparison shows that in contrast with the meanings provided by students, school pupils gave more reports of 'missing parents' (66–76%; compared with students, 30–33%).

A main point of interest is that across all three studies the incidence of reported homesickness was approximately 70% irrespective of whether the decision was made retrospectively (Study 1) or at the time of the experience as in the two diary studies.

The experience of homesickness was found to be associated with increased self report of *non-traumatic ailments*. The effect is not due to reporting bias because there was *no association with the reporting of traumatic ailments*. One obvious interpretation is that unhappiness due to homesickness is associated with increased risk of infection. This would fit with the growing literature suggesting that negative life events are associated with poor physical health (see Fisher, 1985; Totman, 1979). However, it may be the case that minor illnesses *create homesickness* because love and attention of family members if sought by a pupil who feels ill. Fisher *et al.* (1985) found no evidence of increased reporting of non-traumatic ailments in a homesick as compared with non-homesick populations of university students. It is possible that the intensity of the homesick experience is greater in a younger population.

No age or sex differences were found in the reporting of homesickness in the school populations. This confirms results by Fisher *et al.* (1985) with university students. If there are vulnerability factors they do not appear to be age or sex-linked.

Two environmental factors were proposed as likely to be associated with homesickness reporting. The first was geographical distance; it was proposed that increased distance should increase a sense of isolation from home due to the increased difficulty of visits home. Although this was found to be the case for the students studied by Fisher *et al.* (1985), it was not the case for the school pupils in this study. One plausible explanation is that at school, visits home are restricted, therefore the distance factor is not irrelevant.

Perhaps for similar reasons, whereas decisional control was found to be a factor influencing homesickness reporting in students (Fisher *et al.*, 1985), it was not found to be a factor influencing homesickness reporting in the school pupils. Perhaps circumstantial variables act to allow or 'gate out' the influence of such variables; younger children may be used to less decisional involvement in life. Therefore, the absence of decisional control is not operative in the lives of children.

The finding that previous boarding school experience has an ameliorating effect on homesickness may be explained in two ways: Either the child has grown used to the idea of separation from home or has grown used to some of the features of boarding school life so that the features of the new environment are no longer so strange and threatening. Studies on the different profiles of geographical transitions in the lives of boarding school pupils are now needed. Of particular interest is whether all moves away from home, even for holidays, predispose a person to be less homesick or whether only previous experience at a boarding school is influential.

The diary studies although involving relatively small numbers provided information about homesickness experiences. First, the lack of relationship between problem reporting and homesickness (Study 2) argues *against* the view that stressful elements of the new environment are critical factors determining in adverse reaction. This is important in view of the findings of Stokols *et al.* (1983) who reported dissatisfaction with a new environment to be a major factor in poor adjustment to a geographical move in adults. The authors argued that the 'congruence' of the environmental domains (the extent to which they support personal goals) and the temporal context of the move may be critical factors. Fisher *et al.* (1985) also showed that for university students, homesickness reporting was associated with reduced levels of satisfaction with the new environment.

However, against this view; the *strongly homesick* subjects ($N = 3$, in Study 2) were found to report more home orientated problems and more problems in the 'social' category of school orientated problems. They also reported higher worry levels for 'home' problems and higher worry levels for 'social' and 'general' categories of school problems. The results suggest that whereas mild homesickness may not be linked with perceived stress in new environments, in cases of more severe reaction, there is a positive relationship. One possibility is that the severely homesick person becomes homesick and distressed because of problems in the new environment: the lonely, lost person may experience the need for the comforts of home as a focus for distress. The presence of a sibling in the same school might have been expected to reduce feelings of isolation and therefore ameliorate adverse effects. In fact if anything, the presence of a sibling appeared to produce different problems for individuals concerned perhaps because of the increasing competitive demand created.

The pattern of homesickness reporting across the hours of the day proved interesting in that females differed from males in reporting more episodes later in the day, whereas males generally report less. Broadbent (1983) reported (in a quite different context) a sex difference in arousal pattern not inconsistent with the above finding. However, as our data depend on diary recording, it remains possible that males and females adopt different recording techniques.

Very homesick individuals were shown in the third study to have a changed pattern of episodes. Firstly, one of the two respondents reported himself as homesick all day and often endorsed *every available time cell*. This pupil wrote an emotional note to the investigators at the end of the diary indicating his severe homesickness and desperate feelings. The second strongly homesick subject showed an increase in homesickness reporting as the day progressed; a reversal of the previously established pattern for males.

Whereas, in most cases homesickness reporting decreased across the two weeks of the diary studies; this was not true for the strongly homesick subjects who increased their reporting during the diary study. Taken collectively these results suggest that those who are severely homesick have some early 'pathology' in their reporting patterns which can be easily detected.

The association with passivity is worthy of further investigation. A problem is that passivity may be the cause or consequence of being homesick; we cannot recommend that keeping new pupils active is necessarily a means of controlling homesickness. Equally, the pupils' reports of social conditions at the time of homesickness episodes indicate that being alone is not a necessary condition; most episodes were reported as occurring whilst the pupils concerned were in the company of others.

In summary, a two-stage risk model proposed by Fisher *et al.* (1985) continues to provide a working framework; an environmental move is likely to be a necessary but *not a sufficient* condition for a homesickness experience. Of all the factors so far examined it appears that previous experience of boarding school has ameliorating effects and that problems created by the new environment may be augmenting.

Comparisons with the study of university students by Fisher *et al.* (1985) suggests that particular circumstances of the life style of individuals determines which variables influence homesickness reporting. In the case of university students factors such as decisional control over the move to university and the geographical distance involved *were* influential. In the case of the children at boarding school, these factors were not influential, perhaps because of the restrictions of parental and school authorities on the life of a child there is in effect no choice. A child cannot go home because she/he is not allowed to, therefore geographical distance of homesick becomes immaterial. By comparison, a student is free to go home and so the perceived sense of distancing and cost become influential.

A model is needed which allows background and life style circumstances to provide gating devices for determining the *relevant* environmental variables which influence homesickness experiences. It is concluded that a geographical move provides a necessary but not a sufficient condition for homesickness response and that background and life-style circumstances determine or gate out the influence of moderator factors.

References

Broadbent, D. E., Cooper, P. F., Fitzgerald, P. and Parkes, K. R. (1982). The Cognitive Failures Questionnaire (CFQ) and its correlates. *British Journal of Clinical Psychology*, **21**, 1–16.

Broadbent, D. E. (1985). Noise as a public health problem. Turin, Italy, *Fourth International Congress*.

Brown, J. M. and Armstrong, R. (1982). The structure of pupils' worries during transition from junior to secondary school. *British Educational Research Journal*, **8**, 123–131.

Fisher, S. (1984). *Stress and the Perception of Control*. London: Lawrence Erlbaum Associates.

Fisher, S. (1985). *Stress and Strategy*. London: Lawrence Erlbaum Associates.

Fisher, S., Frazer, N. A. and Murray, K. J. (1984). The transition from home to boarding school pupils. *Journal of Environmental Psychology*, **4**, 211–221.

Fisher, S., Murray, K. J. and Frazer, N. A. (1985). Homesickness, health and efficiency in first year students. *Journal of Environmental Psychology*, **5**, 181–195.

Harder, J. J. (1978). *Dissertatio medico de nostalgia oder Heinweh praeside*. Basle: Johannes Heferno. (Quoted by Rather, 1958).

Mandler, G. (1975). *Mind and Emotion*. New York, John Wiley.

Parsons, T. (1955). The school class as a social system: some of its functions in society. *Harvard Educational Review*, **29**, 297–318.

Rather, L. J. (1965). *Diolase, Life and Man*. Selected Essays by Rudolph Virchow, Stanford USA.

Schoggen, P. (1975). An ecological study of children with physical disabilities in school and at home. In R. Weinberg and F. Wood (eds), *Observation of Pupils and Teachers and Special Educational Settings: Alternative Strategies*. Minneapolis: Leadership Training Institute University of Minnesota.

Stokols, D., Shumaker, Sally, A. and Martinez, J. (1983). Residential mobility and personal well-being. *Journal of Environmental Psychology*, **3**, 5–19.

Totman, R. (1979). *Social Causes of Illness*. Souvenir Press (E and A) Ltd.

REPRESENTATION OF FAMILIAR PLACES IN CHILDREN AND ADULTS: VERBAL REPORTS AS A METHOD OF STUDYING ENVIRONMENTAL KNOWLEDGE

GIOVANNA AXIA*,† MARIA ROSA BARONI and ERMINIELDA MAINARDI PERON

* Dipartimento di Psicologia dello Sviluppo e della Socializzazione, University of Padova, Italy and † Dipartimento di Psicologia Generale, University of Padova, Italy

Abstract

This research investigated children's representations of familiar places by means of verbal reporting. Eight-year-old children's representation of the entrance hall and the courtyard of their school was examined under three conditions: free recall; description from memory, intended for a person not acquainted with the place; and direct description, given while looking at the place. Children's descriptions from memory were then compared with those given by a group of teachers in the school. Both contents and organization of verbal reportings were analyzed. Quantitative results showed a difference in recall between places only in children; furthermore, both children and adults mainly remembered the aspects of places which are constant constraints to actions. The importance of considering some qualitative-linguistic indexes was assessed by results showing, for instance, differences in children's discourse organization across conditions, the relevance of cognitive more than physical boundaries of places, and age differences in localizing items.

Introduction

The aim of the present research was to evaluate whether analysis of children's verbal reports would prove successful as a methodological means of studying their environmental knowledge. As the present study is only a first step in a research field ultimately aimed at assessing children's conceptual knowledge of places, we felt it was important to analyze both content and organization of verbal material. For this reason we stressed the importance of qualitative analysis as a means of posing questions to be addressed and hopefully answered in future research.

During the last 10 years children's environmental cognition has been investigated in many studies. However, as Weatherford (1982) points out, it is difficult to compare their results as, among other differences, the places examined vary considerably in size, function, and ecological relevance, ranging from small-scale models (Herman et al., 1982; Newcombe and Liben, 1982) to quite extensive areas such as university campuses (Cohen et al., 1978) and neighbourhoods (Anooshian and Young, 1981). In the present study we therefore felt it advisable to choose places which, although different, could be fairly equated for relevant variables such as familiarity and function, in order to obtain more directly comparable data. We then analyzed how children talk about the entrance hall and courtyard of their school.

Verbal descriptions of spatial networks have already been considered by linguists to be a useful means of studying the relationship between language and thought

---

† To whom correspondence should be addressed at the Dipartimento di Psicologia dello Sviluppo e della Socializzazione, via B. Pellegrino 26, 35137 Padova, Italy.

(see, for example, Linde and Labov, 1975). More recently, Klein (1982, p. 169) stresses how 'route communications are interesting from an interactive, a cognitive and a linguistic point of view', although he concentrates mainly on the third aspect.

Conversely, when used to assess children's large-scale spatial knowledge, verbal reports have generally been employed together with other methods, such as place reconstructions through models (Hart, 1979), children's drawings (Lynch, 1977), and maps and aerial photographs interpretations (Matthews, 1985), while only very recent studies have focused on verbal reports alone (Axia, 1986; Waller, 1986).

The main aim of our research was to investigate children's representations of environments using their verbal reports. In addition to allowing the detection of recalled elements as well as other methods, verbal reports offer linguistic cues which may cast some light on other aspects of children's recall. As we used verbal material, we felt it necessary to check for the possible effects of communicative factors, as the use of verbal reports poses some problems: in particular, it stresses the interactive and communicative aspects of the experimental situations. We hypothesized that instructions stressing different communicative goals would result in the use of different strategies of item presentation. In particular, we expected both quantitative and qualitative differences between free recall of a place, and recall of that same place specifically aimed at describing it to a person who does not know it ('description from memory' condition). A control condition ('direct description'), in which subjects had to describe the place while looking at it, aimed at assessing children's capacity to name environmental elements correctly. The direct description condition can be considered not only a control condition but also an experimental one, in the sense that it allows comparison of descriptions of places anchored to the amount of recall with descriptions which can be freely organized on all the visually present material. Performance in this condition is therefore expected to be quantitatively higher than that in the two memory conditions, as subjects only have to name items, not to remember them. However, children might be overwhelmed by the quantity of perceptual information, and thus use description strategies possibly proving uneconomical in the case of descriptions based on memory. In other words, we expected direct descriptions to differ from descriptions based on memory in a higher number of mentioned items and in the different description strategies adopted.

In order to represent real environments, children have to memorize several kinds of information, in particular spatial information. Children's cognitive mapping of large-scale spaces has been widely studied by analyzing the quantity, kind, and spatial locations of recalled landmarks (Allen *et al.*, 1979). However, as real environments also contain many elements which are not spatial landmarks but which may be relevant for memory in other ways, for instance, for categorizing places, it may be of interest to examine how different kinds of environmental features are selected by subjects when remembering two different familiar places. In the situation considered here the interior (entrance hall) had more movable than fixed elements, while the opposite was true for the exterior (courtyard). Furthermore, while the entrance hall had more precise limits such as walls and doors, the courtyard had perceptually less clear boundaries, e.g. railings which could be seen through, a change in the ground surface, and so on.

The present research aimed at investigating whether such differences between places would appear in subjects' verbal reports.

Environmental elements are distinguished here as 'fixed' and 'movable' items, relating this categorization to the effect of schemata on memory for natural places (e.g., Baroni *et al.*, 1980; Brewer and Treyens, 1981; Schuurmans and Vandierendonck, 1985; Vandierendonck and Schuurmans, 1986). Schemata are structured abstract formats for representing different kinds of knowledge; they are hierarchically organized and can be embedded into larger schemata. For each place schema, three main categories of items may be detected: (a) elements which must be necessarily present in each instance of the schema; (b) elements whose presence is not necessary but compatible with the schema; and (c) elements whose presence will make definition of that given situation as an instance of that schema extremely questionable or even impossible. As an example, let us consider the schema for a cinema: armchairs and a screen are among the schema-expected elements, flowers and a curtain among the schema-compatible elements, and a gas-cooker and a bed among the schema-opposed elements.

In research on episodic memory for interiors it has been assumed that the expected, necessary elements are structural (i.e., floor, walls, ceiling, etc.), while the furniture objects are optional, schema-compatible elements. This distinction has cast some light on the effects of schemata on episodic memory for natural places showing, for instance, that structural elements are recalled better than furniture objects by adults in incidental memory condition, while the reverse is true in intentional memory condition (Salmaso *et al.*, 1983). The distinction made in the present research between 'fixed' (e.g., walls, chandeliers, trees, etc.) and 'movable' items (e.g., clotheshooks, parked cars, etc.) seems to us to reflect the distinction between structural elements and furniture objects already used only for interiors. In addition, this distinction is partially based on Shanon's 'translocability' criterion (Shanon, 1984) and also takes into account the kind of constraints for actions in space. That is, fixed items are constant constraints, while movable ones are not necessarily so.

To assess possible differences between children's and adults' verbal reporting of a familiar interior and exterior, children's performance in the description from memory condition was compared with that of teachers in the same school.

Method

Subjects
Forty-eight children (24 of each sex) and 12 adults (four males and eight females) took part in the experiment. Children's mean age was 8·6 (age range 7·4–9·3). All children were in third grade and they had attended the school where the experiment took place since their first year of school. Only children who, according to their teachers' judgements, showed at least a medium level of linguistic competence took part in the experiment. Children were randomly divided into three groups of 16 subjects each (eight of each sex). Each group was tested in only one of the three experimental conditions, for both interior and exterior.

All adults were teachers in the school. This group was tested only in the description from memory condition for both places. The order of trials was counterbalanced within each group.

Places
One interior and one exterior were considered, respectively: the entrance hall of the subjects' school, and the courtyard in front of it (see Figs 1 to 2).

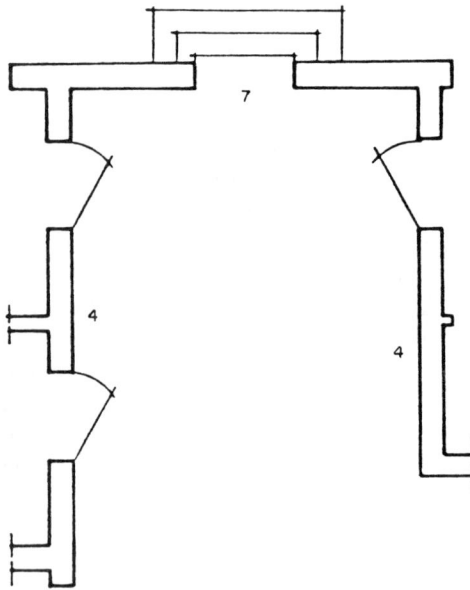

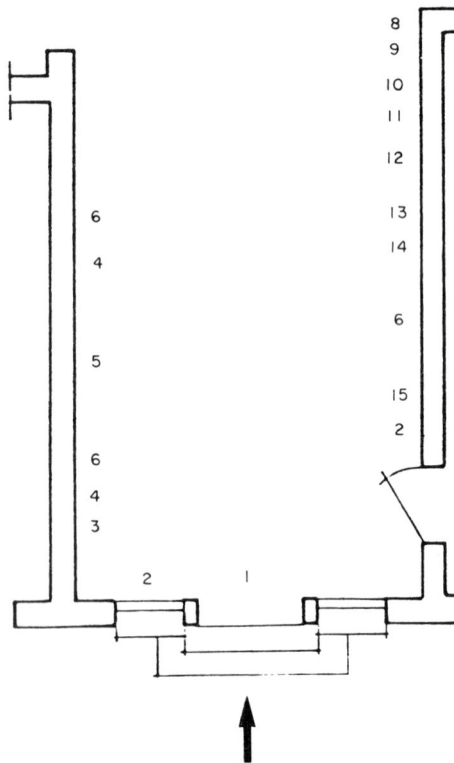

FIGURE 1. Map of the entrance hall. Numbers refer to movable items and their locations. 1, doormats; 2, umbrella stands; 3, fire-hose with pump; 4, clothes-hooks; 5, barometer/thermometer; 6, posters; 7, crucifix; 8, electric bell; 9, wall clock; 10, telephone; 11, table; 12, chairs; 13, newspaper; 14, calendar; 15, notice-board.

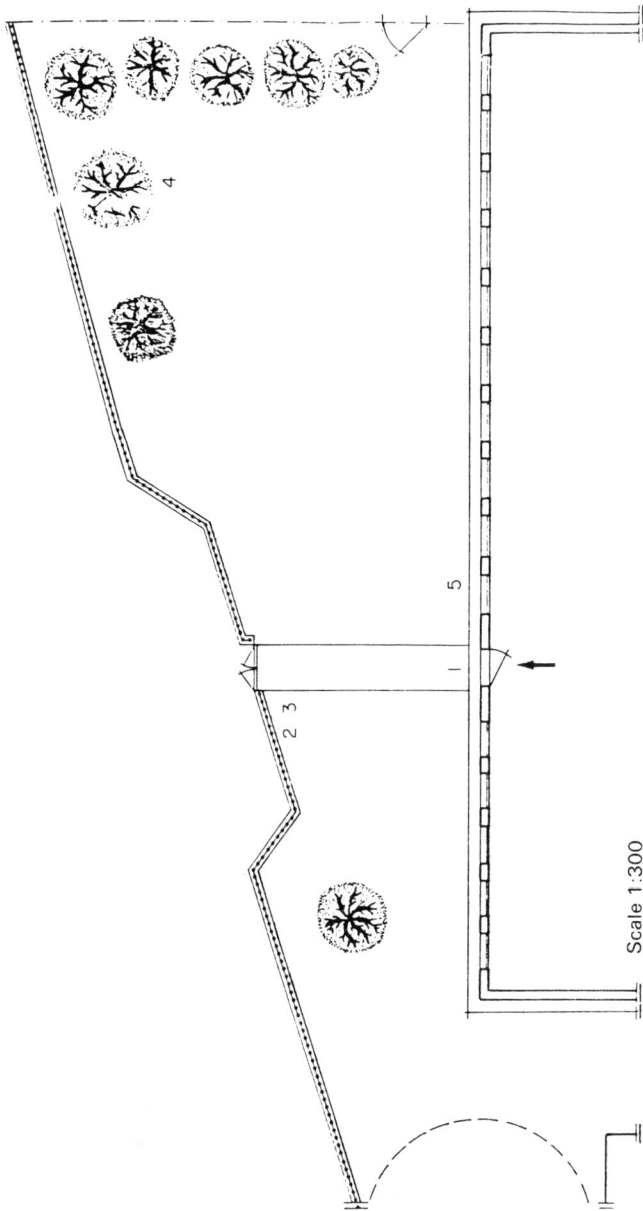

FIGURE 2. Map of the courtyard. Numbers refer to movable items and their locations. 1, doormat; 2, bucket; 3, broom; 4, bicycle rack; 5, parked cars.

Scale 1:300

Both places were equally familiar to the subjects, since they had all known them for the same period of time (three school years). The two places also had similar functions, in that they were both places of transit and of play or rest during recreation. We considered as items only the elements mentioned by subjects in the direct description condition. The rationale for this choice was that, since the three groups of children examined were equated for age and school progress, the items mentioned in the direct description condition could be taken as covering the span of linguistic competence needed for the memory task used here. For each place items were grouped as being either fixed of movable. *Fixed items* were those which were either part of the limits of the place (e.g., doors, walls, fences, gravel) or items whose removal was impossible or very difficult (e.g., concrete path, radiators). *Movable items* were all those which could be fairly easily removed, such as clothes-hooks, posters, cars, bicycle racks, etc. Care was taken to ensure that all the moveable items considered here were usually present in the experimental places, and always in the same locations. This also applied to the items which could easily be moved (e.g., broom, newspaper). When more than one instance from the same category was present, as in the case of walls, clothes-hooks, steps, trees, etc., in order to decide whether to consider them as one single item (e.g., 'clothes-hooks') or several items (e.g. three items each called 'clothes-hook'), we adopted the most frequently chosen utterance as a single item: if the majority of the children just said 'clothes-hooks', we considered them as a single item.

The 10 fixed items in the entrance hall were: main door; windows; radiators; two side corridors; internal doors (to classrooms, toilets, janitor's room); door leading to the back of the school; floor; walls; ceiling; chandeliers. The 15 movable items were: doormats; umbrella stands; fire-hose with pump; clothes-hooks; barometer/thermometer; posters; crucifix; electric bell; wall clock; telephone; table; chairs; newspaper; calendar; notice-board.

The 14 fixed items in the courtyard were: door to school; school walls and school sign; steps; windows; concrete strip round building; school custodian's house; flower-bed; boundary wall and railings; entrance gate; concrete path; trees; wire netting; wire gate; gravel. The five movable items were: doormat; bucket; broom; bicycle rack; parked cars.

Procedure

Three experimental conditions were used: free recall, description from memory, and direct description.

In the *free recall condition* the experimenter and one subject were seated alone in a room in the school building. After the child was familiarized with the experimenter, the latter told the child: 'Tell me all you remember about the entrance hall of your school', or 'Tell me all you remember about the courtyard in front of your school'. Subjects' verbal reports were tape-recorded and later transcribed.

In the *description from memory condition* the instructions for children were: 'Do you see this tape-recorder? A boy is coming to live near here, and he wants to know something about this school. I will record what you say and then let him listen to it. Now describe the entrance hall of your school' (or 'the courtyard in front of your school'). The instructions for adults were: 'Please describe the entrance hall of your school (or, alternatively, 'the courtyard in front of your school') in such a way as to allow a person who has never seen it to have an idea of it from your description'.

In the *direct description condition* instructions were identical to those given in the description from memory condition, but each child now had to describe the place while looking at it. One by one, they stood with their backs to the main entrance to the school and faced either the courtyard or the entrance hall. Care was taken that nobody crossed or stayed in neither of the two experimental places during trials.

The verbal material was analyzed considering *what* subjects said as well as *how* they organized what they said. The first point is called 'analysis of contents' and the second 'analysis of report organization'.

How Children Talk About Familiar Places

Analysis of contents

In this analysis we examined not only subjects' correct reports of places, but also one kind of error which may bear on place identification, that is, mention of items located outside the place ('boundary trespass'). The importance of 'containing features' in favouring reconstruction of familiar places has already been assessed in children's cognitive mapping (Liben *et al.*, 1982; Golbeck, 1985). In addition, Lynch (1960) stressed how important the notion of margin is in the image of a large-scale environment. Borders generally define adjacent areas, are visually prominent, continuous in form, and impenetrable to transversal movement. Transposing and adapting this concept to environments smaller than a city, we assumed that the borders of a courtyard and of an entrance hall coincide with their boundaries, since they satisfy the criteria of being continuous, impenetrable and visually prominent. In our case, except for two open side corridors, the entrance hall boundaries (walls and doors) are opaque, while the courtyard boundaries (railings and netting) are mainly transparent. From a purely perceptual point of view, therefore, it could be argued that the interior is more clearly limited than the exterior. Does this also mean that this place has less definite cognitive boundaries for children? To investigate this question we computed 'boundary trespasses', on the assumption that their presence indicates some difficulty in conceptualizing a place as an environmental unit on its own.

Correctly mentioned items. For each of the two categories of items (fixed, movable), subjects' performance was scored by computing the proportions of items correctly named or unambiguously described on the number of items in each category. Mean proportions of children's performance in the three experimental conditions are reported in Table 1.

All ANOVAs in the present study were carried out on arcsin transformed proportions (Winer, 1970), the rejection region always being $\alpha = 0.05$. A preliminary ANOVA considered the factors mentioned below, plus sex. Since no significant main effect for sex or any significant interaction involving sex as a factor was observed, this variable was discarded. The factors considered in the following ANOVA were thus: *experimental condition* (free recall, description from memory, direct description), *place* (entrance hall, courtyard), and *type of elements* (fixed items, movable items). The sources of variability reaching significance level were: experimental condition $F(2, 47) = 10.07$; place, $F(1, 144) = 4.18$; type of elements, $F(1, 144) = 24.94$; interaction between experimental condition and type of elements, $F(2, 144) = 21.31$; and interaction between place and type of elements, $F(1, 144) = 9.76$. A Newman-Keuls test revealed that subjects' performance was significantly

TABLE 1

Mean proportions of items given by children in three experimental conditions and mean ratios of boundary trespasses

	Experimental conditions		
	Free recall	Description from memory	Direct description
Mean proportion of items			
Entrance Hall			
Fixed Items	0·244	0·194	0·275
Movable Items	0·118	0·112	0·376
Courtyard			
Fixed Items	0·375	0·300	0·325
Movable Items	0·112	0·050	0·450
Mean ratios of boundary trespasses			
Entrance Hall	0·180	0·240	0·003
Courtyard	0·013	0·024	0·002

higher in the direct description condition than in the two memory conditions, which did not differ from each other (mean proportions being 0·21 for free recall, 0·16 for description from memory and 0·35 for direct description).

As for the place factor, subjects performed better in the courtyard than in the entrance hall trials. More fixed than movable items were generally mentioned.

As for the interaction between condition and type of elements, a Newman-Keuls test revealed no difference either for fixed items across conditions or for movable items between the two memory conditions. On the other hand, in the direct description condition, movable items were mentioned significantly more than in the two memory conditions and also more than fixed items in the direct description condition. A Newman-Keuls test revealed that interaction between place and type of elements was due only to the fact that fixed elements were mentioned more frequently in the exterior than in the interior. However, as Table 1 shows, the two memory conditions showed a similar trend for both places (more fixed items being recalled than movable ones), while the opposite trend was observed in the direct description condition.

Boundary trespasses. For all the linguistic aspects of subjects' verbal reports considered here, reports were examined by three independent judges, inter-rater reliability being over 0·80.

For both experimental places boundary trespasses were computed by taking into account the 'direction' of the trespass and not the number of items located outside the experimental places mentioned. Therefore, if subjects described the stairs and the upper floor of the school, this was considered to be *one* boundary trespass, while if they described the inside of a classroom and of the janitor's room (both adjacent to the entrance hall, but one on the left and one on the right of it), *two* boundary trespasses were computed.

Thirteen children gave a total of 24 boundary trespasses for the interior, while only five children gave five boundary trespasses for the exterior.

For the entrance hall, half the trespasses (i.e., 12) concerned children's descriptions of the stairs and the upper floor of the school. It should be noted that the stairs were located on the far end of the side corridors and were therefore not adjacent to the entrance hall itself. An example of this kind of trespass, obtained in the description from memory condition, is

> ... *poi c'è una scala ... e poi s'è qua più avanti, c'è ... le scuole dove si vanno, e poi qua c'è l'ambulatorio e più in fondo ci sono le classi delle quarte, delle quinte ...* (... then there are some stairs ... and then there is further on, there is ... schools where they go' and then here there is the doctor's room and further down there are the fourth-grade, fifth-grade classrooms ...)*.

A less frequent kind of boundary trespass (i.e., 7) in the interior consisted in describing what was inside the rooms adjacent to the entrance hall. For example, in the free recall condition one child said:

> ... *dentro dove lavora la bidella c'è, c'è delle scope, dopo ... e dopo del ... come si dice ... delle cassette dove mette tutte le robe ...* (... inside where the janitor works there is, there is brooms, then ... and then ... how you call it ... some boxes where she puts all the stuff ...).

In five cases children named things outside the school, for instance: ... *una porta che va verso il Nord, che fuori c'è un campo di grano* ('... a door facing north, there is a wheat field outside it'). In describing the courtyard, no child trespassed by mentioning the inside of the school. For a better understanding of the examples of trespasses quoted below, it should be noted that, facing the school, on the left of the courtyard there is a war memorial surrounded by flower-beds; to its left there is a street and then a kindergarten (two children gave boundary trespasses in this direction). Leaving the courtyard entrance gate and crossing a street, there is a square and a church at the end of it (three children gave boundary trespasses in this direction). Some examples of boundary trespasses observed in the description from memory condition are

> ... *vicino a questi alberi c'è un cimitero dei morti della Prima Guerra Mondiale, mi pare, e poi dopo questo c'è l'asilo* (... near these trees there is a cemetery for dead of the First World War, I think, and then after this there is the kindergarten)
>
> *C'è un marciapiede ... um ... c'è il muro di una chiesa ...* (There is a pavement ... um ... there is the wall of a church).

In order to assess children's representations of the places, we thought it necessary to relate boundary trespasses to correct descriptions of the places. The ratios between number of boundary trespasses and number of correctly mentioned items were then computed for each subject. The mean ratios of boundary trespasses for each experimental condition and for both places are given in Table 1. An ANOVA was carried out with the following factors: *experimental condition* (free recall, description from memory, direct description) and *place* (entrance hall, courtyard). The sources of variability reaching significance were experimental condition, $F(2, 45) = 3.59$; place,

---

* Incorrect Italian syntax has been maintained when possible in the English translation.

F (1, 45) = 10·86; and interaction between experimental condition and place, F (2, 45) = 4·42. A Newman-Keuls test revealed that trespasses were significantly lower in the direct description condition than in either memory condition, which did not differ from each other. As for the interaction between experimental condition and place, this test showed that, while almost no trespasses occurred in the direct description condition—thus indicating that children have few doubts about defining the two places and their boundaries—trespasses are significantly more frequent in both memory conditions, and most of all for the entrance hall.

Analysis of report organization

Verbal reports referring to places imply a 'linearization process' because subjects have to order items which are located in a three-dimensional space in only one dimension (their speech). Discourse planning therefore appears to be an index of subjects' capacity to adapt their verbal reports to the communicative aims required by the experimental task. We also investigated whether the experimental conditions used here would give rise to differences in the way in which items were linked to each other. We hypothesized that, in the free recall condition, subjects would tend to list the greatest possible number of items present in the experimental place and would consider mention of features or locations of items fairly unimportant, on the assumption that the experimenter already knows the place in question (Mainardi Peron *et al.*, 1985). In contrast, in the description from memory condition, children were explicitly told that their addressee had no knowledge of the place and would have to get an idea of it from their descriptions. A mere list of items would therefore probably be insufficient, so subjects would probably more often mention various features and/or locations of items. To test this, two indexes were used: 'discourse organization' and 'use of locatives'. In the direct description condition, since instructions were the same as those used for the description from memory condition, it may be argued that this would induce similar performance. However, in the direct description condition the place to be described was perceived: subjects might therefore be overwhelmed by the quantity of perceptual information, and this might result in a more disorganized report.

Another index was 'verbs of motion'. In the development of children's spatial representation of large-scale places, the first level is to rely on body movements (Siegel and White, 1975; Cousins *et al.*, 1983), so we investigated whether the importance of children's movements in space may also be detected in their verbal reports. We assumed that the use of verbs of motion could be taken as an index of the fact that children localize an item according to an imaginary tour, a mental journey in space.

Discourse organization was investigated by dividing subjects' verbal reports into two categories. The first contained reports in which items were given in a disorganized list, one item being linked to the next by utterances of the kind '. . . and then . . .'. Transposing Ullmer-Ehrich's (1982) definition to this situation, we called this the 'enumeration' category. The second category contained verbal reports in which items were mainly given in a more planned context, that is, children either organized the items in a spatial frame or linked them to their previous experience of the place. We called this category 'planned descriptions'. Examples of 'enumeration' reports for the entrance hall and the courtyard, both obtained in the free recall condition, are

Allora, c'è la cassetta meteorologica . . . poi la cassetta dove c'è la pompa, vediamo un po' . . . corridoi, gabinetti, poi, vediamo, telefono, le classi, poi i lampadari . . . i muri che sono di legno (well, there is a meteo-box (barometer and thermometer) . . . and then the box where there is the fire-pump, let me see . . . corridors, toilets, then, let me see, telephone, classrooms, then the table, the chandeliers . . . the walls which are made of wood);

C'è la ghiaia, poi c'è la scuola e . . . e c'è un albero . . . c'è il cancello con tutti i posti per le bici e poi c'è una fila di alberi, poi c'è la casa di Alfredo (there is . . . there is the gravel, then there is the school and . . . and there is a tree . . . there is the entrance gate with lots of places for bikes and there is a row of trees, then there is Alfred's (the custodian's) house).

The following examples of interior and exterior 'planned descriptions' were found in the description from memory condition:

Il cortile ha un sacco di alberi dentro, le maestre possono parcheggiare anche le loro macchine lì, ma il custode non vuole che andiamo là, perchè là ci sono delle margherite e lui ha paura che le raccogliamo . . . (the courtyard has lots of trees in it, the teachers can park their cars there too, but the custodian doesn't want us to go there, because there are some daisies there and he's afraid we'll pick them . . .);

Allora, questo cortile che noi abbiamo davanti alla nostra scuola è molto grande. Da un parte c'è la rete, dall'altra c'è la casa del bidello, e là abita lui con la sua famiglia, poi a terra ci son tutti . . . ci sono i sassi (Well, this courtyard which we have in front of our school is very large. On one side there is the wire netting, on the other side there is the custodian's house, he lives there with his family, then on the ground there are all . . . there are pebbles);

È grande, ci sono dei portaombrelli, un tavolo, e poi in fondo c'è una porta . . . 'It's large, there are umbrella stands, a table, and then at the far end there is a door . . .).

The numbers of planned reports for each experimental place and experimental condition are given in Table 2.

The difference between frequencies of planned descriptions and enumerations in the three experimental conditions was tested for each place by the chi-square test. Results showed a significant difference between conditions and for both places (entrance hall, χ^2 (2, 48) = 6·22; courtyard, χ^2 (2, 48) = 6·38). Furthermore, the McNemar test did not reveal any significant difference between places in any of the three conditions. As can be seen in Table 2, children plan their discourses mainly in the description from memory condition and least of all in the direct description condition.

Use of locatives. The ratio between number of items given with at least one locative and total number of correctly mentioned items was computed for each subject. Mean ratios for each place and experimental condition are given in Table 2. Examples of the use of locatives in the entrance hall are:

c'è una finestra al lato sinistro (there is a window on the left side);

il giornale sopra al tavolino (the newspaper on the small table);

un corridoio per di qua e un corridoio per di qua (a corridor on this side and a corridor on this side (pointing left and right).

TABLE 2
Report organization: results for children's groups

Place	Experimental conditions		
	Free recall	Description from memory	Direct description
Entrance Hall			
Planned descriptions	0·37* (6)	0·44 (7)	0·06 (1)
Locatives	0·30	0·35	0·14
Verbs of motion	0·38	0·40	0·16
Courtyard			
Planned descriptions	0·19 (3)	0·62 (10)	0·25 (4)
Locatives	0·22	0·13	0·07
Verbs of motion	0·04	0·10	0·04

* Numbers in brackets refer to number of subjects in each cell

Examples for the courtyard are:

> ... *davanti alla porta c'è una stradina di* ... *come una strada* (... in front of the door there is a small street of ... like a street);

> ... *e c'è a destra un albero molto grande, davanti alla nostra finestra della classe* (... and on the right there is a very big tree, in front of our classroom window).

An ANOVA examined the factors *experimental condition* (free recall, description from memory, direct description) and *place* (entrance hall, courtyard). The only source of variability reaching significance was place, $F(1, 48) = 9·14$. Results (Table 2) show that locatives prevail in the interior.

Verbs of motion. Examples of the use of verbs of motion in the interior are:

> *Si entra dalla porta, vieni dentro e sorpassi due porte* (You go in at the door, you come in and pass near two doors);

> ... *dopo c'è la porta per andare alla prima* ... (... then there is the door you go through to the first-grade classroom ...).

Examples for the exterior are:

> *Allora, si esce e ci sono tutti i sassi* (Well, you go out and there are all the pebbles);

> ... *C'è il cancello dove i bambini entrano* (... there is the gate where children come in at).

An ANOVA (considering the same factors as in the previous ANOVA) was carried out on the ratios between number of verbs of motion and total number of correctly mentioned items. Mean ratios are given in Table 2. The only source of variability reaching significance was place, $F(1, 48) = 18·33$. Like locatives, verbs of motion were more frequently used in the entrance hall trials, means being 0·314 for the entrance hall and 0·064 for the courtyard. Since the two places were equally familiar to the subjects and also involved in similar daily routines, this result cannot be attributed to children's different experiences of motion in the two places.

Comparison with Adults

Adults' memory for the previously described places was compared with the performance of the eight-year-olds only in the description from memory condition, as only 12 teachers fulfilled the requirement of having the same degree of familiarity with the experimental places as the children. Adults were obviously expected to perform quantitatively better than children for both places. Furthermore, adults' recall for the interior, was not expected to differ significantly from that of the exterior, since the two places were equated for function and familiarity. Adults were also expected to plan their discourse more than children, with a clear majority of planned descriptions and more item specifications. Assuming that there is a developmental trend in representation from contextualized knowledge of the world to more general and categorical knowledge (Nelson, 1983), we expected to find a more limited use of verbs of motion in adults than in children, as this index is related to the description of movements in space. As for the boundary trespasses index, if adults' and children's data did not differ, this would suggest that, in human memory for natural familiar places, boundaries are defined more psychologically than physically. Instead, if adults did not trespass in either place, this interpretation could only apply to eight-year-olds' memory.

Analysis of contents
Correctly mentioned items. As all items mentioned by teachers were also mentioned by children in the direct description condition, the range of items was the same. For the adults' group the mean proportions of remembered items were: in the entrance hall: fixed items, 0·45; movable items, 0·14, and in the courtyard: fixed items, 0·40; movable items, 0·14.

An ANOVA was carried out on proportions of items remembered by children and by adults in the description from memory condition. The factors considered were: *age* (adults, children), *place* (entrance hall, courtyard), and *type of elements* (fixed items, movable items). The sources of variability reaching significance were age, $F(1, 26) = 4·52$; type of elements, $F(1, 26) = 61·96$; and interaction between age, place, and type of elements, $F(1, 26) = 6·52$. Adults' performance was always higher than that of children. Furthermore, as revealed by the Newman-Keuls test, adults remembered the interior and the exterior equally well, and in both places remembered more fixed than movable items. This confirms the results obtained by Baroni *et al.* (1985) which suggest that, when remembering familiar places, adults mainly rely on the fixed aspects. The significant differences revealed by this analysis thus confirm differentiated recall of places only in children.

Boundaries trespasses. The mean ratios of boundary trespasses for the adults' group were 0·275 for the entrance hall, and 0·178 for the courtyard.

An ANOVA was carried out on the following factors: *age* (adults, children), and *place* (entrance hall, courtyard). The only source of variability reaching significance was place, $F(1, 26) = 7·28$. As no age difference was revealed, it may be argued that a certain type of place induces boundary trespasses.

Analysis of report organization
Results referring to all indexes in the analysis of report organization for the adults' group are given in Table 3.

TABLE 3
Report organization: results for adults' group

Descriptions from memory	
Entrance Hall	
Planned descriptions	1·00 * (12)
Locatives	0·52
Verbs of motion	0·11
Courtyard	
Planned descriptions	0·75 (9)
Locatives	0·46
Verbs of motion	0·04

* Numbers in brackets refer to number of subjects in each cell

Discourse organization. The difference between frequencies of planned descriptions and of enumerations in the two age groups was computed for each place by the chi-square test, employing Yates's corrections. Results show that adults give significantly more planned descriptions than children in the entrance hall (planned descriptions compared to enumerations for two age levels: χ^2 (1, 28) = 7·51), while no significant difference was observed in the courtyard.

Use of locatives. An ANOVA was carried out on the following factors: *age* (adults, children), and *place* (entrance hall, courtyard). The sources of variability reaching significance were age, F (1, 26) = 7·54, and place, F (1, 26) = 5·16. When remembering familiar places, adults tend to localize environmental elements much more than children.

Verbs of motion. An ANOVA with the same factors considered for the use of locatives was employed here. The sources of variability reaching significance were age, F (1, 26) = 4·67, and place, F (1, 26) = 14·51. These results tend to confirm that, when recalling familiar environments, children describe actions more frequently than adults.

Discussion

Our results offer some evidence of children's knowledge of places which involves aspects other than spatial information. Research in this field is still at a very early stage, but some promising lines for future studies are sketched out by our data. The most intriguing results of the present study is that eight-year-old children's recall was better for a familiar exterior than for an equally familiar interior. Instead, adults recalled the same quantity of environmental elements in both places, thus indicating that children's better recall of the exterior may mainly be attributed to their performance and not to place differences *per se*. Children's behavior may be better explained by focusing on some results obtained here.

It appears that, in both free recall and description from memory conditions, eight-year-old children base their recall mainly on the fixed elements of the two places, thus suggesting that their recall of highly familiar environments relies mainly on items which are strictly connected with the place schema and which are fixed constraints to action in space. As a similar trend of data was also observed for the adult group, this may be a general tendency in representation of familiar places.

Children generally recalled fewer interior than exterior items, and in particular they recalled fewer fixed elements in the former. These results may be further clarified by the data on boundary trespasses. Only interior boundaries appear to be fuzzy in memory performance, in both children and adults. The interior and exterior considered here do not therefore appear to stand in memory on the same categorical level as place units. While the courtyard tends to be considered a place on its own, the entrance hall is embedded in a larger internal place-unit, i.e., the school building. This does not affect adults' quantitative memory performance for the two places, but it may account for the significant difference in children's quantitative memory between the two places. These data also bear on the organizational principles of environmental knowledge, in the sense that they suggest the different effect on place representation of considering an environment either as a place on its own or as a part of a larger unit.

In our opinion, the importance of taking into account some aspects of the organization of verbal reporting also stems from the above considerations. For instance, joint consideration of locatives and verbs of motion shows two interesting phenomena. First, children tend to name movements in space while adults give mainly items locations, thus suggesting that children's representation of places may be connected with the activity in space that places allow, as Cohen and Cohen (1985) have argued. Second, children tend to localize items by means of locatives and to mention verbs of motion mainly when remembering the entrance hall, i.e. the place remembered least well. It should also be noted that adults' performance does not differ for the two places, either in quantity of recalled items or in ratios of locatives. One possible explanation for our children's data is offered by the work of Nelson and Ross on the development of long-term memory in young children, in which it is noted that '... location of important objects is important for younger children, while episodes or events are more important to the older ones' (Nelson and Ross, 1980, p. 94). Transposing these observations to better- or worse-recalled places, a similar trend is observed in our data, in the sense that localizations appear mainly at the lowest levels of memory, as if children revert to developmentally earlier kinds of memory when having difficulty in remembering.

The communicative aspects of our experimental situations affected children's performance both quantitatively and qualitatively, as expected. In fact, eight-year-old children show better discourse organization in the description from memory than in the free recall condition, thus indicating their capacity to adapt verbal reports to the social aim requested. However, the instruction effect is less important than the effective presence of the material, as children scarcely plan their discourses in the direct description condition. The difference in discourse organization between the two memory conditions might be related to the quantity of recall, in the sense that enumerations and slightly higher recall are observed in the free recall condition, while planned descriptions and slightly lower recall occur in the description from memory condition.

Considered overall, the present results support the analysis of verbal reports as a useful methodological means for better understanding of children's representation of natural familiar environments. In addition, various aspects of a place must be separately considered in assessing place representation. On this issue, our proposed distinction between fixed and movable items proved to be relevant. Furthermore, in the present approach quantitative data can usefully be integrated by joint analyses of linguistic indexes used here. In particular, report organization and boundary trespasses should be further investigated at different age levels and in various kinds of familiar and unfamiliar places, to clarify their effects on children's knowledge of places.

Acknowledgements

The authors wish to thank Tommy Gärling and Remo Job for their useful comments and suggestions on a previous version of this paper.

References

Allen, G. L., Kirasic, K. C., Siegel, A. W. and Herman, J. F. (1979). Developmental issues in cognitive mapping: the selection and utilization of environmental landmarks. *Child Development*, **50**, 1062–1070.

Anooshian, L. J. and Young, D. (1981). Developmental changes in cognitive maps of a familiar neighborhood. *Child Development*, **52**, 341–348.

Axia, G. (1986). *La mente ecologica: la conoscenza dell'ambiente nel bambino.* Firenze: Giunti Ed.

Baroni, M. R., Job, R., Mainardi Peron, E. and Salmaso, P. (1980). Memory for natural settings: role of diffuse and focused attention. *Perceptual and Motor Skills*, **51**, 883–889.

Baroni, M. R., Job, R., Mainardi Peron, E. and Salmaso, P. (1985). Recalling interiors and outdoors: effects of familiarity. Paper presented at the 16th Environmental Design Research Association Conference, New York, June 10–13, 1985.

Brewer, W. F. and Treyens, J. C. (1981). Role of the schemata in memory for places. *Cognitive Psychology*, **13**, 207–230.

Cohen, R., Baldwin, L. M. and Sherman, R. C. (1978). Cognitive mapping of a naturalistic setting. *Child Development*, **49**, 1216–1218.

Cohen, S. L. and Cohen, R. (1985). The role of activity in spatial cognition. In R. Cohen (ed.), *The Development of Spatial Cognition*. Hillsdale, NJ: Lawrence Erlbaum Associates, pp. 199–224.

Cousins, J. H., Siegel, A. W. and Maxwell, S. E. (1983). Way finding and cognitive mapping in large-scale environments: a test of a developmental model. *Journal of Experimental Child Psychology*, **35**, 1–20.

Golbeck, S. L. (1985). Spatial cognition as a function of environmental characteristics. In R. Cohen (ed.), *The Development of Spatial Cognition*. Hillsdale, NJ: Lawrence Erlbaum Associates, pp. 225–255.

Hart, R. (1979). *Children's Experience of The Place*. New York: Irvington.

Herman, J. F., Roth, S. F., Miranda, C. and Getz, M. (1982). Children's memory for spatial locations: the influence of recall perspective and type of environment. *Journal of Experimental Child Psychology*, **34**, 257–273.

Klein, W. (1982). Local deixis in route directions. In R. Jarvella and W. Klein (eds) *Speech, Place and Action* Chichester: Wiley, pp. 162–182.

Liben, L., Moore, M. and Golbeck, S. (1982). Preschoolers' knowledge of their classroom environment: evidence from small-scale and life-size spatial tasks. *Child Development*, **53**, 1275–1284.

Linde, C. and Labov, W. (1975). Spatial networks as a site for the study of language and thought. *Language*, **51**, 924–939.

Lynch, K. (1960). *The Image of The City*. Cambridge, MA: M.I.T. Press.

Lynch, K. (1977). *Growing Up In Cities*. Cambridge, MA: M.I.T. Press.

Mainardi Peron, E., Baroni, M. R., Job, R. and Salmaso, P. (1985). Cognitive factors and communicative strategies in recalling unfamiliar places. *Journal of Environmental Psychology*, **5**, 325–333.

Matthews, M. H. (1985). Young children's representations of the environment: a comparison of techniques. *Journal of Environmental Psychology*, **5**, 261–278.

Nelson, K. (1983). The derivation of concepts and categories from event representations. In E. R. Scholnick (ed.), *New Trends in Conceptual Representation: Challenges to Piaget's Theory?* Hillsdale, NJ: Lawrence Erlbaum Associates, pp. 129–149.

Nelson, K. and Ross, G. (1980). The generalities and specifics of long-term memory in infants and young children. In M. Perlmutter (ed.), *Children's Memory.* San Francisco: Jossey-Bass, Inc., pp. 87–101.

Newcombe, N. and Liben, L. S. (1982). Barrier effects in the cognitive maps of children and adults. *Journal of Experimental Child Psychology,* **34**, 46–58.

Salmaso, P., Baroni, M. R., Job, R. and Mainardi Peron, E. (1983). Schematic information, attention, and memory for places. *Journal of Experimental Psychology: Learning, Memory and Cognition,* **9**, 263–268.

Schuurmans, E. and Vandierendonck, A. (1985). Recall as communication: Effects of frame anticipation. *Psychological Research,* **47**, 119–124.

Shanon, B. (1984). Room Descriptions. *Discourse Processes,* **7**, 225–255.

Siegel, A. W. and White, S. H. (1975). The development of spatial representation of large-scale environments. In H. W. Reese (ed.), *Advances in Child Development and Behaviour,* Vol. X. New York: Academic Press, pp. 10–55.

Ullmer-Ehrich, V. (1982). The structure of living place descriptions. In R. J. Jarvella and W. Klein (eds), *Speech, Place and Action.* Chichester: Wiley, pp. 219–249.

Vandierendonck, A. and Schuurmans, E. (1986). Interaction of incidental and intentional learning with frame usage. *Cahiers de Psychologie Cognitive,* **6**, 405–418.

Waller, G. (1986). The development of route knowledge: multiple dimensions? *Journal of Environmental Psychology,* **6**, 109–119.

Weatherford, D. L. (1982). Spatial cognition as a function of size and scale of the environment. In R. Cohen (ed.), *Children's Conceptions of Spatial Relationships.* San Francisco: Jossey-Bass, Inc., pp. 5–18.

Winer, B. J. (1970). *Statistical Principles in Experimental Design.* London: McGraw-Hill.

INDEX